Johannes M. Zanker • Jochen Zeil (Eds.)

Motion Vision

Springer

Berlin
Heidelberg
New York
Barcelona
Hong Kong
London
Milan
Paris
Singapore
Tokyo

Johannes M. Zanker · Jochen Zeil (Eds.)

Motion Vision

Computational, Neural, and Ecological Constraints

With 90 Figures

 Springer

EDITORS

Johannes M. Zanker* and Jochen Zeil
Centre for Visual Sciences
Research School of Biological Sciences
Australian National University
PO Box 475
Canberra, ACT 2601
Australia

Present address:
Department of Psychology
Royal Holloway
University of London
Egham
Surrey TW20 OEX
England

The publication of this book was assisted by a grant from the Publications Committee of the Australian National University

ISBN 3-540-65166-7   Springer-Verlag Berlin Heidelberg New York

Library of Congress Cataloging-in-Publication Data
Motion vision: computational, neural, and ecological constraints / Johannes M. Zanker, Jochen Zeil
(eds.). p. cm. Includes bibliographical references and index. ISBN 3540651667
1. Motion perception (Vision) I. Zanker, Johannes M., 1957- II. Zeil, Jochen, 1949-
QP493 .M685 2000 612.8'4--dc21   00-059557

Springer-Verlag Berlin Heidelberg New York
a member of BertelsmannSpringer Science+Business Media GmbH
© Springer-Verlag Berlin Heidelberg 2001
Printed in Germany

Cover illustration based on „Bush Berries", tryptich by Josie Petrick Kemarre, by courtesy of the Anmatyerre people, with support from the Australian National University
Cover Design: *design & production*, Heidelberg
Typesetting: Camera-ready by the editors

SPIN: 10665014 31/3130xz – 5 4 3 2 1 0 – Printed on acid free paper

Preface

This book originated from a small workshop on the question of how image motion is processed under natural conditions that was held in 1997 at the Institute of Advanced Studies of the Australian National University in Canberra. Most of the contributing authors of this book were participants of the workshop. We intended to bring together a multidisciplinary group of researchers to discuss the neural, computational, and ecological constraints under which visual motion processing has evolved. We took this as an opportunity to assess how far we have come in our understanding of neural processing mechanisms in relation to natural scenes. The workshop was characterized by vivid discussions, tough debates about methods and approaches, thorough reviews of our knowledge base, and the discovery of kindred spirits amongst colleagues from very different areas of expertise. The intense atmosphere not only motivated the participants to contribute to "yet another" book, it also kept us as editors going through the sometimes tedious and tough editorial process, which, not surprisingly, took much longer than intended.

To reflect the discursive nature of the workshop, this book has an unusual structure. Each part consists of a keynote paper, usually written by authors from different research areas, who introduce our six main topics, which is followed by two or three companion articles that provide a comment to the keynote, add alternative views, or expand on some of the issues involved in the topic. We do not cover the whole field comprehensively or in a balanced manner in this way, but rather intend to mark out some critical issues that in our and the authors' mind will be important to tackle if we are to understand the design of the neural processing mechanisms underlying motion vision.

Many people contributed in the background to the completion of this book. Our first thanks naturally go to the authors, for bearing with us through a painstakingly complex process of reviewing which was dictated by the book's ambitious structure. We secondly would like to thank the referees as the real heroes behind this book. They contributed in the background with their effort, patience, diligence, and dedication substantially to the quality of the book. Last, but not least, we thank Waltraud Pix for her meticulous work during the preparation of the final typeset manuscript, and our publisher, in particular Ursula Gramm, for her patience and her support throughout the preparation of the book. For the workshop and for editing this book we received financial support from the Research School of Biological Sciences, the Centre for Visual Sciences, and the Department of Industry, Science and Tourism. Finally, we promise our families that we will think twice before we should ever be tempted again to edit another book.

Johannes Zanker & Jochen Zeil, Sunny Canberra Winter 2000

Contents

Part IV: Motion Vision in Action

Part V: Neural Coding of Motion

Part VI: Motion in Natural Environments

Contributors

Janette Atkinson
Visual Development Unit, Department of Psychology, University College
London, Gower Street, London WC1E 6BT, England
j.atkinson@ucl.ac.uk

Crista L. Barberini
Howard Hughes Medical Institute and Department of Neurobiology, Stanford
University, Sherman Fairchild Bldg Rm D209, Stanford CA 94305-5401
crista@monkeybiz.Stanford.EDU

William Bialek
NEC Research Institute 4 Independence Way Princeton, New Jersey 08540, USA
Bialek@research.nj.nec.com

Alexander Borst
ESPM-Division of Insect Biology, University of California, 201 Wellman Hall,
Berkeley, CA 94720-3112
borst@nature.berkeley.edu

Oliver Braddick
Visual Development Unit, Department of Psychology, University College
London, Gower Street, London WC1E 6BT, England
o.braddick@ucl.ac.uk

Simon J. Cropper
Department of Physiology University of Melbourne, Victoria, 3010, Australia
s.cropper@physiology.unimelb.edu.au

Hans-Jürgen Dahmen
Lehrstuhl fuer Kognitive Neurowissenschaften, Auf der Morgenstelle 28, D-
72076 Tübingen, Germany
hansjuergen.dahmen@uni-tuebingen.de

Dawei W. Dong
Complex Systems & Brain Sciences, FAU 3091, 777 Glades Road, Boca Raton,
FL 33431-0991, USA
dawei@dove.ccs.fau.edu

John K. Douglass
Arizona Research Laboratories, Division of Neurobiology, University of Arizona, Tucson, AZ 85721, USA
jkd@neurobio.arizona.edu

Michael P. Eckert
University of Technology, Sydney, School of Electrical Engineering, PO Box 123, Briadway, NSW 2007, Australia
meckert@ee.uts.edu.au

Martin Egelhaaf
Lehrstuhl fuer Neurobiologie, Fakultaet fuer Biologie, Universitaet Bielefeld, Postfach 10 01 31, 33501 Bielefeld, Germany
martin.egelhaaf@biologie.uni-bielefeld.de

Matthias O. Franz
Image Understanding (FT3/AB) DaimlerChrsyler AG, Research & Technology, P.O. Box 23 60, D-89013 Ulm, Germany
matthias.franz@daimlerchrysler.com

Shigang He
Vision, Touch and Hearing Research Centre, The University of Queensland, Brisbane QL 4072, Australia
s.he@vthrc.uq.edu.au

Gregory D. Horwitz
Howard Hughes Medical Institute and Department of Neurobiology, Stanford University, Sherman Fairchild Bldg Rm D209, Stanford CA 94305-5401
horwitz@salk.edu

Michael Ibbotson
Centre for Visual Sciences, RSBS, Australian National University, G.P.O. Box 475, Canberra, A.C.T. 2601, Australia
ibbotson@rsbs.anu.edu.au

Holger G. Krapp
Lehrstuhl fuer Neurobiologie, Fakultaet fuer Biologie, Universitaet Bielefeld, Postfach 10 01 31, D-33501 Bielefeld
holger.krapp@biologie.uni-bielefeld.de

Michael F. Land
Centre for Neurosciences, University of Sussex, Brighton BN1 9QG, England
M.F.Land@sussex.ac.uk

William R. Levick
Division of Psychology, The Australian National University, Canberra, Australia
William.Levick@anu.edu.au

Ted Maddess
Centre for Visual Sciences, RSBS, Australian National University, G.P.O. Box 475, Canberra, A.C.T. 2601, Australia
maddess@rsbs.anu.edu.au

Frederick A. Miles
Laboratory of Sensorimotor Research, Building 49, Room 2A50, 49 Convent Drive, Bethesda MD 20892-4435, USA
fam@lsr.nei.nih.gov

William T. Newsome
Howard Hughes Medical Institute and Department of Neurobiology, Stanford University, Sherman Fairchild Bldg Rm D209, Stanford CA 94305-5401
bill@monkeybiz.Stanford.EDU

David C. O'Carroll
Department of Zoology, University of Washington, Seattle, Box 351800 WA 98195, U.S.A.
davidoc@u.washington.edu

Francesco Panerai
Laboratoire de Physiologie de la Perception et de l'Action, Collège de France, 75005 Paris, France
francesco.panerai@college-de-france.fr

John A. Perrone
Psychology Department, University of Waikato, Private Bag 3105, Hamilton, New Zealand
jpnz@waikato.ac.nz

Ning Qian
Centre for Neurobiology and Behaviour, Columbia University, 722 West 168th Street, # 730A, New York NY 10032, USA
nq6@columbia.edu

Rob de Ruyter van Steveninck
NEC Research Institute 4 Independence Way Princeton, New Jersey 08540, USA
ruyter@research.nj.nec.com

Giulio Sandini

Laboratory for Integrated Advanced Robotics, Department of Communication Computers and Systems Science, University of Genova, Genova, Via Opera Pia 13, 16145 Genova, Italy
sandini@dist.unige.it

Nicholas J. Strausfeld

Arizona Research Laboratories, Division of Neurobiology, University of Arizona, Tucson, AZ 85721, USA
flybrain@manduca.neurobio.arizona.edu

Mandyam V. Srinivasan

Centre for Visual Sciences, RSBS, Australian National University, G.P.O. Box 475, Canberra, A.C.T. 2601, Australia
m.srinivasan@anu.edu.au

W. Rowland Taylor

John Curtin School of Medical Research, The Australian National University, Canberra, Australia
rowland.taylor@anu.edu.au

David I. Vaney

Vision, Touch and Hearing Research Centre, The University of Queensland, Brisbane QL 4072, Australia
vaney@vthrc.uq.edu.au

Anne-Katrin Warzecha

Lehrstuhl fuer Neurobiologie, Fakultaet fuer Biologie, Universitaet Bielefeld, Postfach 10 01 31, D-33501 Bielefeld
AK.Warzecha@biologie.uni-bielefeld.de

Johannes M. Zanker

Centre for Visual Sciences, RSBS, Australian National University, G.P.O. Box 475, Canberra, A.C.T. 2601, Australia
johannes@rsbs.anu.edu.au

Jochen Zeil

Centre for Visual Sciences, RSBS, Australian National University, G.P.O. Box 475, Canberra, A.C.T. 2601, Australia
zeil@rsbs.anu.edu.au

Processing Motion in the Real World

Johannes M. Zanker and Jochen Zeil

Visual Sciences Group, Research School of Biological Sciences, Australian National University, Canberra, ACT, Australia

1. Observing motion

As Nick Wade illustrates in wonderful detail in his book on the history of visual science (1998), it was the development of experimental sciences that converted scholars from outdoor observers into laboratory experimenters. This novel approach opened new opportunities to study nature, and since the 17th century advanced knowledge in visual sciences as in so many other areas with impressive speed. Experimental sciences, combined with the theory of evolution, opened a new perspective by comparing design principles for similar sensory and behavioural functions across different species. Carrying out critical, theory-driven experiments under highly controlled conditions is so powerful, compared to pure observation of nature, that euphoria about precise and clean evidence made scientists sometimes forget the conditions under which such evidence is collected. The visual stimuli used in laboratory experiments are by necessity rather "abstract", simple, and barely representative of the natural operating conditions under which biological systems have evolved and continue to operate. The focus on laboratory studies has dominated visual science ever since, although in the work of some scientists observation and experimentation had always been part of a unified approach. All the way along there have been continuing attempts for instance, to understand of human perception in its real life context (a number of "classical" examples can be found in Helmholtz 1864; Metzger 1975; Gibson 1979).

This book tries to pull together research from very different areas of visual sciences with the aim of evaluating whether the developments in knowledge and technology during the last decades allow us to consider visual motion processing under conditions that have been relevant in evolution. What is the framework for such a paradigm shift from laboratory-based investigations of visual systems to their analysis in the real world?

- *Technological advances*. Recording devices with high resolution in time and space allow us now to monitor behavioural patterns in the field with hitherto unknown precision. The availability of digital cameras and powerful storage devices at a reasonable price furthermore opens the possibility to record the structure and dynamics of natural habitats at a large scale, with the option to analyse the spatial distribution of light together with its spectral composition and its polarization.
- *Theoretical approaches*. Motion detector theory and simulation devices have been developed to a stage at which responses to complex inputs can be analysed. At the same time, theoretical approaches to understanding neuronal coding strategies have reached a solidity and sophistication that encourage us to move away from interpreting average responses to average stimulus conditions. Understanding responses to a single stimulus presentation may offer the chance to trace a behavioural decision in a visuo-motor task down to the level of the individual neurones in a complex neuronal circuit.
- *Neurophysiology and Behaviour*. The extensive knowledge of the computational properties of individual neurones and the interactions in neuronal circuits in the visual pathways provides us with a solid basis to ask how the image processing of natural scenes is limited by the structure and function of real brains. There are recent attempts to record nerve cell activity under more and more natural conditions, leading eventually to field electrophysiology. One of the goals that may be achieved in the nearby future is to study the intricate relation between motor and sensory patterns emerging during active vision in its natural context.
- *Robotics*. Advances in systems control theory, opto-electronics and micromechanics have led to rapid developments in robotics so that we can now test neuroethological concepts in the closed sensory-motor loop. At the same time neurobiological knowledge can be used to develop robotics, as demonstrated by the emerging discipline of "biomimetics". Implementations of the principles of visual information processing that have been identified in biological systems can be put to test in artificial systems in a variety of platforms and environments.

In this book, we try to illustrate how far we have come in our understanding of motion vision, and where the essential advances could be expected in the future. Motion vision is special because it is a non-trivial visual processing task and has a high "information content" for any organism (Borst and Egelhaaf 1989). Motion information is used to control orientation and movement, to relate to other animals of the same or other species, and to extract information on the three-dimensional structure of the environment (Nakayama 1985). It is thus not surprising that motion vision is highly developed in basically all diurnal creatures. Consequently, motion vision has been extensively studied in neuroethology (e.g. Miles and Wallman 1993) and psychophysics (e.g. Ullman 1979; Braddick and Sleigh 1983; Smith and Snowden 1994; Watanabe 1998), and continues to be a challenge for machine vision (Marr 1982; Aloimonos 1993; Srinivasan and Venkatesh 1997). Recognizing the recent achievements of multidisciplinary research on motion vision, this book brings together authors from various disciplines ranging from engineering and biology to psychology.

2. Processing constraints

We begin our survey with the fundamental neuronal mechanisms of motion detection and the integration of local motion information. On this basis we discuss a number of essential control problems that are solved by using motion information, and ask how the motion signals necessary for such control systems can be encoded with neuronal processing elements that have limited bandwidth and reliability. Three types of constraints, which limit the performance of biological signal processing systems, are considered in this book.

- *Computational:* The visual system has to extract relevant information about egomotion, about the three-dimensional layout of the environment, and about moving objects, from complex, dynamic, two-dimensional images (Gibson 1979). In each case it has to cope with highly ambiguous data. Well known examples of such ambiguities are the so-called "aperture" or "correspondence" problems which arise at the elementary level of motion detection (Marr 1982; Hildreth and Koch 1987). They demonstrate that the basic computational problems in motion vision are mathematically ill-posed.
- *Neural:* Biological systems perform computations with neurones which suffer from a number of severe processing limitations. Neurones possess a comparatively small dynamic range for representing intensities and temporal changes with analogue neuronal signals or spike trains (Barlow 1981), they can only approximate exact mathematical operations (Torre and Poggio 1978), and suffer from internal noise (Bialek and Rieke 1992; Laughlin 1998).
- *Ecological:* Visual systems operate in concrete and often very specific worlds, which are characterized by differences in the structure of behaviour and the topography of the environment (Lythgoe 1979; Dusenbery 1992). Given that motion processing mechanisms have evolved under selective pressure in specific visual habitats and in the context of specific lifestyles (O'Carroll et al. 1996), the systematic analysis of visual environments and visual tasks should help us to understand the functional and adaptive properties of neural processing strategies. At the moment we have surprisingly little to say regarding the question that arises repeatedly in this book: what are the actual motion signals visual systems have to work with?

3. Steps to analyse a complex system

The book deals with six major topics to investigate the significance of these constraints for motion vision. Each part is organized in a keynote chapter that introduces a topic and the crucial concepts, which are expanded, complemented or juxtaposed by shorter companion articles that provide additional or alternative views on the same topics. This format naturally does not cover the field comprehensively, but hopefully offers the reader a multi-facetted insight into a set of questions that need to be addressed when we try to assess our knowledge of visual motion processing in natural environments.

(I) The first part of the book deals with the biological basis of motion detection from a physiological and anatomical point of view. It lays out what we know about how motion detection is implemented with neurones and synapses, and how their properties can be related to theoretical models of elementary motion detection, which is fundamental to all consecutive processing. How do neurones perform the basic mathematical operations that are necessary to extract directional selective signals from the spatial and temporal changes of image intensity (Reichardt 1987)? The review by Vaney et al. describes in great detail the neuronal machinery of such a spatiotemporal correlation mechanism for the rabbit retina, which now has been studied over decades. Although such cellular models of connectivity are now described down to the level of the biochemistry of the synapses, it is surprising to see that some essential questions are still a matter of debate. The neuroanatomical and functional structure of directionally selective ganglion cells in the rabbit is compared in the companion chapters with two very different biological systems. Ibbotson illustrates how specific models of motion detection can be discriminated by careful experimentation in the marsupial, a comparatively distant relative of the rabbit, in which – like in many higher mammals, including primates – the elementary steps of motion detection are not carried out in the retina, but in the cortex. Despite a completely different localization within the visual processing stream, which involves different classes of neurones, the computational structure of the local motion detecting process is strikingly similar in the rabbit and the wallaby, apart from some specific variations in synaptic connectivity. The theme of functionally equivalent processing by very different neuronal elements is further developed in the contribution by Douglass and Strausfeld who review the anatomical knowledge of motion detection networks in the visual system of flies. Neurones in animals with widely different phylogenetic history have such similar functional properties in the context of motion detection that they have even been assigned similar labels such as "magnocellular stream". We have thus to appreciate that computational needs have recruited very different neuronal substrates in the course of evolution to solve one and the same task.

(II) Although we know the fundamental principles and the biological realization of the initial stages of motion processing quite well, we are far from understanding how behaviourally relevant information is extracted. The reason being that local motion information, as it is extracted by elementary motion detectors, is noisy, ambiguous, or even misleading (Egelhaaf et al. 1989). The crucial information is often only carried by the whole distribution of motion signals (Koenderink 1986). The first processing stage in which such distributions can be extracted involves two fundamental classes of operations: image segmentation and spatio-temporal integration (Braddick 1997). The fact that we are able to perceive two motion signals that differ in direction or speed simultaneously within the same region of the visual field – a phenomenon called "transparency" – is often regarded as critical for understanding the competitive demands of integration and segmentation. Braddick and Qian discuss this topic from the viewpoints of both human psychophysics and primate electrophysiology. The authors address the question of how local motion signals are pooled across space and time while retaining sensitivity to different motion directions and indicate at which levels in the cortical processing stream the two mechanisms need to be localized. Braddick and Qian suggest that there must be an intermediate integration stage that has not yet been identified in terms of neurones. This role of motion opponency – being an essential part of local motion detection – for motion transparency is considered further by Zanker, who develops a computational model that accounts for specific properties of motion transparency and segmentation that are found in psychophysical studies. This model converges with the physiological considerations raised in the keynote paper, and predicts spatial constraints of separating motion signals. A more fundamental approach is taken by Cropper in his companion article to discuss the question how useful different kinds of motion stimuli are to study "global" motion percepts. He scrutinizes the variety of local features that can be used in a segmentation or pooling process, and asks to what extent our current experimental paradigms fail to address the question of how stimulus feature combinations are represented in cortical processing. These critical questions remind us that in trying to understand the interaction between motion signals we must be aware that brains are extremely powerful in combining information across modalities.

(III) More complex motion signal distributions, which have an extraordinary significance in everyday life, are the optic flow fields experienced by a moving observer. Extracting reliable information from optic flow is a crucial task for any mobile organism, because vision is required for the control of locomotion (Gibson 1979). The task of estimating egomotion parameters from optic flow (Koenderink 1986) is discussed in this part of the book from both biological and theoretical perspectives, in an attempt to characterize the operating principles. Dahmen et al. identify the principal limitations in comprehensive simulation experiments and then ask to what degree visual systems are optimized to extract egomotion parameters from optic flow. Simulations demonstrate how the fundamental algorithms can be realized by matched filters. By making assumptions about the statistical structure of the world and about typical patterns of locomotion, Dahmen et al. derive matched filters, which turn out to resemble the distribution of directional sensitivity of large field integrating neurones in the insect visual system. In his companion paper, Perrone draws attention to the fact, that despite all their sophistication, current models continue to be hampered by the aperture problem that can lead to significant misjudgements of local motion direction. On the other hand, Srinivasan discusses how some invertebrates could use "quick and dirty short-cuts" to overcome some of the difficult problems of egomotion estimation. It is clear from the papers presented in this part that comparatively simple algorithms can be designed with pragmatic assumptions to analyse certain aspects of optic flow for a range of conditions, and that biological systems are experts in doing so. But the accuracy of the local motion information limits in various ways the precision that can be achieved, and under many conditions the visual system may need to work around typical pitfalls by using "rules-of-thumb".

(IV) The next part takes a closer look at the intricate connection between the control of locomotion and the motion signals that are to be processed for this purpose. What are the fundamental computational strategies that are involved in gaze stabilization and tracking eye movements, and how is performance limited by the constraints imposed by the elementary motion detection process, and by the neural implementation? Sandini et al. focus in their keynote paper on the coordination of two eyes and the binocular integration of motion information that is needed to keep an object in the centre of the visual field. This task is complicated by geometrical aspects, such as translational components that result from excentric rotations, and particularities of motor dynamics. The use of mechanosensory cues can be very helpful in this context, and Sandini at al. demonstrate how both biological and artificial systems make use of such information. This relation between different sensory cues and the design of motor systems makes us aware of the need to consider the cross-modal context in which animals normally operate. One particularly interesting aspect is how a growing and learning organism adapts to the changes in perspective and the size and shape of its own sensory organs. The world seen by a newborn is not the world seen by a 20 year old and this again differs from people in their sixties. The visual field, for instance, has been reported to expand from a more ventral region to a more dorsal one, when human infants work their way up from a predominantly horizontal to an upright posture (Mohn and van Hof-van Duin 1991). We are far from understanding, however, the general patterns of perceptual adaptations that reflect changes in the visual environment. The chapter of Atkinson and Braddick addresses this issue by taking a developmental perspective of how the human visual system through growth, maturation and learning finally reaches the finely tuned visuo-motor coordination that we all rely on. Visual control of motor activity clearly has to be acquired when it comes to driving vehicles, and motion information is likely to play a crucial role in this task (Lee 1976). Land however provides a contrasting view to our mantra about the importance of motion information. He shows that during steering a car the control of gaze is not determined exclusively by the analysis of optic flow, but can be described in terms of simple geometric operating rules. Under natural operating conditions nervous systems thus exploit and combine sources of information that are useful and reliable.

(V) Up to this point we have treated motion information as if it was represented in some more or less instantaneous manner and independent of the immediate history of the sensory signals. Given the rapidly changing visual input in natural environments, this approach has two major limitations. It fails to take into account firstly, how adaptative properties of neurones change the instantaneously available information, and secondly with what precision the neural system can encode rapid changes. In biological systems, information is coded by neurones which have a limited bandwidth, in particular in the temporal domain. So what are the limits to the precision with which motion is represented by neurones? We had to discover in the preparation of this part of the book, that this question is by no means settled and therefore present highly controversial opinions – backed by elaborate experiments and sophisticated mathematical analysis – side by side. On the one hand, Warzecha and Egelhaaf argue in their contribution that in the motion sensitive neurones of the fly little or no information is carried by the exact timing of individual action potentials. Flies rely on temporal averages within a window of about 40 ms. On the other hand, de Ruyter van Steveninck et al. make a case that the same class of neurones exhibit extremely high precision in the timing of spikes. These opposing interpretations are derived from surprisingly similar experimental and theoretical approaches, in which, however, a crucial question remains unresolved, namely what exactly natural stimuli are. It is left to the reader to evaluate this pointed scientific discourse, in which cutting edge experimental and theoretical techniques are put to a biological reality check[1]. A comparative component is added by the paper of Barberini et al., demonstrating that MT neurones employ similar coding strategies to those of fly visual interneurones, which takes us back to the theme of analogous implementations in different branches of the animal kingdom that has been touched upon in the first part. A final word of caution is added by Maddess who notes that adaptation of neuronal activity, on a variety of time scales, can change the information content of spike trains considerably. We thus have to realize that the attempt to assess "motion vision in the real world" needs to include the consideration of signal processing dynamics and coding limitations, which despite substantial theoretical and experimental advances in recent times are far from being understood.

(VI) Under natural operating conditions, an animal is not only exposed to the neural and computational constraints of motion vision, but also to those imposed by its habitat and lifestyle. Two aspects of natural operating conditions have to be considered: (i) the spatial and temporal distribution of biologically relevant signals in a given visual habitat and ethological context, and (ii) the structure of locomotion which to a large extent determines the pattern of motion signals an observer experiences. What do we need to know about lifestyle and the dynamic structure of the environment, and how can we relate such knowledge to

---

[1] The note added "in proof" (i.e., after the end of the refereeing process) by the authors of the keynote paper in response to the companion article illustrates the intensity of the continuing discussion of the issues covered by this part of the book

the neural and computational constraints? In their keynote paper, Eckert and Zeil make an attempt to compile a preliminary inventory of the relevant questions that need to be asked and of the available facts about the motion signals that animals experience under natural conditions. The paper emphasizes the fact that the major part of image motion is generated by animals themselves, so that the analysis of behaviour will play a crucial role in understanding the conditions under which motion processing normally operates. A challenging theme emerging from this analysis is that of characteristic motion habitats and their statistical properties: Even if environments are statistically self-similar, different animals have to attend to different relevant events, and – depending on their way of locomotion – will experience different "motion environments". A more formal analysis of the spatiotemporal structure of image sequences is added by Dong in his companion chapter, who suggests characteristic coding strategies that resemble those proposed for achieving optimal representation of static images (Olshausen and Field 1996). Our last chapter provides neurophysiological evidence for motion processing being adapted to particular behavioural and ecological niches. By analysing large-field motion sensitive insect neurones, O'Carroll demonstrates how the structure of the sensory organs, the style of locomotion and the coding properties of neurones in insects reflect specific environments and lifestyles.

4. Conclusion

We are thus at a point where it becomes feasible, from a technical point of view, to describe and interpret biological visual systems in the context of their natural operating conditions. However, we have to acknowledge that we are just beginning to understand how neural, computational, and environmental constraints have driven the evolution of neuronal information processing mechanisms. It is only with a clear knowledge of these constraints that we can hope to develop smart machines which are as versatile, robust, competent, and flexible as the most humble animals evidently are.

References

Aloimonos Y (1993) Active perception (Computer Vision). Erlbaum, Hillsdale

Barlow HB (1981) Critical limiting factors in the design of the eye and visual cortex. Proc Roy Soc Lond 212: 1-34

Bialek W, Rieke F (1992) Reliability and information transmission in spiking neurons. Trends Neurosci 15: 428-434

Borst A, Egelhaaf M (1989) Principles of visual motion detection. Trends Neurosci 12: 297-306

Braddick OJ (1997) Local and global representations of velocity: transparency, opponency and global direction perception. Perception 26: 995-1010

Braddick OJ, Sleigh AC, eds (1983) Physical and biological processing of images. Springer Verlag, New York

Dusenbery DB (1992) Sensory ecology. How organisms acquire and respond to information. Freeman and Co, New York

Egelhaaf M, Borst A, Reichardt W (1989) Computational structure of a biological motion-detection system as revealed by local detector analysis in the fly's nervous system. J Opt Soc Am A 6: 1070-1087

Gibson JJ (1979) The ecological approach to visual perception. Lawrence Erlbaum Assoc, Hillsdale, New Jersey

Helmholtz HLF von (1864) Treatise on physiological optics. English translation from German (1962), Southall JPC (ed) Dover Publications, Dover

Hildreth E-C, Koch C (1987) The analysis of visual motion: From computational theory to neuronal mechanisms. Ann Rev Neurosci 10: 477-533

Koenderink JJ (1986) Optic flow. Vision Res 26: 161-180

Laughlin SB (1998) Observing design with compound eyes. In: Weibel ER, Taylor CR, Bolis L (eds) Principles of animal design. Cambridge University Press, Cambridge, pp 278-287

Lee DN (1976) A theory of visual control of braking based on information about time-to-collision. Perception 5: 437-459

Lythgoe JN (1979) The ecology of vision. Clarendon Press, Oxford

Marr D (1982) Vision: A computational investigation into the human representation and processing of visual information. Freeman and Co, San Francisco

Metzger W (1975) Gesetze des Sehens. W Kramer Verlag, Frankfurt/Main

Miles FA, Wallman J, eds (1993) Visual motion and its role in the stabilization of gaze. Elsevier, Amsterdam

Mohn G, van Hof-van Duin J (1991) Development of spatial vision. In: Regan D (ed) Vision and visual dysfunction 10. Spatial vision. Macmillan Press, London, pp 179-211

Nakayama K (1985) Biological image motion processing: A review. Vision Res 25: 625-660

O'Carroll D, Bidwell NJ, Laughlin SB, Warrant EJ (1996) Insect motion detectors matched to visual ecology. Nature 382: 63-66, 1996

Olshausen BA, Field DJ (1996) Emergence of simple-cell receptive field properties by learning a sparse code for natural images. Nature 381: 607-609

Reichardt W (1987) Evaluation of optical motion information by movement detectors. J Comp Physiol A 161: 533-547

Smith AT, Snowden RJ (1994) Visual detection of motion. Academic Press, London

Srinivasan MV, Venkatesh S, eds (1997) From living eyes to seeing machines. Oxford University Press, Oxford

Torre V, Poggio T (1978) A synaptic mechanism possibly underlying directional selectivity to motion. Proc Roy Soc Lond B 202: 409-416

Ullman S (1979) The interpretation of visual motion. MIT Press, Cambridge

Wade NJ (1998) A Natural history of vision. MIT Press, Cambridge MA

Watanabe T (1998) High-level motion processing. Computational, neurobiological, and psychophysical perspectives. MIT Press, Cambridge

Part I

Early Motion Vision

Direction-Selective Ganglion Cells in the Retina
David I. Vaney, Shigang He, W. Rowland Taylor and William R. Levick

Identification of Mechanisms Underlying Motion Detection in Mammals
Michael Ibbotson

Pathways in Dipteran Insects for Early Visual Motion Processing
John K. Douglass and Nicholas J. Strausfeld

Direction-Selective Ganglion Cells in the Retina

David I. Vaney[1], Shigang He[1], W. Rowland Taylor[2] and William R. Levick[3]

[1]Vision, Touch and Hearing Research Centre, The University of Queensland, Brisbane, Australia; [2]John Curtin School of Medical Research, [3]Division of Psychology, The Australian National University, Canberra, Australia

Contents

1. Abstract

The first stages in the neuronal processing of image motion take place within the retina. Some types of ganglion cells, which are the output neurones of the retina, are strongly stimulated by image movement in one direction, but are inhibited by movement in the opposite direction. Such direction selectivity represents an early level of complex visual processing which has been intensively studied from morphological, physiological, pharmacological and theoretical perspectives. Although this computation is performed within two or three synapses of the sensory input, the cellular locus and the synaptic mechanisms of direction selectivity have yet to be elucidated.

The classic study by Barlow and Levick (1965) characterized the receptive-field properties of direction-selective (DS) ganglion cells in the rabbit retina and established that there are both inhibitory and facilitatory mechanisms underlying the direction selectivity. In each part ("subunit") of the receptive field, apparent-motion experiments indicated that a spatially asymmetric, delayed or long-lasting inhibition "vetoes" excitation for movement in one direction (the "null" direction), but not for movement in the opposite direction (the "preferred" direction). In addition, facilitation of excitatory inputs occurs for movement in the preferred direction.

Subsequently, pharmacological experiments indicated that a GABAergic input from lateral association neurones (amacrine cells) may inhibit an excitatory cholinergic input from other amacrine cells and/or a glutamatergic input from second-order interneurones (bipolar cells). An added complication is that the cholinergic amacrine cells also synthesize and contain GABA, raising the possibility that these "starburst" cells mediate both the excitation and inhibition underlying direction selectivity (Vaney et al. 1989).

This review focuses on recent studies that shed light on the cellular mechanisms that underlie direction selectivity in retinal ganglion cells. He and Masland (1997) have provided compelling evidence that the cholinergic amacrine cells mediate the facilitation elicited by motion in the preferred direction; however, it now appears that the cholinergic facilitation is non-directional, although the null-direction facilitation is normally masked by the directional inhibitory mechanism. The null-direction inhibition may act presynaptically on the excitatory input to the DS ganglion cell; in this case, the release of transmitter from the excitatory neurone would itself be direction selective, at least locally. Alternatively, the null-direction inhibition may act postsynaptically on the ganglion cell dendrites, probably through the non-linear mechanism of shunting inhibition.

In the rabbit retina, there are two distinct types of DS ganglion cells which respond with either On-Off or On responses to flashed illumination; the two types also differ in their specificity for stimulus size and speed and their central projections. The On-Off DS cells comprise four physiological subtypes, whose preferred directions are aligned with the horizontal and vertical ocular axes, whereas the On DS cells comprise three physiological subtypes, whose preferred directions correspond to rotation about the best response axes of the three semicircular canals in the inner ear. The On DS cells, which project to the accessory optic system, appear to respond to global slippage of the retinal image, thus providing a signal that drives the optokinetic reflex. The On-Off DS cells, which are about ten times more numerous than the On DS cells, appear to signal local motion and they may play a key role in the representation of dynamic visual space or the detection of moving objects in the environment.

2. Functional organization

Visual information undergoes a sophisticated coding process in the retina, culminating in the diverse output of many types of ganglion cells to the brain. There is massive convergence of retinal interneurones onto the ganglion cells and thus the optic nerve is effectively the information bottleneck in the visual system. The operations performed by ganglion cells, as reflected in their receptive-field properties, represent the outcome of retinal strategies for compressing the representation of the visual scene (Levick and Thibos 1983). In the mammalian retina, there are about 20 distinct types of ganglion cells, which respond preferentially to different features of the visual image, such as local contrast, colour, and the speed and direction of image movement. It appears that each ganglion cell population achieves complete and efficient coverage of the retina (Wässle et al. 1981; DeVries and Baylor 1997) and that the information provided by these 20 congruent maps is sent to the brain in parallel (Rodieck 1998).

Most types of retinal ganglion cells respond to the temporal modulation of luminance within their receptive fields and, consequently, they may be activated by images moving across the receptive field, resulting either from an object moving in visual space or from self-motion of the animal. In particular, ganglion cells that respond transiently to step changes in illumination are likely to be involved in motion processing. A subset of these cells is differentially responsive to the direction of image motion and this appears to be the essence of their function (see Section 4). Although such direction-selective (DS) ganglion cells have been found in the retina of all vertebrate classes (references in Wyatt and Daw 1975), they have been encountered only rarely in the primate retina (Schiller and Malpeli 1977; DeMonasterio 1978), where their presence is probably masked by the great excess of midget and parasol ganglion cells (Rodieck 1988, 1998).

The majority of studies on DS ganglion cells have used the rabbit retina as a model system, reflecting both the wide availability of this laboratory animal and the importance of the classic study by Barlow and Levick (1965). Some of the key findings in the rabbit retina have been confirmed in the turtle retina (Marchiafava 1979; Ariel and Adolph 1985; Rosenberg and Ariel 1991; Kittila and Granda 1994; Smith et al. 1996; Kogo et al. 1998), indicating that similar mechanisms may underlie the generation of direction selectivity in diverse vertebrate retinas. In the rabbit retina, there are two distinct types of DS ganglion cells (Barlow et al. 1964). The commonly encountered On-Off DS cells are excited by objects that are lighter or darker than the background and they respond over a wide range of stimulus velocities; the rarer On DS cells are excited by objects that are lighter than the background and they respond optimally to slow movements (Oyster 1968; Wyatt and Daw 1975). Throughout this chapter, references to DS ganglion cells should be taken to mean the On-Off DS cells of the rabbit retina, unless otherwise specified. The numerous physiological and morphological studies on vertebrate DS ganglion cells have been most recently reviewed by Amthor and Grzywacz (1993a), who placed special emphasis on the spatiotemporal characteristics of the excitatory and inhibitory inputs to the On-Off DS cells.

Although the actual neuronal circuitry that underlies the generation of direction selectivity in the retina has yet to be elucidated, the diverse models that have been proposed over the last 35 years provide guideposts for future experiments (Barlow and Levick 1965; Torre and Poggio 1978; Ariel and Daw 1982; Koch et al. 1982; Grzywacz and Amthor 1989; Vaney et al. 1989; Oyster 1990; Vaney 1990; Borg-Graham and Grzywacz 1992; Grzywacz et al. 1997; Kittila and Massey 1997). These models are judged primarily by their ability to account for the detailed functional properties of the DS ganglion cells, but this is only one of the requirements. Morphological and biophysical constraints also pose hurdles for candidate mechanisms. For example, it would not be appropriate to require a higher density of a particular neuronal type than is known to exist. Nor would it be sound to postulate highly localised synaptic interactions on dendritic segments where the electrotonic properties indicate more extensive interactions. Finally, the developmental requirements need to be kept in mind: it should be possible to achieve the appropriate specificity in the neuronal connections by such mechanisms as Hebbian-type synaptic modification or the selective expression of marker molecules.

In this Section, we briefly describe the fundamental receptive-field properties of the DS ganglion cells, but defer until Section 5 discussion of recent physiological and neuropharmacological studies that examine the cellular mechanisms of direction selectivity.

2.1 Classical receptive field

When the receptive field of an On-Off DS cell is mapped with a small flashing spot, transient excitatory responses are usually elicited at both the Onset and Offset of illumination (Barlow et al. 1964). Some regions near the edge of the receptive field may respond to only one phase of illumination (Barlow and Levick 1965) but the responses to a flashing spot do not reveal any receptive-field substructure that can be correlated with the directional responses produced by moving stimuli. Although quantitative one-dimensional mapping of the responses across the receptive field of the On-Off DS cells produced spatial profiles that are both flat topped and steep edged (Yang and Masland 1994), suggesting a uniform weighting of local excitatory input throughout the dendritic field (Kier et al. 1995), more detailed two-dimensional mapping revealed some troughs in responsiveness even in the middle of the receptive field (He 1994), presumably corresponding to the indentations apparent in the irregular dendritic fields of the On-Off DS cells (see Section 3.1).

The receptive fields mapped with moving or flashing stimuli follow the shape and size of the dendritic field. Yang and Masland (1994) demonstrated this directly by mapping the receptive fields of the On-Off DS cells in an isolated preparation of the rabbit retina and then injecting the recorded cells with Lucifer yellow to reveal the dendritic morphology. For the majority of cells, the receptive-field border lay just beyond the reach of the most distal dendrites, suggesting that the classical receptive field of the ganglion cell represents the summed excitatory input from presynaptic neurones with small receptive fields: they are presumably bipolar cells that terminate in the same strata of the inner plexiform layer as the dendrites of the DS ganglion cells (see Section 5.2). The bipolar cells touching the distal dendrites of the ganglion cell would receive some input from photoreceptors located beyond the edge of the dendritic field, but this increases the receptive-field width by only 6%.

Yang and Masland (1994) reported that the receptive field of many On-Off DS cells was shifted relative to the dendritic field and this shift was always towards the side that is first encountered by a stimulus moving in the preferred direction (the "preferred side"). The shift was ≥10% of the field width for 42% of the cells and ≥20% of the field width for 6% of the cells. The receptive-field shift was not associated with a particular preferred direction, but was exhibited by examples from each of the four subtypes of On-Off DS cells. In cells with displaced fields, the sizes of the dendritic field and receptive field remained closely matched, and thus this phenomenon does not represent an asymmetric enlargement of the receptive field, as predicted by some models of direction selectivity (Vaney 1990). The possible origin of the receptive-field shift is discussed in Section 5.1.

2.2 Direction selectivity

The characteristic behaviour of DS ganglion cells appears when moving stimuli are used. There is a path across the receptive field for which motion in one direction (the preferred direction) elicits the strongest response while motion in the opposite direction (the null direction) yields essentially no discharge of impulses. This asymmetry for oppositely directed motion is robust: it persists despite reversal of stimulus contrast and despite changes in the size, shape, velocity and trajectory of the moving stimulus (Barlow and Levick 1965).

Oyster and Barlow (1967) demonstrated that there are four subtypes of On-Off DS ganglion cells in the rabbit retina, each with a different preferred direction that roughly corresponds to object movement in one of the four cardinal ocular directions (upwards/superior, backwards/posterior, downwards/inferior and forwards/anterior, for the laterally pointing rabbit eye). By contrast, there are only three subtypes of On DS cells, with preferred directions that correspond to either anteriorly directed object movement, upwards movement with a posterior component, or downwards movement with a posterior component. These data relate to receptive fields that are centrally located (0-40° elevation in the visual field).

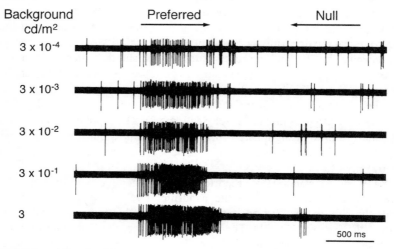

Fig. 1 Responses of an On-Off DS ganglion cell to a light spot moved through the receptive field in the preferred direction and then back in the null direction, over a 10,000-fold range of background illumination; in each case, the stimulating spot was 10× brighter than the background; an artificial pupil of 3 mm diameter was used. The DS responses are maintained even under scotopic conditions (3×10^{-4} cd/m^2). (WR Levick unpublished)

The On-Off DS cells retain their direction selectivity over large changes in the mean illumination of the visual field (Fig. 1), indicating that the signals from the rods and the cones feed into a common neuronal mechanism. The cone bipolar cells act as both second-order neurones in the cone-signal pathway and as third- or fourth-order neurones in the rod-signal pathway: the rod signal is thought to pass from rods → rod bipolar cells → AII amacrine cells → cone bipolar cells under scotopic conditions, and from rods → cones → cone bipolar cells under mesopic conditions (Nelson 1977; Kolb and Nelson 1984; Smith et al. 1986). However, DeVries and Baylor (1995) found intriguing evidence in the rabbit retina that the latter pathway may also be effective under scotopic conditions: at stimulus intensities that were two log units lower than required to produce a pure rod response in the ganglion cells, the responses of the On-Off DS cells were resistant to blockade of the rod → rod bipolar cell synapse. It has yet to be tested whether the DS ganglion cells are responsive under low scotopic conditions, when the rod signal appears to be channelled only through the rod bipolar pathway.

The concept of the DS subunit has been central to understanding the mechanism of direction selectivity since it was first introduced by Barlow and Levick (1965). They reported that the smallest movements that produced directional responses varied from 0.1-0.4° in different On-Off DS cells and thus concluded that "the complete mechanism for direction selectivity is contained within a subunit of the receptive field extending not much more than 0.25° in the preferred-null axis. Since the result does not depend critically upon the position of the [stimulus] within the receptive field, it looks...as if the sequence-discriminating mechanism must be reduplicated perhaps a dozen or more times to cover the whole receptive field." The subsequent quest for the cellular mechanism of the DS subunit has been the holy grail of research on direction selectivity in the retina (see Section 6).

2.3 Non-directional zone

When mapping the responses of the On-Off DS cells to a moving target in different parts of the receptive field, Barlow and Levick (1965) observed that there is an "inhibition-free" zone located on the preferred side of the roughly circular field. When this zone was stimulated with a moving 0.5° spot, displacements of about 1° or 25% of the receptive-field diameter produced equivalent responses in the null and preferred directions (Barlow and Levick 1965).

The non-directional zone was analysed quantitatively by He (1994) using visual stimuli confined by a slit aperture, whose long side was twice the diameter of the receptive field and whose short side covered one tenth of the receptive field along the preferred-null axis. The response profile of the receptive field was first mapped by flashing the slit On and Off; the resulting flat-topped profile showed a similar decline in responsiveness on the null and preferred sides. These responses were then compared with those to moving stimuli, which were produced by drifting a grating behind the slit aperture (Fig. 2). Local movement in the preferred direction elicited robust responses over much of the classic receptive field, although both the null and preferred edges of the receptive field were comparatively unresponsive to preferred-direction motion. The response profile to preferred-direction motion was much more dome-shaped than the response profile to flashing stimuli and, in the centre of the receptive field, the responses to moving stimuli greatly exceeded the responses to flashing stimuli. Local movement in the null direction elicited no response from the middle and the null side of the receptive field, but elicited small stable responses from the preferred side of the receptive field, in agreement with the original findings of Barlow and Levick (1965).

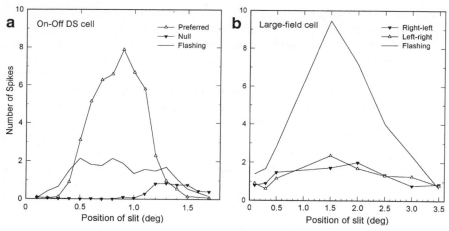

Fig. 2 Comparison of the responses of two types of rabbit retinal ganglion cells to a flashing slit and to an edge moving within the slit, which covered 0.1-0.2 of the width of the receptive field. **a** The On-Off DS cell shows a fairly uniform sensitivity to flashing stimuli over most of the receptive field and these responses are greatly facilitated in the central receptive field by preferred-direction movement. Null-direction movement within the narrow slit abolishes the responses over most of the receptive field whereas, on the preferred side of the receptive field, there is a non-directional zone in which the null-direction responses may approach or even exceed the preferred-direction responses. **b** Although the large-field ganglion cell has classically been regarded as a non-directional movement detector, it actually gives much greater responses to the flashing stimuli than to the moving stimuli. (After He 1994)

It is not known why the excitation produced by null-direction motion in the non-directional zone is not subject to the null-direction inhibition present in other parts of the receptive field and, therefore, the implications of the zone for the mechanism of direction selectivity are still unclear (see Section 5.2).

3. Neuronal architecture

The wealth of physiological investigations on the receptive-field properties of DS ganglion cells has been complemented by diverse morphological investigations on the dendritic architecture of these cells and their presynaptic neurones, with the result that the functional studies are powerfully informed by the structural studies, and vice versa. There was a gap of 21 years between the first physiology paper (Barlow and Hill 1963) and the first morphology paper (Amthor et al. 1984), and the two approaches have been significantly integrated in only a few recent studies (Yang and Masland 1992, 1994; Amthor and Oyster 1995; He and Masland 1997).

3.1 On-Off DS ganglion cells

Dendritic morphology. Intracellular dye injection into physiologically identified On-Off DS cells in rabbit eyecup preparations revealed that these cells have a characteristic bistratified dendritic morphology (Amthor et al. 1984, 1989b; Oyster et al. 1993; Yang and Masland 1994; Amthor and Oyster 1995). Although there are several types of bistratified ganglion cells in the rabbit retina (Amthor et al. 1989a, 1989b; Vaney 1994a), the type 1 bistratified (BiS1) morphology of the On-Off DS cells is particularly distinctive, enabling these neurones to be identified from their morphology alone, both in the adult retina (Famiglietti 1987, 1992b, 1992c; Vaney et al. 1989; Vaney 1994b) and in the developing retina (Wong 1990; Vaney 1994a).

The On-Off DS cells stratify narrowly at 20% and 70% depth of the inner plexiform layer, where 0% depth and 100% depth correspond to the outer and inner borders, respectively. Thus the outer dendritic stratum in sublamina *a* of the inner plexiform layer receives input from Off-centre interneurones (depolarized by decreasing illumination), whereas the inner dendritic stratum in sublamina *b* receives input from On-centre interneurones (depolarized by increasing illumination; Bloomfield and Miller 1986). Moreover, the On-Off DS cells costratify precisely with the cholinergic (starburst) amacrine cells (Famiglietti 1987, 1992c; Vaney et al. 1989), which provide direct excitatory drive to these ganglion cells (Masland and Ames 1976; Ariel and Daw 1982). The Off-centre starburst cells branching in sublamina *a* have their somata in the inner nuclear layer, whereas the On-centre starburst cells branching in sublamina *b* have their somata in the ganglion cell layer (Famiglietti 1983; Tauchi and Masland 1984; Vaney 1984).

On-Off DS cells typically give rise to three or four primary dendrites which may branch to the 10th order or more. The dendrites in sublamina *a* can arise from dendrites of any order in sublamina *b*, but only rarely do dendrites in sublamina *b* arise from sublamina *a*. The branching systems arising from each primary dendrite tile the dendritic field with minimal overlap in each sublamina. Dendrites of all orders give rise to thin terminal branches, creating a space-filling lattice (Oyster et al. 1993; Kier et al. 1995; Panico and Sterling 1995) whose mesh-like appearance is enhanced by the retroflexive dendrites, which sometimes appear to form "closed loops" within the dendritic tree.

Careful examination of the dendritic morphology of the On-Off DS cells reveals no asymmetry that can be correlated with the preferred direction of the cell (Amthor et al. 1984) and, therefore, the anisotropic responses would seem to arise from asymmetries in the synaptic inputs. Nevertheless, some distinctive features of the dendritic morphology may underlie key aspects of the receptive-field organization. For example, the space-filling dendrites would provide a homogeneous substrate for locally generating direction selectivity throughout the receptive field. Moreover, the short terminal dendrites provide sites away from the dendritic trunks where excitatory and inhibitory inputs may interact locally, as required by postsynaptic models of direction selectivity (see Section 5.2).

Territorial dendritic fields. The sublamina *a* dendritic field may differ greatly in size, shape and relative position from the sublamina *b* dendritic field and, thus, the two fields are not coextensive (Oyster et al. 1993; Vaney 1994b). In extreme cases, the dendrites are largely confined to sublamina *a* and, interestingly, both Barlow et al. (1964) and Oyster (1968) reported finding Off DS cells. These essentially monostratified cells do not comprise a separate type of ganglion cell because they form part of a regular array of bistratified cells (see below). The morphological observations support physiological evidence (Amthor and Grzywacz 1993b; Cohen and Miller 1995; Kittila and Massey 1995) that the directional responses do not require interaction between the On and Off inputs and, therefore, each sublamina on its own must contain the neuronal circuitry for generating direction selectivity.

The four directional subtypes of On-Off DS cells appear to tile the retina in a similar manner. A synthesis of three complementary studies (Vaney 1994b; Amthor and Oyster 1995; DeVries and Baylor 1997) indicates that each On-Off DS cell is surrounded by a ring of 4-6 cells with the same preferred direction, whose somata are generally located beyond the dendritic field of the central cell. Such an array of cells provides complete coverage of the retina, with minimal overlap of the dendritic fields in each sublamina. This mirrors, on a larger scale, the territorial organization of branching systems within the dendritic tree of individual cells. Moreover, the dendrites at the edge of the dendritic field often form tip-to-shaft or tip-to-tip contacts with dendrites from neighbouring cells of the same subtype, thus appearing to form dendritic loops that resemble those found within the dendritic tree. Consequently, the combined dendrites of the interlocking cells of each subtype are distributed regularly and economically across the retina (Vaney 1994b; Panico and Sterling 1995). Thus the sensitivity profile of the summed receptive fields of each subtype may be rather uniform across the retina (DeVries and Baylor 1997). For some On-Off DS cells, the dendritic tree in sublamina *a* is not unbroken but forms a major arborization and one or more minor arborizations (Famiglietti 1992b; Oyster et al. 1993); these dendritic "islands" occupy holes in the dendritic trees of neighbouring DS cells of the same subtype (Vaney 1994b).

There is little variation in the dendritic-field size of On-Off DS cells at each retinal eccentricity, suggesting that the different subtypes have similar spatial distributions (Vaney 1994b; Yang and Masland 1994). The dendritic-field area in each sublamina ranges from ~8,000 μm^2 (100 μm diameter) in the visual streak to ~130,000 μm^2 (400 μm diameter) in the far periphery. The combined area of the two dendritic fields is fairly constant, so that an extra-large field in one sublamina is compensated by an extra-small field in the other sublamina.

It has been possible to show directly that there is a reciprocal trade-off between increasing dendritic-field area and decreasing cell density, by taking advantage of the unexpected finding that some On-Off DS cells show homologous tracer coupling when injected intracellularly with Neurobiotin, a gap-junction permeant tracer (Vaney 1991, 1994b). When two or three overlapping On-Off DS cells, presumably comprising subtypes with different preferred directions, are injected with Neurobiotin, only one cell of the group shows tracer coupling. This suggests that the tracer-coupled subtype corresponds to one of the physiological subtypes, but its preferred direction has yet to be established. The tracer coupling reveals the local somatic array of 5-20 cells of the coupled subtype, whose density ranges from 8 cells/mm^2 in the far periphery to 145 cells/mm^2 in the peak visual streak, thus accounting for 3% of the ganglion cells. If each of the non-coupled subtypes is present at the same density as the coupled subtype, there would be a total of 40,000 On-Off DS cells in the rabbit retina.

Dendritic fasciculation. Although the somatic arrays of the four subtypes of On-Off DS cells appear to be spatially independent, their dendritic trees are not randomly superimposed but commonly run together in loose fascicles of 2-4 dendrites (Fig. 3a); this was demonstrated by injecting dye into overlapping On-Off DS cells with closely spaced somata (Vaney 1994b), comprising subtypes with different preferred directions (Amthor and Oyster 1995). In each sublamina, the On-Off DS cells follow the dendritic meshwork of the starburst amacrine cells (Vaney et al. 1989; Vaney and Pow 2000), whose widely overlapping dendrites are also strongly fasciculated, forming cords of dendrites surrounding lacunae of 10-50 µm diameter (Tauchi and Masland 1985; Brandon 1987; Famiglietti and Tumosa 1987). Each ganglion cell dendrite is invariably associated with a star-burst fascicle and, conversely, there are few starburst fascicles that do not contain at least one dendrite from an On-Off DS cell (Fig. 3b).

The striking dendritic fasciculation suggests that a small neuronal assem-blage, comprising a bundle of ganglion cell dendrites running a "gauntlet" of presynaptic amacrine boutons and bipolar terminals (Brandon 1987; Famiglietti 1991), may contain all the neuronal wiring that is necessary for extracting image motion in four orthogonal directions. The fasciculation would enable the starburst amacrine cells to contact several subtypes of On-Off DS cells simultaneously, thus providing an efficient substrate for isotropic cholinergic input to the ganglion cells, as proposed by He and Masland (1997). It is not known how the dendritic fasciculation develops, but the starburst amacrine plexus may provide the initial scaffold, because the dendrites of On-Off DS cells commonly run along the sides of the starburst dendritic fascicles rather than down the middle (Vaney and Pow 2000; cf. Famiglietti 1991).

3.2 On DS ganglion cells

The dendritic morphology of the On DS ganglion cells in the rabbit retina has been characterized using two complementary approaches. First, intracellular dye injection into physiologically identified On DS cells revealed that they have large monostratified dendritic trees in sublamina *b* of the inner plexiform layer (Amthor et al. 1989b; He and Masland 1998). Second, intracellular dye injection into somata that were retrogradely labelled from either the medial terminal nucleus (MTN; Buhl and Peichl 1986) or the nucleus of the optic tract (NOT; Pu and Amthor 1990) revealed ganglion cells with the same morphology.

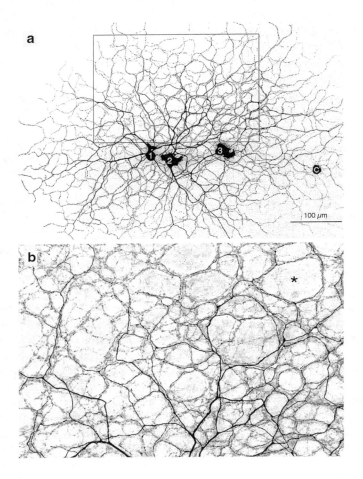

Fig. 3 Confocal micrographs showing the dendritic fasciculation of DS ganglion cells and starburst amacrine cells. **a** Intracellular Neurobiotin injection of three overlapping type 1 bistratified ganglion cells, presumably comprising On-Off DS cells with different preferred directions, reveals that their dendrites cofasciculate in both the On and Off sublaminae of the inner plexiform layer; cell #1 was tracer coupled to surrounding cells of the same subtype, one of which (C) is included in the field. **b** Higher power micrograph of the On sublamina, showing the dendrites of the Neurobiotin-filled ganglion cells (black) and the co-stratified population of On starburst amacrine cells (grey). The fasciculated DS dendrites follow the cords of starburst dendrites, which surround large lacunae that are relatively free of dendrites (asterisk). (after Vaney and Pow 2000)

In several respects, the dendritic trees of On DS cells appear to be scaled-up versions of the sublamina *b* arborization of On-Off DS cells, with both types of ganglion cells giving rise to terminal dendrites of relatively uniform length throughout the dendritic tree. At each retinal eccentricity, the dendritic field of the On DS cells is 2-3 times wider than that of adjacent On-Off DS cells, ranging in size from 300 μm diameter in the peak visual streak to over 800 μm diameter in the inferior periphery (Buhl and Peichl 1986; Pu and Amthor 1990; Famiglietti 1992b). Although the two cell types contain similar numbers of dendritic branches in sublamina *b*, the On DS cells differ from the On-Off DS cells in that their arborization appears less regular, perhaps reflecting a more patchy distribution of the terminal dendrites. A minority of On DS cells give rise to a restricted arborization in sublamina *a*, accounting for only 5-15% of the cell's total dendritic length (Buhl and Peichl 1986; Famiglietti 1992c; He and Masland 1998).

Widely overlapping On DS cells with closely spaced somata, which presumably comprise cells with different preferred directions, show pronounced fasciculation of their dendrites (He and Masland 1998), comparable to that shown by different subtypes of On-Off DS cells. By contrast, On DS cells with somata separated by about the width of a dendritic field, which are more likely to comprise cells with the same preferred direction, show economical coverage of the retina with few dendritic crossings (Buhl and Peichl 1986). The territorial organization of the On DS cells thus appears to be similar to that of the On-Off DS cells, including the presence of tip-to-tip dendritic contacts which make it difficult to assign some dendrites to their cell of origin. The On DS cells, which stratify around 75% depth of the inner plexiform layer, lie adjacent to the sublamina *b* arborization of the On-Off DS cells and the displaced starburst amacrine cells, which stratify around 70% depth of the inner plexiform layer (Famiglietti 1992c). This study also showed that the dendrites of a Golgi-stained On DS cell generally did not follow the dendrites of an overlapping On-Off DS cell. However it is possible that the On DS cells cofasciculate with the adjacent starburst amacrine cells, because only part of the meshwork of starburst dendrites is followed by the dendrites of a single On-Off DS cell.

Simpson and colleagues (1979) proposed that the On DS cells provide the primary retinal input to the terminal nuclei of the accessory optic system (AOS; see Section 4.1) and this hypothesis has been confirmed directly by electrophysiological recordings from MTN-projecting ganglion cells (Brandon and Criswell 1997). In the rabbit retina, about 2000 ganglion cells can be retrogradely labelled from the MTN (Giolli 1961; Oyster et al. 1980) and they presumably comprise the two subtypes of On DS cells with near-vertical preferred directions. There appear to be no MTN-projecting cells within 4-5 mm of the superior and inferior edges of the rabbit retina and, correspondingly, almost all On DS cells encountered electrophysiologically have been located within 40° of the visual streak (Oyster 1968; Vaney et al. 1981a). This contrasts with the ubiquitous distribution of the On-Off DS ganglion cells. The On DS cells with anterior preferred directions probably project to the dorsal terminal nucleus (DTN) of the AOS

(Soodak and Simpson 1988) and perhaps also to the NOT (Collewijn 1975; Pu and Amthor 1990), which is not clearly delineated from the adjacent DTN. The MTN-projecting cells do not project to the superior colliculus (SC), unlike 99% of the ganglion cells in the rabbit retina, but it is not known whether this is the case for the DTN-projecting cells (Vaney et al. 1981b; Buhl and Peichl 1986).

Although the On DS cells probably comprise less than 1% of all ganglion cells in the rabbit retina, they are encountered experimentally at frequencies of 5-7% in central retina, reflecting the fact that they have larger somata than 90% of the ganglion cells in the visual streak (Vaney et al. 1981b). It should be noted that if the population of On DS cells in the rabbit retina had relatively small somata and slowly conducting axons, like their counterparts in the cat retina (Cleland and Levick 1974; Farmer and Rodieck 1982), they would be encountered very infrequently indeed.

4. Functions of DS ganglion cells

In assessing what the rabbit's eye tells the rabbit's brain, we need to be cautious in assuming that the function of a ganglion cell is inherent in its "trigger feature" (Barlow et al. 1964), particularly given mounting evidence that a local ensemble of ganglion cells may code visual information that is not apparent in the responses of single cells (Meister et al. 1995). For example, the brisk transient nature of the alpha ganglion cells makes them particularly responsive to visual motion, but these cells may also serve a variety of other functions including coarse form perception and the rapid mobilization of most primary visual areas (Ikeda and Wright 1972; Levick 1996). Nevertheless, a considered analysis of direction selectivity, which is the archetypal trigger feature, leads to the inescapable conclusion that the essential function of DS ganglion cells is to code visual motion. At the single cell level, direction selectivity is a remarkably stereotyped and robust phenomenon (Barlow and Levick 1965), requiring the evolution of specific neuronal circuits that would only provide selective advantage if used to code moving stimuli.

4.1 On DS cells signal global motion

The On DS ganglion cells are single-minded in their purpose. Whereas all other ganglion cells in the rabbit retina project to the superior colliculus, with most sending a collateral projection to the lateral geniculate nucleus, the On DS cells appear to do neither (Oyster et al. 1971; Buhl and Peichl 1986; cf. Vaney et al. 1981a). Rather, the ON DS cells may provide the sole retinal input to the three terminal nuclei of the accessory optic system (AOS): the medial terminal nucleus (MTN), the lateral terminal nucleus (LTN), and the dorsal terminal nucleus (DTN). Most of the neurones in the AOS are direction selective, with preferred directions that broadly match the three subtypes of On DS ganglion cells (Soodak and Simpson 1988). The MTN and LTN units responded best to near vertical movements, either upwards with a posterior component or downwards with a posterior component. Although the majority of MTN units were excited by upwards movement and the majority of LTN units were excited by downwards movement, there is morphological evidence that the MTN receives a direct retinal projection from at least two subtypes of On DS ganglion cells (Buhl and Peichl 1986). The neurones in the DTN and the adjacent nucleus of the optic tract (NOT) have preferred directions towards the anterior, and are thus excited by temporal to nasal movement (Collewijn 1975; Soodak and Simpson 1988). Most of the DS units in the AOS responded optimally to slow stimulus velocities of 0.1-1°/s, comparable to the velocity tuning of the On DS ganglion cells (Oyster et al. 1972).

Despite the direction and speed similarities, the receptive-field properties of the DS units in the AOS differ from those of the On DS ganglion cells in several important respects (Soodak and Simpson 1988). First, the AOS units in the rabbit were maximally stimulated by textured patterns 30° square or larger, corresponding to 25 mm^2 on the retina; this would cover the receptive fields of more than 100 MTN-projecting ganglion cells, suggesting a massive convergence of the On DS ganglion cells. Thus the AOS units were readily stimulated by movement of the whole visual field but were unresponsive to localized movement of targets smaller than 5° in diameter. Second, in the MTN and LTN, the preferred and null directions were not collinear, in that both directions commonly showed a posterior component. This suggests that these cells receive both excitation from one subtype of vertically tuned On DS cell (upwards or downwards) and inhibition from the other subtype, perhaps mediated by inhibitory connections between the MTN and LTN (Soodak and Simpson 1988; see also Kogo et al. 1998).

Under natural conditions, the movement of a large portion of the image of the visual field would usually reflect self-motion, resulting from movement of the eye or the head. Rotational head movement activates the semicircular canals, which trigger vestibulo-ocular reflexes that stabilize the directions in which the eyes are pointing. However, these compensatory eye movements are imperfect for several reasons, producing residual slippage of the global image on the retina, which would readily activate the AOS units. With the head still, the AOS units may also be responsive to the small involuntary eye movements (about 0.25° amplitude) that persist when the gaze is fixed, again providing a retinal slip signal. The visual and vestibular signals converge in the vestibulo-cerebellum, leading to common motor pathways for stabilizing eye position relative to the visual scene. The preferred directions of the three subtypes of On DS cells appear to correspond geometrically to rotation about the best response axes of the three semicircular canals, which would allow signals of rotational head motion from two different sensory modalities to be combined in a common coordinate system (Simpson 1984; Simpson et al. 1988).

Under artificial conditions, the contribution of the visual input to image stabilization can be studied in isolation by keeping the head fixed and moving the visual field, thus producing a relatively pure optokinetic reflex that is uncontaminated by vestibulo-ocular and vestibulocollic reflexes (Wallman 1993). In the rabbit, Collewijn (1969) measured the velocity of the slow phase of optokinetic nystagmus as a function of visual field velocity under open loop conditions (with the stimulated eye immobilized and the measurements made on the yoked movements of the unstimulated eye). The eye movement velocity was maximal at a visual field velocity of 0.4°/s, dropping to 10% of the maximum response at 0.01°/s and 3.5°/s: this response profile closely matched that of the On DS ganglion cells to a similar whole-field stimulus, providing support for the hypothesis that these cells are primarily responsible for driving the optokinetic reflex (Oyster et al. 1972). Visual field velocities up to 20°/s continued to elicit optokinetic nystagmus, perhaps driven by the On-Off DS cells (see Section 4.2), but the gain of the response (eye velocity/stimulus velocity) was very low under these open loop conditions.

Many of the properties of the On DS ganglion cells can be interpreted in the framework that these cells signal slippage of the retinal image, resulting either from small involuntary eye movements or from residual deficits in the vestibulo-ocular reflexes that compensate for head movements. The retinal image slip will be quite small even for rapid head movements, and thus a system that responds best to slow velocities is optimal for this purpose. The low density and large receptive fields of the On DS cells do not limit the performance of the system because the signals from many cells are pooled in the AOS to provide information about movement of the whole visual field. Global rotation about any of the three axes defined by the semicircular canals should be signalled reliably by the On DS cells in the central retina and, therefore, any input from peripheral retina may be redundant. This may account for the apparent absence of On DS cells in the superior and inferior retina, but the deficit may also reflect the fact that these regions offer reduced scope for signalling horizontal image motion, corresponding to rotation about the vertical axis of the lateral semicircular canals. The sensitivity bias in favour of brightening (On) stimuli correlates neatly with a similar bias in the rabbit's optokinetic reflex when luminance changes of different polarity are presented (Rademaker and Ter Braak 1948).

At first glance, the three sets of preferred directions of the On DS cells appear to preclude the possibility of an organization in terms of antagonistic pairings, as in the case of the On-Off DS cells (see Section 4.2). However, the symmetry of an antagonistic arrangement is recovered by linking the cells from both eyes and considering the responses in terms of rotations rather than translations of the visual field. Thus, the On DS cells of one eye preferring the forward direction of image motion could be matched to those with similar sensitivity in the fellow eye, to make an antagonistic pair with respect to rotations of the visual field about a vertical axis. The same arrangement would also apply to the two vertical sets of preferred directions, and a similar pattern holds for the "functional polarizations" of primary vestibular afferents (Goldberg and Fernandez 1971).

4.2 On-Off DS cells signal local motion

Although the On DS cells appear to play the leading role in signalling global rotation of the visual field, the On-Off DS cells may also contribute to optokinetic reflexes. In particular, the On-Off DS cells appear to project directly to the NOT (Pu and Amthor 1990), which is reciprocally connected with the AOS nuclei (Simpson 1984). In the rabbit, the units in the NOT resembled those in the adjacent DTN, in that they had very large receptive fields (up to 40° × 150°), which responded selectively to movement in the anterior direction; the NOT units responded to a wider range of velocities (0.01-20°/s), perhaps reflecting the additional input from the On-Off DS cells. Most NOT units covered much of the velocity range of the optokinetic reflex under open loop conditions (Collewijn 1969) and electrical stimulation of the NOT elicited vigorous horizontal nystagmus (Collewijn 1969, 1975). It is not known whether there is a particular pretectal pathway mediating vertical optokinetic nystagmus, which might involve the vertically tuned On-Off DS cells.

Notwithstanding the findings in the NOT, a number of the properties of the On-Off DS cells indicate that these ganglion cells are specialized for signalling local motion rather than global motion. First, the number of On-Off DS cells exceeds the number of On DS cells by an order of magnitude and it does not make sense from the point of neuronal economy to assign some 40,000 On-Off DS cells simply to provide a global retinal slip signal. This view discounts the importance of the demonstration that the four preferred directions of the On-Off DS cells appear to be aligned with the four rectus muscles (Oyster and Barlow 1967), without abrogating the related concept that the four subtypes of On-Off DS ganglion cells enable higher order DS neurones to make use of excitation and inhibition from spatially superimposed mirror-symmetric subtypes (anterior/posterior or upwards/downwards; Levick et al. 1969). Second, the On-Off DS cells are present throughout the whole of the retina, providing a substrate for signalling the motion direction of both light and dark objects in all parts of the visual field. By contrast, the On DS system is only responsive to light objects in the central visual field, which appears sufficient to signal global motion reliably. Third, the effect of the surround beyond the excitatory receptive field may be fundamentally different in the two types of DS ganglion cells. In On-Off DS cells, the response to preferred direction movement is greatly reduced by concurrent stimulation of the silent inhibitory surround (Barlow and Levick 1965; Oyster et al. 1972; Wyatt and Daw 1975). In On DS cells, by contrast, the response to preferred direction movement appears to be unaffected or even facilitated when the stimulus extends beyond the classical receptive field (Oyster 1968). Fourth, there is only limited convergence of On-Off DS ganglion cells onto DS units in the rabbit lateral geniculate nucleus (LGN; Levick et al. 1969), indicating that the comparatively fine spatial sampling of this system is conserved in higher visual centres.

The activity of the DS units in the LGN is modulated about a maintained firing rate that is much higher than that of the On-Off DS ganglion cells: the LGN units are strongly excited by movement in the preferred direction and completely inhibited by movement in the opposite direction. The DS LGN units are excited over a narrower range of directions than the On-Off DS ganglion cells, and the LGN units respond poorly to stationary flashed spots. Taken together, these properties are consistent with the hypothesis that each DS unit in the LGN receives excitatory input from a subtype of On-Off DS ganglion cell and inhibitory input from the mirror-symmetric subtype, presumably mediated by an inhibitory interneurone (Levick et al. 1969). Thus the DS units in the LGN may be comparable in complexity to the fully opponent version of the correlation-type movement detector (Hassenstein and Reichardt 1956; Reichardt 1961; Borst and Egelhaaf 1989, 1990), with the two convergent subtypes of On-Off DS ganglion cells corresponding to the mirror-symmetrical subunits in the Reichardt model.

As detectors of local motion, the On-Off DS cells may serve two rather distinct functions, both concerned with image segmentation (see Part II of this volume). First, the On-Off DS cells would detect moving objects such as prey, predators and conspecifics, and this information may be used to inform spatial-attention mechanisms. The many On-Off DS cells located in the inferior rabbit retina image the sky and they are likely to be important for detecting birds of prey. Second, the On-Off DS cells would be sensitive to the faster relative motion of the images of foreground objects arising from translational movements of the head (motion parallax). Thus they may play a lower level role in complex visual performance such as depth perception and figure-ground discrimination (Miles 1993). The translational movements of forward locomotion are associated with a pattern of graded retinal image motion over much of the superior retina, arising from the texture of the ground plane of the terrain. A correct interpretation of three-dimensional scene structure in terms of the pattern of optic flow depends critically on a precise local analysis of the direction and speed of movement of textural details in the retinal image.

5. Cellular mechanisms of direction selectivity

Any neuronal model of direction selectivity should take account of the specific physiological and morphological properties of retinal interneurones. For example, the cotransmission model of direction selectivity (Vaney 1990) and a closely related computational model (Borg-Graham and Grzywacz 1992) predicted how asymmetrical responses could be derived from symmetrical starburst amacrine cells, based on the radial asymmetry of their input and output synapses (Famiglietti 1991). These cellular models are not supported by the results of He and Masland (1997) but there are currently no alternatives that account for direction selectivity in terms of identified types of retinal neurones. We are thus forced to deal with more generic models, such as the postsynaptic inhibitory scheme shown in figure 7 below (see Section 6). While such simple models may account plausibly for some basic experimental findings, they will undoubtedly lack the sophistication to explain the more subtle features. Moreover, it is probable that several distinct mechanisms contribute to the generation of direction selectivity in retinal ganglion cells (Grzywacz et al. 1997), which greatly complicates the experimental dissection of the phenomenon.

The mechanism underlying the inhibition by movement in the null direction is likely to be fundamentally different from the mechanism underlying facilitation by movement in the preferred direction (Grzywacz and Amthor 1993). Because of the interplay between inhibition and facilitation, it is not possible to map the spatial organization of each of these components in isolation, particularly if the cell's response is gauged only from the axonal firing rate. Consequently, it cannot be assumed that both the null-direction inhibition and the preferred-direction facilitation are anisotropic, as outwardly appears to be the case. Other scenarios are also possible: for example, the cotransmission model proposed that the directional tuning of On-Off DS cells arises because anisotropic facilitation shapes isotropic inhibition (Vaney 1990). In fact it now seems likely that the opposite is the case, with anisotropic inhibition shaping isotropic facilitation (He and Masland 1997). In order to distinguish between these scenarios, it is necessary to use a range of tools that selectively probe the different mechanisms, as described below.

Barlow and Levick's influential study (1965) on the receptive-field properties of On-Off DS cells indicated that null-direction inhibition is the key mechanism underlying direction selectivity in the retina. This conclusion was subsequently supported by evidence that γ-aminobutyric acid (GABA) antagonists block the direction selectivity of both the On-Off DS and On DS ganglion cells (Wyatt and Daw 1976; Caldwell et al. 1978; Ariel and Daw 1982; Massey et al. 1997). However, the interpretation of these pharmacological experiments is problematic because the null-direction inhibition may differentially affect the excitatory inputs from the glutamatergic bipolar cells and the cholinergic starburst cells. Moreover, there are dozens of different types of GABAergic amacrine cells (Vaney 1990), whose individual actions cannot be dissected by pharmacological means except at a rather coarse receptor level (Massey et al. 1997). Both the bipolar cells and the starburst cells are subject to direct inhibition from GABAergic amacrine cells that may be different from those that mediate the null-direction inhibition (Linn and Massey 1992; Zhou and Fain 1995). Shunting inhibition could play a crucial role in differentiating the responses of starburst cells to centripetal and centrifugal motion (Borg-Graham and Grzywacz 1992; Peters and Masland 1996; Grzywacz et al. 1997), so even preferred-direction facilitation may depend indirectly on an inhibitory GABAergic mechanism.

5.1 Spatial asymmetries

The basic requirement for the generation of direction selectivity is an asymmetric nonlinear interaction between spatially separate inputs (Barlow and Levick 1965;

Poggio and Torre 1981; Borst and Egelhaaf 1989). The spatial asymmetry is usually represented schematically as a lateral process extending in either the null direction (for inhibitory interactions) or the preferred direction (for facilitatory interactions). Thus, in a simplified inhibitory model (Fig. 4a), the excitatory input at each position (*D*) is vetoed by prior inputs that are spatially offset towards the null side of the receptive field (*A*, *B*, *C*). In a simplified excitatory model (Fig. 4b), the input at each position (*D*) is facilitated by prior inputs that are spatially offset towards the preferred side (*E*, *F*, *G*). The identification of the underlying spatial asymmetry is a necessary prerequisite for establishing the cellular basis of direction selectivity in the retina, because this will constrain both the locus and the synaptic mechanism of the nonlinear interactions.

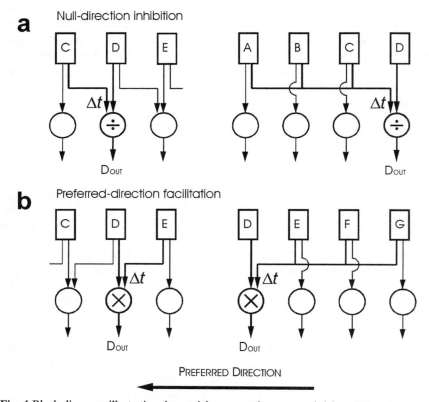

Fig. 4 Block diagrams illustrating the spatial asymmetries, temporal delays (Δt) and non-linear interactions that could underlie the generation of direction selectivity by either divisive-like ($\div$) null-direction inhibition (**a**) or multiplicative-like ($\times$) preferred-direction facilitation (**b**), with reference to a DS subunit labelled *D*. In classic representations of the DS subunit, the direct excitatory component interacts with a spatially asymmetric input arising from an adjacent subunit (left) but, in reality, the excitatory component may interact with multiple asymmetric inputs (right).

In the past, it has been postulated that the spatial asymmetry arises from retinal interneurones that either are elongated, give rise to short intraretinal axons, or have asymmetrically located somata (Mariani 1982). For example, the axons of B-type horizontal cells provide a possible substrate for asymmetric inhibitory interactions (Barlow and Levick 1965), but there are compelling arguments that direction selectivity is not computed in the outer retina (see Section 5.2). In the inner retina, dye-injection studies and classical Golgi studies have failed to identify candidate interneurones that could serve each of the preferred directions coded by the DS ganglion cells: although one unusual type of amacrine cell in the rabbit retina has "dorsally directed" processes, there appear to be no corresponding types that could code for the other three cardinal directions (Famiglietti 1989).

Starburst amacrine cells: inhibitory interactions. In the last decade, several related models of direction selectivity have hypothesized that the different preferred directions are generated by shared interneurones (Vaney et al. 1989; Vaney 1990; Werblin 1991; Borg-Graham and Grzywacz 1992; Poznanski 1992; He and Masland 1997). Attention has been focused primarily on the starburst amacrine cells, which costratify narrowly with the On-Off DS cells in both sublaminae of the inner plexiform layer (Perry and Walker 1980; Famiglietti 1983, 1992c). The starburst cells receive synapses from bipolar cells and amacrine cells over the whole dendritic tree but they contact ganglion cells only in a varicose distal zone (Famiglietti 1991). The proximo-distal segregation of the input and output synapses could underlie the spatial asymmetry necessary for direction selectivity (Vaney and Young 1988), provided that dendrites on different sides of the starburst cell contact different subtypes of DS ganglion cells. The starburst amacrine cells contain both acetylcholine and GABA (Masland et al. 1984; Brecha et al. 1988; Vaney and Young 1988) and, therefore, they could mediate either the preferred-direction facilitation or the null-direction inhibition (Vaney et al. 1989). If starburst cells provide the spatial asymmetry necessary for direction selectivity, then each DS ganglion cell should receive either GABAergic input from starburst cells located on the null side of its dendritic field or cholinergic input from starburst cells located on the preferred side.

He and Masland (1997) tested directly whether the starburst amacrine cells provide anisotropic input by laser ablating the cells in small retinal patches that were located on different sides of On-Off DS cells (Fig. 5). They targeted the On starburst cells in the ganglion cell layer of the isolated rabbit retina, thus avoiding damage to the rest of the amacrine cells located in the inner nuclear layer. This preserved the neuronal circuitry underlying the generation of direction selectivity in the Off sublamina, which provided a built-in control for the experimental manipulations confined to the On sublamina. Selective ablation of starburst cells located on the null side of the On-Off DS cell had little effect on the responses of the cell to null-direction movement, indicating that the starburst cells do not mediate the null-direction inhibition (Fig. 5a and c). What then is the function of the GABA in the starburst amacrine cells? The synapses that starburst cells make with each other (Millar and Morgan 1987) are probably GABAergic rather than cholinergic (Zhou et al. 1993; Brandstätter et al. 1995; Baldridge 1996), and consequently they may play a role in generating the symmetrical inhibitory surround of the starburst cells (Taylor and Wässle 1995; Peters and Masland 1996).

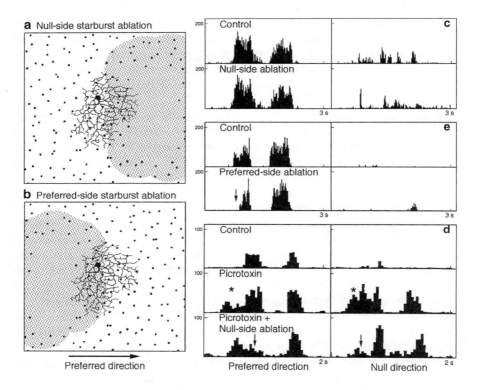

Fig. 5 Starburst amacrine cell ablation experiments. **a, b** Schematic representation of the laser ablation of On starburst cells (black dots) on either the null side (**a**) or the preferred side (**b**) of an On-Off DS cell, leading to reduced starburst coverage over one side of the ganglion cell's dendritic tree (shading). **c-e** Effects of starburst cell ablation on the responses (spikes/s) of On-Off DS cells to the leading edge (On response) and the trailing edge (Off response) of a light bar (500 µm wide on the retina) moved through the receptive field in the preferred direction and then back in the null direction. **c** Null-side starburst ablation had little effect on the responses to null-direction movement, indicating that the starburst cells do not provide the null-direction inhibition. **d** Preferred-side starburst ablation reduced the On response to preferred-direction movement in that portion of the receptive field overlapped by the ablated cells (arrow). **e** A GABAergic antagonist (50 µM picrotoxin) abolished direction selectivity and increased the size of the receptive field activated by the leading edge of the stimulus (asterisks); subsequent null-side starburst ablation reduced the On response to both preferred-direction and null-direction movement in that portion of the receptive field overlapped by the ablated cells (arrows). Taken together, (d) and (e) indicate that the starburst amacrine cells provide an isotropic facilitatory input to the DS ganglion cells. ((c) and (d) after He and Masland 1997; (e) S He and RH Masland, unpublished).

Starburst amacrine cells: facilitatory interactions. Laser ablation of On starburst cells located on the preferred side of the On-Off DS cell dramatically reduced the On response to targets moving in the preferred direction (Fig. 5b and d; He and Masland 1997). The effects of ablating the starburst amacrine cells were very similar to those elicited by blocking doses of nicotinic cholinergic antagonists, which reduced the preferred-direction responses by about half but did not affect the null-direction inhibition (Ariel and Daw 1982; Grzywacz et al. 1997; He and Masland 1997; Kittila and Massey 1997). Although the responses of the On-Off DS cells were unaffected by ablation of starburst cells on the null side, this does not mean that the cholinergic input from the starburst cells is anisotropic. The excitatory drive elicited by null-direction movement would normally be hidden by the null-direction inhibition but it can be unmasked if the inhibition is blocked by addition of a GABA antagonist; under these circumstances, the DS ganglion cells responded equally well to movements in all directions. Moreover, these responses were symmetrically reduced with the further addition of a nicotinic antagonist, indicating that the cholinergic input accounts for about half the excitatory drive in both the null and preferred directions under these conditions (He and Masland 1997).

The question remained whether the null-direction cholinergic input actually arises from null-side starburst cells rather than preferred-side starburst cells, and thus whether the cholinergic starburst input to the On-Off DS ganglion cells is anatomically symmetrical. He and Masland (unpublished) examined this question by blocking the null-direction inhibition with a GABA antagonist, and then comparing the responses before and after laser ablation of null-side starburst cells. For both null-direction and preferred-direction movements, the On response of the ganglion cell was greatly reduced in the portion of the receptive field overlapped by the ablated starburst cells (Fig. 5e). This indicates that the null-side starburst cells provide substantial excitatory drive to On-Off DS cells. By comparison, laser ablation of the null-side starburst cells had no effect on the null-direction inhibition (He and Masland 1997). The starburst amacrine cells were the first neurones shown to colocalize classical excitatory and inhibitory transmitters (Vaney and Young 1988), but there is currently no evidence that individual synapses made by starburst cells onto either ganglion cells or amacrine cells are both excitatory and inhibitory in function.

It thus appears that the starburst amacrine cells provide isotropic cholinergic input which is spatially shaped by anisotropic GABAergic inhibition from some other type of amacrine cell (He and Masland 1997). It is likely that the starburst amacrine cells mediate the preferred-direction facilitation, which operates over a 100-200 μm range near the visual streak (Grzywacz and Amthor 1993). Moreover, preferred-direction facilitation is elicited over a larger area than the classical receptive field (Amthor et al. 1996) and this resolves the paradox that the classical receptive field is much smaller than the spatial extent of the interneurones that potentially provide excitatory input (Yang and Masland 1992). It appears that the classical receptive field of On-Off DS ganglion cells is delimited by the dendritic envelope of the afferent bipolar cells (Yang and Masland 1994), whereas the facilitatory receptive field may be delimited by the dendritic envelope of the afferent starburst amacrine cells (He and Masland 1997). It remains to be shown directly that preferred-direction facilitation is abolished when the input from starburst cells is blocked with cholinergic antagonists.

Like the more extensive inhibitory surround, the facilitatory surround stimulated on its own is normally silent as far as ganglion cell firing is concerned; both surrounds probably act in concert to modulate the direct excitatory drive from the bipolar cells (see Section 6). The facilitatory surround may be unmasked by GABA antagonists (Fig. 5e), but this effect has not been analysed systematically. However, a natural manifestation of preferred-direction facilitation may be the 10-30% shift of the classical receptive field towards the preferred side of the dendritic field observed in about 40% of the On-Off DS cells (Yang and Masland 1994), but it is not known why the majority of cells do not show this asymmetrical organization. Moreover, it is difficult to account for the finding that cells with a shifted receptive field do not respond to a flashing spot located over the null side of the dendritic tree: the unmasking of cholinergic facilitation should result in enlargement of the receptive field, rather than inhibition of the bipolar cell input on the null side. The preferred side of the receptive field contains the non-directional zone (see Section 2.3) and it would be interesting to know whether the extent of the receptive-field shift is correlated with the width of the non-directional zone, which might indicate that the zone is an epiphenomenon of the mechanism that produces motion facilitation rather than being associated with the mechanism that produces direction selectivity.

He and Masland (1997) concluded that the function of starburst amacrine cells is to potentiate generally the responses of retinal ganglion cells to moving stimuli, regardless of the direction of motion. Three features of the neuronal architecture of starburst cells can be rationalized in this context. First, the proximo-distal segregation of the input and output synapses of starburst cells (Famiglietti 1991) provides the spatial offset that is a prerequisite for motion facilitation, just as it is also necessary for direction selectivity. Second, the large dendritic fields of the starburst cells ensures that the facilitatory mechanism is responsive to both small and large displacements (Fig. 4, right). The extensive dendritic-field overlap is not redundant because each starburst cell that provides input to a local region of a ganglion cell's receptive field would be most responsive to a different vector of motion (Vaney 1990). Third, the dendritic fasciculation of the starburst amacrine cells and the On-Off DS cells enables each terminal dendrite of a starburst cell to contact several subtypes of ganglion cells (Vaney 1994b).

Although the facilitatory input from the starburst amacrine cells appears to be isotropic, the strong null-direction inhibition ensures that the facilitation is demonstrable only for movements with a component in the preferred direction (Grzywacz and Amthor 1993). Consequently, the difference between the preferred- and null-direction responses is significantly enhanced by the cholinergic input from starburst amacrine cells. Thus the On-Off DS cells respond much more strongly to moving stimuli than to stationary flashed stimuli (Fig. 2), and this effect is even more pronounced in higher order DS neurones, which appear to receive excitation and inhibition from mirror-symmetric subtypes of DS ganglion cells (Levick et al. 1969).

As a counterpoint to the above conclusion that the cholinergic input to On-Off DS cells is isotropic, a recent study has proposed that there are pronounced differences in the neuronal circuitry underlying the directional signals elicited by moving edges and drifting gratings or textures (Grzywacz et al. 1998). The basic finding was that drifting gratings activated an anisotropic cholinergic input to the On-Off DS cells which was independent of the null-direction inhibition. However, the drifting gratings elicited very low firing rates and this raises the question of whether the observed effect is functionally significant. Moreover, null-direction spikes appeared when the cholinergic input was blocked with tubocurarine and this unexpected result is not consistent with the authors' own model.

GABAergic amacrine cells. The available evidence indicates that the null-direction inhibition is mediated by GABAergic amacrine cells that are different from the starburst amacrine cells. It is likely that the On-Off DS cells are served by separate On and Off types of amacrine cells, because apparent-motion experiments indicated that the interaction between excitation and null-direction inhibition is largely segregated between the On and Off pathways (Amthor and Grzywacz 1993b). Correspondingly, the Off responses retained their direction selectivity when the On component of the bipolar cell input was blocked (Kittila and Massey 1995). Given that most types of GABAergic amacrine cells are widely overlapping unistratified neurones (Vaney 1990), it would be predicted that the null-direction inhibition of the On-Off DS cells is mediated by a complementary pair of amacrine cell types, stratifying around 20% and 70% depth of the inner plexiform layer in the rabbit retina.

In a recent study on the rabbit retina, MacNeil et al. (1999) systematically surveyed the dendritic morphology of 261 randomly labelled amacrine cells, which they classified into at least 27 cell types. Interestingly, none of these types appear to meet the criteria outlined above, apart from the cholinergic/GABAergic starburst amacrine cells. Indeed, the mirror-symmetric populations of starburst cells would appear to provide an ideal substrate for mediating null-direction inhibition, because several characteristics of the starburst cells that are important for cholinergic motion facilitation may also be necessary for GABAergic motion inhibition, including the widely overlapping dendritic fields and the spatial offset between the input and output synapses. Until candidate amacrine cells are identified, we are forced to model the DS mechanism with a hypothetical amacrine cell that differs from the starburst cell in providing an asymmetric input to the DS ganglion cell (see Section 7).

It is still unclear whether GABA-mediated null-direction inhibition could account for all of the direction selectivity in retinal ganglion cells. Earlier in vivo studies showed that systemic infusion of the GABA antagonist picrotoxin, which acts on both $GABA_A$ and $GABA_C$ receptors, greatly reduced but did not abolish the direction selectivity of both the On-Off DS cells and the On DS cells (Wyatt and Daw 1976; Caldwell et al. 1978; Ariel and Daw 1982). Recent in vitro studies provide contradictory evidence: Grzywacz et al. (1997) reported that saturating doses of picrotoxin left a residual directionality in most cells tested, whereas Massey et al. (1997) reported that the direction selectivity of On-Off DS cells was eliminated by either picrotoxin or low concentrations of the selective $GABA_A$ antagonists, bicuculline and SR-95531, indicating that $GABA_A$ receptors may account for all of the null-direction inhibition. The localization of $GABA_C$ receptors in bipolar cells of the mammalian retina (Enz et al. 1996) does not preclude the possibility that the null-direction inhibition acts directly on the bipolar cells, because $GABA_C$ receptors account for only about 20% of the GABA-induced currents in cone bipolar cells, with the bulk of the current mediated by $GABA_A$ receptors (Euler and Wässle 1998).

5.2 Locus of null-direction inhibition

The null-direction inhibition could act presynaptically on the excitatory inter-neurones or postsynaptically on the DS ganglion cell itself (Torre and Poggio 1978; Ariel and Daw 1982; Koch et al. 1982, 1983). Both the cholinergic input and the glutamatergic input are subject to null-direction inhibition (Kittila and Massey 1997), so it is generally assumed that a purely presynaptic mechanism would require the asymmetric inhibitory inputs to make selective contact with both the starburst cells and the bipolar cells (but see Section 5.2). Thus a post-synaptic locus for the null-direction inhibition may provide the most parsimonious circuitry for implementing direction selectivity. However, we emphasize in advance of our detailed arguments that there is presently no direct evidence that establishes or refutes either a presynaptic model or a postsynaptic model and, moreover, it is quite possible that direction selectivity is implemented by a combination of presynaptic and postsynaptic mechanisms.

Cholinergic input from starburst amacrine cells. Although the spatial asymmetry that underlies direction selectivity does not appear to arise from the starburst amacrine cells (see Section 5.1 and Fig. 5), the release of acetylcholine onto an individual DS ganglion cell would be direction selective if the null-direction inhibition acts presynaptically on the starburst cells. The starburst cells show a 25- to 70-fold overlap of their dendritic fields (Tauchi and Masland 1984; Vaney 1984) but there is no morphological or physiological evidence that they comprise four subtypes, each providing input to one of the four subtypes of On-Off DS cells (cf. Amthor and Grzywacz 1994). Patch-electrode recordings from starburst somata indicated that the cells are not direction selective, in that they responded isotropically to moving stimuli (Peters and Masland 1996). However, the starburst cells appeared to respond more strongly to centrifugal movements than to centripetal movements and thus the release of acetylcholine from a terminal dendrite may be direction sensitive; that is, terminals on the left side of the starburst cell may be depolarized more strongly by leftwards movement than rightwards movement (see also Borg-Graham and Grzywacz 1992). In order to tap into this directional signal, a ganglion cell would have to make preferential contact with dendrites located on one side of the starburst cells, but the study by He and Masland (1997) indicates that this is not the case.

A presynaptic mechanism may still be feasible if the direction selectivity is coded at a very local level in the starburst cell, with adjacent terminal dendrites receiving different asymmetric inputs from the processes that mediate null-direction inhibition; movement in one direction would then inhibit only a subset of the terminals distributed around the dendritic tree. In order to tap into this directional signal, a ganglion cell would have to make preferential contact with these terminals, which would be a complex task developmentally. It would also be necessary for the null-direction inhibition to be effective only locally, although this caveat also applies to any postsynaptic mechanism (see Section 5.2).

It has been argued that two pharmacological experiments provide evidence that the null-direction inhibition modulates the cholinergic input postsynaptically rather than presynaptically. First, the depolarizing effects of nicotinic cholinergic agonists on DS ganglion cells were suppressed by null-direction movements, suggesting that the null-direction inhibition acts postsynaptically on the ganglion cells (Ariel and Daw 1982; Kittila and Massey 1997). Second, the direction selectivity of the On-Off DS cells was greatly reduced by the cholinergic potentiator, physostigmine, which is not consistent with the cholinergic input to the ganglion cell being direction selective, as required by a presynaptic mechanism (Ariel and Daw 1982; Grzywacz et al. 1997). However, physostigmine prevents the rapid breakdown of acetylcholine by acetylcholinesterase, and it is possible that the ganglion cell is being excited by acetylcholine released at synapses on other DS ganglion cells with different preferred directions. Thus the physostigmine result can be variously interpreted as supporting either the presynaptic model or the postsynaptic model.

Application of saturating concentrations of nicotinic antagonists, which completely blocked the response of the cells to exogenous nicotinic agonists, reduced the preferred-direction responses of the On-Off DS cells by about half, but did not affect the strength of the null-direction inhibition (Ariel and Daw 1982; Grzywacz et al. 1997; Kittila and Massey 1997). This important result indicates that the cholinergic input is not essential for direction selectivity and that the null-direction inhibition also acts on the other excitatory inputs, which presumably arise from bipolar cells.

Glutamatergic input from cone bipolar cells. The ganglion cells that are postsynaptic to the starburst amacrine cells receive direct synaptic input from cone bipolar cells (Brandon 1987; Famiglietti 1991), but unravelling the bipolar contribution to the generation of direction selectivity is problematic because the same bipolar cells also drive both the starburst amacrine cells and probably the GABAergic amacrine cells that are presumed to provide the null-direction inhibition. However, the bipolar cell input to these amacrine cells is mediated naturally by glutamate receptors that are sensitive to α-amino-3-hydroxy-5-methyl-4-isoxazolepropionic acid (AMPA) and kainate (KA) and, correspondingly, the effects of AMPA/KA antagonists are closely mimicked by the combined application of cholinergic and GABA antagonists (Fig. 6). AMPA/KA antagonists abolished the null-direction inhibition of On-Off DS cells (Cohen and Miller 1995; Kittila and Massey 1997) and this effect appears to be due to blockade of GABA release rather than blockade of acetylcholine release (Linn et al. 1991), because cholinergic antagonists alone did not reduce the direction selectivity.

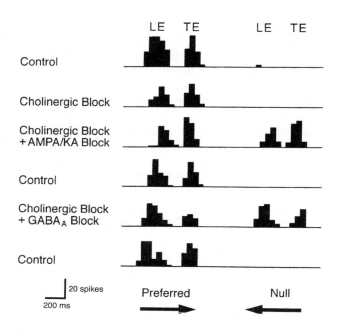

Fig. 6 Effect of neurotransmitter antagonists on the responses of an On-Off DS ganglion cell to the leading edge (LE: On response) and the trailing edge (TE: Off response) of a light bar (about 500 μm wide on the retina) moved through the receptive field in the preferred direction and then back in the null direction. The direction selectivity apparent in the control responses was maintained when the cholinergic inputs were blocked with nicotinic and muscarinic antagonists (100 μM hexamethonium bromide and 2 μM atropine), although the preferred-direction responses were reduced. When the cholinergic antagonists were combined with a $GABA_A$ antagonist (2 μM SR95531), the direction selectivity was completely abolished, indicating that the underlying GABAergic mechanism is independent of the cholinergic input from the starburst amacrine cells. The same effect was produced by combining the cholinergic antagonists with an AMPA/KA antagonist (10 μm NBQX), suggesting that NBQX blocks the bipolar cell glutamatergic drive to the GABAergic amacrine cells that mediate null-direction inhibition. (After Kittila and Massey 1997, with permission)

Whereas glutamatergic transmission from bipolar cells to amacrine cells appears to be mediated largely by AMPA/KA receptors, the direct transmission from bipolar cells to DS ganglion cells appears to be dominated by glutamatergic receptors that are sensitive to N-methyl-D-aspartic acid (Massey and Miller 1990; Cohen and Miller 1995; Kittila and Massey 1997). NMDA antagonists reduced the preferred-direction responses of the On-Off DS cells by about 31% but, unlike AMPA/KA antagonists, had no discernible effect on the null-direction inhibition. This is qualitatively similar to the effects of cholinergic antagonists, which reduced the preferred-direction responses by about 54% (Kittila and Massey 1997). Simultaneous blockade with NMDA and cholinergic antagonists reduced the responses by about 92%; the residual 8% responsiveness probably represents the direct bipolar input that is mediated by AMPA/KA receptors. Thus the NMDA receptors would account naturally for 70-80% of the glutamatergic input to On-Off DS cells, which is much greater than reported for other types of ganglion cells (Massey and Miller 1990; Cohen and Miller 1994; Taylor et al. 1995).

If the null-direction inhibition acts presynaptically on the cone bipolar cells, then the release of glutamate onto individual ganglion cells would be direction selective. This could be achieved most simply by having subtypes of bipolar cells dedicated to each preferred direction, and recordings from the turtle retina have revealed both DS bipolar cells and DS amacrine cells (DeVoe et al. 1989). The cone bipolar cells that costratify with the On starburst amacrine cells in sublamina *b* of the inner plexiform layer can be selectively labelled by CD15 immunocyto-chemistry in the juvenile rabbit retina (Brown and Masland 1999), and they have a similar density distribution to other identified populations of cone bipolar cells, which terminate at different levels in the inner plexiform layer (Mills and Massey 1992; Massey and Mills 1996). The CD15-immunoreactive bipolar cells are mor-phologically homogeneous and show unitary coverage of both their dendritic fields in the outer plexiform layer and their axonal fields in the inner plexiform layer (Brown and Masland 1999): if they comprised four physiological subtypes, there would be significant gaps in the retinal coverage by each subtype. For the same reason, it is unlikely that the null-direction inhibition would be mediated by GABAergic horizontal cells in the outer retina, because this would also require dedicated subtypes of bipolar cells for each preferred direction.

The possibility remains that direction selectivity is coded at a very local level in the axonal tree of the cone bipolar cells, with different terminal endings receiving different asymmetric inputs from the processes that mediate null-direc-tion inhibition. As discussed in Section 5.2 for the starburst amacrine cells, such a presynaptic scheme would require that the inhibitory interactions in adjacent branches are relatively isolated from each other, which may be difficult given that the cable properties of the bipolar cells ensure that graded potentials from the dendrites are transmitted reliably to all parts of the compact axonal tree. Never-theless, the scheme presumes that a null-direction inhibitory input would be located directly upon the terminal ending providing excitatory output to the corre-sponding subtype of DS ganglion cell and, therefore, it would be off the current path leading to other terminal endings coding different preferred directions (see Section 5.2).

A simplified presynaptic model. Although a flashing target located beyond the preferred edge of the classical receptive field produces no spikes in an On-Off DS cell, this stimulus augments the subsequent response of the cell to a flashing target located inside the classical receptive field (Amthor et al. 1996). The interpretation of this result is that stimulation of starburst amacrine cells straddling the receptive-field edge facilitates the response elicited by stimulation of bipolar cells contacting the ganglion cell directly. Thus the cholinergic input from the starburst amacrine cells, on its own, does not appear to produce a suprathreshold excitatory response and is effectively silent. While the nature of the non-linear interaction is not clear (Grzywacz and Amthor 1993), the glutamatergic input from the bipolar cells appears somehow to gate the cholinergic input from the starburst amacrine cells, although it is difficult to imagine how this could be implemented biophysically.

If there is such an interaction between the glutamatergic and cholinergic inputs then it is possible that the effective cholinergic input to the DS ganglion cell would appear to be direction selective even if the null-direction inhibition acts only on the bipolar cells and not directly on the starburst cells or the DS ganglion cells. That is, null-direction motion may stimulate the release of acetylcholine from the starburst amacrine cells, but this would be ineffective if no glutamate was released from the bipolar cells because of null-direction inhibition. The model is not challenged by the finding that the direction selectivity is unaffected by NMDA antagonists, because they block only about 80% of the glutamatergic input from the bipolar cells to the DS ganglion cells (Cohen and Miller 1995; Kittila and Massey 1997). Thus it may be premature to favour a postsynaptic model of direction selectivity simply because of its parsimony.

Physiological experiments. In the turtle retina, patch-electrode recordings from DS ganglion cells revealed that the direction selectivity was not affected when the inhibitory inputs were blocked intracellularly with an electrode solution lacking Mg^{2+} and ATP; this suggested that the excitatory input to these ganglion cells was direction selective, thus requiring the null-direction inhibition to act presynaptically (Borg-Graham and Grzywacz 1992). These results contrast with earlier findings in the turtle retina that were obtained with sharp electrodes: Marchiafava (1979) reported that DS ganglion cells showed a strong EPSP with superimposed action potentials to preferred-direction movement and a reduced EPSP to null-direction movement. On its own, the reduced null-direction response could reflect either a reduced input from excitatory interneurones (presynaptic inhibition), or shunting of the excitatory input by an inhibitory conductance with a reversal potential near the resting potential (postsynaptic inhibition). In the latter case, the difference between the preferred- and null-direction responses would depend on the membrane potential, because it arises from the postsynaptic interaction of conductances with different reversal potentials. Marchiafava (1979) demonstrated that the null-direction EPSP was converted to an IPSP when a steady-state depolarizing current was injected into the soma, thus suggesting that the null-direction inhibition is mediated by a shunting conductance located on the ganglion cell.

In the rabbit retina, Amthor et al. (1989b) reported that null-direction movement produces an EPSP without spikes in On-Off DS cells, comparable to that observed in the turtle retina by Marchiafava (1979). The effects of current injection were not tested, but an earlier study with On-Off DS cells reported that null-direction movement elicits IPSPs during injury-induced depolarization (Miller 1979). Taken together, these observations suggest that the null-direction inhibition may act postsynaptically through a shunting conductance, at least in part. Other evidence favouring a postsynaptic mechanism of direction selectivity is largely indirect, and primarily revolves around several findings indicating that the release of acetylcholine onto individual DS ganglion cells is not direction selective (see Section 5.2). Definitive proof will require a systematic series of patch-electrode experiments under different current-clamp and voltage-clamp conditions, using strategies to isolate the GABAergic inputs.

The presence of a non-directional zone on the preferred side of the receptive field of On-Off DS cells (He 1994; see Section 2.3) would appear to provide indirect support for the hypothesis that the null-direction inhibition acts postsynaptically on the ganglion cell. An exclusively presynaptic implementation of direction selectivity should not be dependent on the dendritic architecture of the postsynaptic ganglion cells: if direction selectivity is implemented uniformly across the array of presynaptic elements, then ganglion cells sampling that array should have uniform response properties throughout the receptive field. The observation that some visual stimuli do not produce directional responses in a restricted part of the receptive field suggests that this property may be a consequence of implementing direction selectivity postsynaptically in the dendritic tree of the DS ganglion cell, rather than presynaptically in the bipolar cells or starburst amacrine cells.

Barlow and Levick (1965) proposed that the existence and location of the non-directional zone can be rationalized in terms of the spatial asymmetry that underlies null-direction inhibition, as outlined in the following argument. Consider the two subunits D and E in figure 4a: if subunit D connects to the preferred side of a DS ganglion cell and subunit E connects to the null side of an adjacent ganglion cell with the same preferred direction, then restricted movement of a target in the null direction from D to E will excite the first ganglion cell but not the second, because the output from subunit D is not subject to null-direction inhibition. That is, the first ganglion cell will exhibit a non-directional zone on the preferred side of its receptive field. However, now consider what would happen if subunits D and E were connected to branching systems arising from different primary dendrites of the same ganglion cell, rather than being connected to different ganglion cells. By the same logic, an individual branching system located in the middle of the dendritic field should also show a non-directional zone, because the excitation from subunit D would be transmitted to the soma unimpeded by any inhibition from subunit E on the adjacent branching system. In principle, this logic can also be extended to adjacent terminal dendrites, provided that the excitation on one dendrite is not shunted by inhibition on the other dendrite (see below). Thus the original rationalization of the non-directional zone simply in terms of asymmetric inhibition leads to the prediction that the specific stimuli that produce null-direction responses in the zone should produce null-direction responses anywhere in the receptive field. The fact that they do not suggests that the anomalous generation of null-direction responses does not have such a straight-forward explanation.

Topology of cellular interactions. Because direction selectivity is implemented locally throughout the ganglion cell's receptive field (Barlow and Levick 1965), any postsynaptic mechanism of direction selectivity requires that the inhibitory inputs to a local region (subunit) of the dendritic tree should veto only the excitatory inputs to that region, without affecting the excitatory inputs to other parts of the dendritic tree (see Section 6). If the excitatory and inhibitory inputs to a DS ganglion cell were simply summed linearly at the axon hillock to produce a spike train, the integrated responses to preferred- and null-direction movements might well be the same (although the responses may differ in shape), regardless of the spatial asymmetries in the inputs.

With a postsynaptic mechanism of direction selectivity, as exemplified by the Torre-Poggio-Koch model, the outcome of any visual stimulation represents the sum of net longitudinal currents along dendrites, which converge through the dendritic tree and soma of the ganglion cell to depolarize the spike-trigger zone near the axon hillock (Torre and Poggio 1978; Koch et al. 1982, 1983). The model requires that DS responses are generated locally in a small part of the dendritic tree, with the responses from different parts of the dendritic field being additive in their effect at the spike-trigger zone. Both the local processing constraint and the pooling of excitatory responses are largely handled by implementing the null-direction inhibition through a shunting conductance, which has an equilibrium potential close to the resting potential, so that the effect on currently inactive parts of the dendritic tree is small. Excitatory synaptic current is short-circuited through nearby inhibitory synapses, and is thus lost from the longitudinal dendritic loop through the spike-trigger zone.

Nevertheless, in the circuit diagram of Torre and Poggio (1978), longitudinal resistances are provided in the paths linking the sites of local interaction to the spike-trigger zone because they are necessary to retain independent additivity of effects. The inclusion of these resistances implicitly invokes the detailed topology of the dendritic tree. The consequence is that null-direction inhibition would need to be excluded from more proximal dendritic segments that channel currents from more distal sites. This arrangement is necessary to avoid on-the-path inhibition producing aberrations in the responses to some moving stimuli, including the premature truncation of the ganglion cell's discharge to a preferred-direction stimulus moving across the null side of the dendritic tree. Moreover, this must also be the case to prevent On inhibition from vetoing Off excitation in On-Off DS cells (Amthor and Grzywacz 1993b), because the Off dendrites in sublamina a can arise from On dendrites of any order in sublamina *b*.

Koch et al. (1982, 1983, 1986) used one-dimensional cable theory to model the postsynaptic implementation of direction selectivity and they concluded that the excitatory input on one dendrite would not be shunted significantly by the inhibitory input on a neighbouring dendrite, provided that the inhibitory input was located 10-20 µm off the direct path to the soma in the case of a graded depolarization, or 5 µm in the case of a dendritic spike. Although such figures are critically dependent on both the actual dimensions of the dendrites and the biophysical properties of the cell membrane and the cytoplasm, the conclusion that excitation will be reduced significantly by on-the-path inhibition appears to be robust.

The topological issue also arises with a presynaptic mechanism of direction selectivity, but the context is quite different. The number of bipolar cells that terminate in the appropriate stratum of the inner plexiform layer appears to be insufficient to allow separate subtypes for each of the four preferred directions of motion. In this case, it would be necessary to suppose that separate terminal branches of an individual bipolar cell would be dedicated to different preferred directions. The issue then arises as to how synapses on different branches could be capable of independent activity in response to common-sourced excitatory drive, which arises in the outer plexiform layer from the cone photoreceptor input to the bipolar cell dendrites. In the inner plexiform layer, the current change divides into the branches of the axon terminal and, in this presynaptic model, each branch would be subject to laterally sourced inhibition by way of a direct amacrine to bipolar synapse (Raviola and Raviola 1967). The topological requirements are that the local synaptic complex should be located well away from the axonal trunk and that the shafts of the terminal branches should be thin. This would allow presynaptic inhibition to produce at least small departures from isopotentiality in a local part of the axon terminal. To amplify the differential effects, it would be necessary to suppose that transmitter release is a steeply accelerating function of terminal membrane potential (Eccles 1964) and that the steady-state resting potential of the bipolar cell is maintained close to the threshold for transmitter release. Detailed modelling would be required to substantiate the validity of the foregoing conjectures.

6. Direction-selective subunits

In Section 5, we outlined the cellular mechanisms that may underlie the generation of direction selectivity; in this Section, we consider the spatial and temporal characteristics of these mechanisms. The size of the DS subunit was originally equated with the smallest movement that produced directional responses (Barlow and Levick 1965). Subsequent neuropharmacological studies indicated that local excitatory input from glutamatergic bipolar cells participates in both inhibitory and facilitatory interactions with wider ranging GABAergic and cholinergic amacrine cells, respectively, and the combined ensemble can be regarded as comprising the DS subunit (Fig. 7). This definition of the subunit emphasizes its lateral extent rather than its spacing, and takes account of the distinction made in Section 5.2 between the classical receptive field of a DS ganglion cell on the one hand, and the broader inhibitory and facilitatory fields on the other.

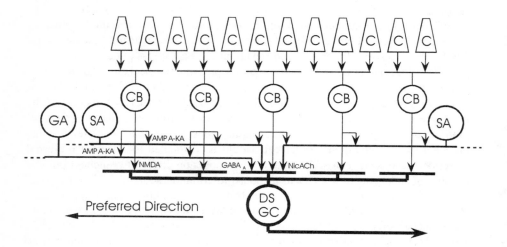

Fig. 7 Schematic representation of the neuronal circuitry that may underlie direction selectivity in the retina. Five subunits in the DS ganglion cell (DS GC) receive direct excitatory inputs from overlying cone bipolar (CB) cells, primarily through NMDA glutamate receptors. The lateral inputs that interact with the direct excitatory input to the central subunit are shown, including the hypothetical GABAergic amacrine (GA) cell that provides null-direction inhibition through GABA_A receptors, and the starburst amacrine (SA) cells that provide symmetrical facilitation through nicotinic cholinergic (NicACh) receptors. These amacrine cells receive cone bipolar input through AMPA-KA glutamate receptors. (The amacrine cells that provide symmetrical surround inhibition are not shown for the sake of clarity.) In this postsynaptic scheme, the null-direction inhibition acts on a terminal dendrite of the DS ganglion cell; in the corresponding presynaptic scheme, the GABAergic input would simply be directed towards the adjacent processes of the cone bipolar cell and the starburst amacrine cells. C = cone photoreceptor.

6.1 Bipolar cell array and subunit grain

Given that the envelope of cone bipolar cells converging on the dendritic tree almost matches the classical receptive field mapped with flashing spots (Yang and Masland 1994), then it is attractive to suppose that each bipolar cell provides the direct excitatory component of a notional subunit of the receptive field. This raises the question: Are the spatial properties of the bipolar cell array reflected in either the receptive-field organization of DS ganglion cells or the properties of individual subunits? The CD15-immunoreactive bipolar cells, which costratify with the On starburst cells, would have an intercellular spacing of 30-50 µm in the adult rabbit retina (Brown and Masland 1999), and this broadly matches the stimulus spacings that produced maximum directionality in apparent-motion (two-slit) experiments. Thus Barlow and Levick (1965) reported that On-Off DS cells were maximally inhibited by a null-direction spacing of 0.13-0.2° (20-30 µm), and that the cells were maximally facilitated by a preferred-direction spacing of 0.28° (40 µm). Similarly, Amthor and Grzywacz (1993b) reported that the null-direction inhibition increased as the displacement was reduced to 50 µm, and that the pre-ferred-direction facilitation increased as the displacement was reduced to 80 µm (Grzywacz and Amthor 1993), but they did not quantify the responses to smaller displacements. These results indicate that the strength of the null-direction inhibi-tion (or the preferred-direction facilitation) on the excitatory drive from a bipolar cell increases as the stimulus approaches the subunit centre, but then decreases for displacements below some critical value.

It needs to be admitted that there is no evidence that the DS subunits are literally discrete entities, with the territorial limits of individual cone bipolar cells forming the boundaries of the direct excitatory component of DS subunits. Although the issue has not been thoroughly tested, it is generally accepted that the DS property is smoothly distributed throughout the receptive field. This can be reconciled with the discreteness of the CD15 bipolar cell array by noting that the dendritic field of each bipolar cell encompasses about 10 cone photoreceptors, which have a spacing of 8-14 µm in the rabbit retina (Young and Vaney 1991). This would permit incremental movements to be smoothly signalled as the stimu-lus crossed neighbouring bipolar cells. Similar considerations would apply to the amacrine cells that underlie null-direction inhibition and preferred-direction facilitation because they are driven presumably by the same bipolar cells that provide the direct excitatory input. Image motion will produce distributed responses in neighbouring cells, which will be coded by the ganglion cell as though the stimulus had activated a DS subunit with characteristics corresponding to the weighted sum of the activated bipolar cells. In this way, any point in the receptive field of a DS ganglion cell could be the notional centre of a DS subunit, whether located over the centre of a bipolar cell or in between bipolar cells.

6.2 Directional acuity

The foregoing considerations need to be combined with an appreciation of the limitations inherent in visual stimulation. Retinal images are inevitably blurred versions of real objects; transduction by the photoreceptor array leads to further blurring by virtue of both the limited angular acceptance properties of individual photoreceptors and the presence of oblique rays crossing the outer segments. In block diagrams of the DS subunit (Fig. 4), there is implicit representation of the minimum extent of receptive field that would include sufficient lateral inhibitory input and direct excitatory drive to produce DS responses to moving stimuli. The original experiments of Barlow and Levick (1965) showed that the minimum displacement required to produce directional responses depends strongly on the form of the testing stimuli. Thus, a continuous-motion (single-slit) experiment indicated a minimum of 0.25° (40 µm), whereas an apparent-motion (two-slit) experiment on the same cell indicated a minimum of 0.1° (15 µm) or less.

Grzywacz et al. (1994) devised a special stimulus configuration in which a spatial jump of a long stimulus edge (400 µm) presented within a wide slit (100 µm) consistently produced significant directional responses in On-Off DS cells for displacements of 4-8 µm on the retina, and some cells showed directional responses for displacements of only 1.1 µm. However, the changes in illumination resulting from such a small jump would have been spread over a retinal region of substantially greater width, estimated to be 7-45 µm along the preferred-null axis, and this needs to be taken into account when interpreting such "directional hyper-acuity". Although small preferred-direction displacements caused an incremental response not seen with null-direction displacements, the experiments do not establish whether this asymmetry arises through preferred-direction facilitation or null-direction inhibition. If this excitatory effect resulted from cholinergic facilitation it might be isotropic (He and Masland 1997), suggesting that it was not detected during null-direction displacements because it was masked by null-direction inhibition.

6.3 Spatiotemporal properties

In this Section we examine the relations between the directional mechanisms and non-directional mechanisms that affect the spatial-temporal properties of the subunits. The measurement of the spatial extent of either the null-direction inhibition or the surround facilitation may be confounded by their interactions with each other and with the symmetrical surround inhibition. Robust facilitation of On-Off DS cells was produced by preferred-direction displacements that covered less than half the width of the excitatory receptive field (Grzywacz and Amthor 1993), which corresponds to about half the dendritic-field diameter of overlapping starburst amacrine cells. This suggests that the excitation generated on one side of the starburst cell is conducted effectively to nearby terminal dendrites, but does not spread appreciably through the soma to the other side of the starburst cell. However, the full extent of the preferred-direction facilitation may be masked normally by the surround inhibition, with the result that preferred-direction displacements covering more than half the receptive field produced net inhibition (Barlow and Levick 1965, Table 3; Grzywacz and Amthor 1993).

Null-direction apparent motion produced net inhibition over a wide range of displacements covering much of the receptive field, but the inhibition from larger displacements appeared to have a faster time course than that from shorter displacements, suggesting that larger displacements predominantly activate the surround inhibition, which has a faster rise time and a more transient time course than the null-direction inhibition (Amthor and Grzywacz 1993b; Merwine et al. 1995). In addition, the difference in time course may reflect the fact that shorter null-direction displacements may also activate the facilitatory mechanism, the dynamics of which would be compounded with those of the two inhibitory mechanisms in a complicated manner. Thus it cannot be assumed that null-direction inhibition operates over a wider range than preferred-direction facilitation, and an open mind should be kept regarding the morphological substrate that underlies null-direction inhibition.

If a moving object stimulates two spatially separate inputs, the signal from the first input must be delayed or spread out so as to interact with the signal from the second input (Hassenstein and Reichardt 1956; Barlow and Levick 1965). In principle, the temporal difference (Δt) and the extent of the spatial offset ($\Delta\varphi$) determine the optimal velocity ($\Delta\varphi/\Delta t$) of the DS detector and its dynamic range. For a given temporal delay, broad velocity tuning requires a variable spatial offset with a small minimum offset, and physiological experiments on the On-Off DS cells provide clear evidence that the null-direction inhibition is indeed wide-ranging (Barlow and Levick 1965; Wyatt and Daw 1975; Amthor and Grzywacz 1993b).

For a given spatial offset, broad velocity tuning requires short latency, long duration interactions, but physiological experiments provide equivocal evidence as to whether the null-direction inhibition is particularly long lasting. When Wyatt and Daw (1975) used an apparent-motion stimulus consisting of two flashing bars separated by 65-140 µm, the On-Off DS cells showed directional responses for time intervals of 10-50 ms only. This seems consistent with pharmacological evidence that the null-direction inhibition is mediated by $GABA_A$ receptors (Massey et al. 1997), which have an activation time constant of about 10 ms (Feigenspan and Bormann 1998). By contrast, when Amthor and Grzywacz (1993b) used an apparent motion stimulus consisting of short displacements (about 40 µm) of the slit stimulus, the null-direction inhibition reached 50% of the peak value within 20 ms and remained elevated for over 2 000 ms. Although this was a much more sustained response than the suprathreshold excitatory response elicited by the first slit, the result only provides evidence that the null-direction inhibition is sustained rather than long lasting, because the first slit was On for the whole period (Amthor and Grzywacz, personal communication).

7. Future Directions

This review has focused on recent advances in our understanding of the structural and functional properties of DS ganglion cells and the cellular mechanisms that may generate direction selectivity in the retina. In this final Section, we briefly outline some of the most important questions that need to be answered.

Morphology. The most pressing issue concerns the identity of the inhibitory inter-neurones. For the bistratified On-Off DS cells, the current working hypothesis is that the null-direction inhibition is mediated by two matching populations of GABAergic amacrine cells, which would provide independent input to the On and Off dendritic strata. Moreover, each population is expected to show at least a four-fold dendritic-field coverage, commensurate with the presumed coverage of the four subtypes of On-Off DS cells. A recent survey of rabbit retinal amacrine cells by MacNeil et al. (1999) did not identify cell types that meet these criteria, apart from the matching populations of starburst amacrine cells. This result suggests that our identikit picture of the "most-wanted" amacrine cell may be misleading in some important respects.

It is apparent, however, that there are not enough amacrine cells in the rabbit retina for each DS subunit to have its own inhibitory amacrine cell, as shown in block diagrams of the DS mechanism. A population of amacrine cells with the density of the CD15 bipolar cells would account for about 10% of the amacrine cells (Brown and Masland 1999); this is equivalent to the proportion of the AII amacrine cells, which are by far the most common type of amacrine cell (Vaney 1990; MacNeil and Masland 1998). Matching populations of inhibitory inter-neurones for the On and Off dendritic strata would thus require 20% of the amacrine cells, and dedicated populations for each of the four subtypes of On-Off DS cells would require 80% of the amacrine cells. Although it has been suggested that the program of identification and classification of retinal neurones begun by Cajal is nearing completion (MacNeil and Masland 1998), much work remains to be done in characterizing the GABAergic wide-field amacrine cells, including those with spatially disparate dendritic and axonal fields (Dacey 1988; Vaney et al. 1988; Famiglietti 1992a), which could provide the input-output offset neces-sary for generating motion sensitivity (Werblin 1991).

Physiology. The functional properties of the amacrine cells that mediate the null-direction inhibition may provide important clues to the mechanism of directional selectivity. Diverse pharmacological agents have been shown to affect direction selectivity in the rabbit retina and it will be interesting to test their effects on can-didate inhibitory amacrine cells. For example, specific omega-conotoxins that block Q-type calcium channels abolish the direction selectivity of the On-Off DS cells without blocking the excitatory responses (Jensen 1995). Low concentrations of NBQX have a similar effect and this AMPA antagonist is thought to block the glutamatergic input to amacrine cells from bipolar cells (Cohen and Miller 1995).

Paired recordings between an inhibitory amacrine cell and a DS ganglion cell would be very instructive. The On-Off DS cells can already be identified in vitro with a high success rate (Vaney 1994b; Yang and Masland 1994) and the challenge will be to selectively label the inhibitory amacrine cells. This may be possible with present transgenic techniques because the target amacrine cells might selectively express a particular type of calcium channel, and this would provide a basis for selectively labelling the cells with green fluorescent protein. Without resorting to paired recordings, laser ablation of the fluorescent amacrine cells under microscopic control could demonstrate that the cells provide the null-direction inhibitory input to the DS ganglion cells, in the same way that He and Masland (1997) demonstrated that the starburst amacrine cells provide non-directional facilitatory input.

References

Amthor FR, Grzywacz NM (1993a) Directional selectivity in vertebrate retinal ganglion cells. In: Miles, FA, Wallman J (eds) Visual motion and its role in the stabilization of gaze. Elsevier, Amsterdam, pp 79-100

Amthor FR, Grzywacz NM (1993b) Inhibition in ON-OFF directionally selective ganglion cells of the rabbit retina. J Neurophysiol 69: 2174-2187

Amthor FR, Grzywacz NM (1994) Morphological and physiological basis of starburst-ACh amacrine input to directionally selective (DS) ganglion cells in rabbit retina. Soc Neurosci Abstr 20: 217

Amthor FR, Oyster CW (1995) Spatial organization of retinal information about the direction of image motion. Proc Natl Acad Sci USA 92: 4002-4005

Amthor FR, Oyster CW, Takahashi ES (1984) Morphology of on-off direction-selective ganglion cells in the rabbit retina. Brain Res 298: 187-190

Amthor FR, Takahashi ES, Oyster CW (1989a) Morphologies of rabbit retinal ganglion cells with concentric receptive fields. J Comp Neurol 280: 72-96

Amthor FR, Takahashi ES, Oyster CW (1989b) Morphologies of rabbit retinal ganglion cells with complex receptive fields. J Comp Neurol 280: 97-121

Amthor FR, Grzywacz NM, Merwine DK (1996) Extra-receptive-field motion facilitation in on-off directionally selective ganglion cells of the rabbit retina. Vis Neurosci 13: 303-309

Ariel M, Adolph AR (1985) Neurotransmitter inputs to directionally sensitive turtle retinal ganglion cells. J Neurophysiol 54: 1123-1143

Ariel M, Daw NW (1982) Pharmacological analysis of directionally sensitive rabbit retinal ganglion cells. J Physiol 324: 161-185

Baldridge WH (1996) Optical recordings of the effects of cholinergic ligands on neurons in the ganglion cell layer of mammalian retina. J Neurosci 16: 5060-5072

Barlow HB, Hill RM (1963) Selective sensitivity to direction of motion in ganglion cells of the rabbit's retina. Science 139: 412-414

Barlow HB, Levick WR (1965) The mechanism of directionally selective units in rabbit's retina. J Physiol 178: 477-504

Barlow HB, Hill RM, Levick WR (1964) Retinal ganglion cells responding selectively to direction and speed of image motion in the rabbit. J Physiol 173: 377-407

Bloomfield SA, Miller RF (1986) A functional organization of ON and OFF pathways in the rabbit retina. J Neurosci 6: 1-13

Borg-Graham LJ, Grzywacz N (1992) A model of the direction selectivity circuit in retina: transformations by neurons singly and in concert. In: McKenna T, Davis J, Zornetzer SF (eds) Single neuron computation. Academic Press, San Diego, pp 347-375

Borst A, Egelhaaf M (1989) Principles of visual motion detection. Trends Neurosci 12: 297-306

Borst A, Egelhaaf M (1990) Direction selectivity of blowfly motion-sensitive neurons is computed in a two-stage process. Proc Natl Acad Sci USA 87: 9363-9367

Brandon C (1987) Cholinergic neurons in the rabbit retina: dendritic branching and ultra-structural connectivity. Brain Res 426: 119-130

Brandon C, Criswell MH (1997) Rabbit retinal ganglion cells that project to the medial terminal nucleus are directionally-selective. Soc Neurosci Abstr 23: 1023

Brandstätter JH, Greferath U, Euler T, Wässle H (1995) Co-stratification of GABA$_A$ receptors with the directionally selective circuitry of the rat retina. Vis Neurosci 12: 345-358

Brecha N, Johnson D, Peichl L, Wässle H (1988) Cholinergic amacrine cells of the rabbit retina contain glutamate decarboxylase and gamma-aminobutyrate immunoreactivity. Proc Natl Acad Sci USA 85: 6187-6191

Brown SP, Masland RH (1999) Costratification of a population of bipolar cells with the direction-selective circuitry of the rabbit retina. J Comp Neurol 408: 97-106

Buhl EH, Peichl L (1986) Morphology of rabbit retinal ganglion cells projecting to the medial terminal nucleus of the accessory optic system. J Comp Neurol 253: 163-174

Caldwell JH, Daw NW, Wyatt HJ (1978) Effects of picrotoxin and strychnine on rabbit retinal ganglion cells: lateral interactions for cells with more complex receptive fields. J Physiol 276: 277-298

Cleland BG, Levick WR (1974) Properties of rarely encountered types of ganglion cells in the cat's retina and an overall classification. J Physiol 240: 457-492

Cohen ED, Miller RF (1994) The role of NMDA and non-NMDA excitatory amino acid receptors in the functional organization of primate retinal ganglion cells. Vis Neurosci 11: 317-332

Cohen ED, Miller RF (1995) Quinoxalines block the mechanism of directional selectivity in ganglion cells of the rabbit retina. Proc Natl Acad Sci U S A 92: 1127-1131

Collewijn H (1969) Optokinetic eye movements in the rabbit: input-output relations. Vision Res 9: 117-132

Collewijn H (1975) Direction-selective units in the rabbit's nucleus of the optic tract. Brain Res 100: 489-508

Dacey DM (1988) Dopamine-accumulating retinal neurons revealed by in vitro fluorescence display a unique morphology. Science 240: 1196-1198

DeMonasterio FM (1978) Properties of ganglion cells with atypical receptive-field organization in retina of macaques. J Neurophysiol 41: 1435-1449

DeVoe RD, Carras PL, Criswell MH, Gur RB (1989) Not by ganglion cells alone: directional selectivity is widespread in identified cells of the turtle retina. In: Weiler R, Osborne NN (eds) Neurobiology of the inner retina. Springer, Berlin, pp 233-246

DeVries SH, Baylor DA (1995) An alternative pathway for signal flow from rod photoreceptors to ganglion cells in mammalian retina. Proc Natl Acad Sci USA 92: 10658-10662

DeVries SH, Baylor DA (1997) Mosaic arrangement of ganglion cell receptive fields in rabbit retina. J Neurophysiol 78: 2048-2060

Eccles JC (1964) The physiology of synapses. Springer, Berlin

Enz R, Brandstätter JH, Wässle H, Bormann J (1996) Immunocytochemical localization of the GABA$_C$ receptor rho subunits in the mammalian retina. J Neurosci 16: 4479-4490

Euler T, Wässle H (1998) Different contributions of GABA$_A$ and GABA$_C$ receptors to rod and cone bipolar cells in a rat retinal slice preparation. J Neurophysiol 79: 1384-1395

Famiglietti EV (1983) 'Starburst' amacrine cells and cholinergic neurons: mirror-symmetric on and off amacrine cells of rabbit retina. Brain Res 261: 138-144

Famiglietti EV (1987) Starburst amacrine cells in cat retina are associated with bistratified, presumed directionally selective, ganglion cells. Brain Res 413: 404-408

Famiglietti EV (1989) Structural organization and development of dorsally-directed (vertical) asymmetrical amacrine cells in rabbit retina. In: Weiler R, Osborne NN (eds) Neurobiology of the inner retina. Springer, Berlin, pp 169-180

Famiglietti EV (1991) Synaptic organization of starburst amacrine cells in rabbit retina: analysis of serial thin sections by electron microscopy and graphic reconstruction. J Comp Neurol 309: 40-70

Famiglietti EV (1992a) Polyaxonal amacrine cells of rabbit retina: PA2, PA3, and PA4 cells. Light and electron microscopic studies with a functional interpretation. J Comp Neurol 316: 422-446

Famiglietti EV (1992b) New metrics for analysis of dendritic branching patterns demonstrating similarities and differences in ON and ON-OFF directionally selective retinal ganglion cells. J Comp Neurol 324: 295-321

Famiglietti EV (1992c) Dendritic co-stratification of ON and ON-OFF directionally selective ganglion cells with starburst amacrine cells in rabbit retina. J Comp Neurol 324: 322-335

Famiglietti EV, Tumosa N (1987) Immunocytochemical staining of cholinergic amacrine cells in rabbit retina. Brain Res 413: 398-403

Farmer SG, Rodieck RW (1982) Ganglion cells of the cat accessory optic system: morphology and retinal topography. J Comp Neurol 205: 190-198

Feigenspan A, Bormann J (1998) GABA-gated Cl⁻channels in the rat retina. Prog Retinal Eye Res 17: 99-126

Giolli RA (1961) An experimental study of the accessory optic tracts (transpeduncular tracts and anterior accessory optic tracts) in the rabbit. J Comp Neurol 121: 89-108

Goldberg JM, Fernandez C (1971) Physiology of peripheral neurons innervating semicircular canals of the squirrel monkey. I. Resting discharge and response to constant angular accelerations. J Neurophysiol 34: 635-660

Grzywacz NM, Amthor FR (1989) A computationally robust anatomical model for retinal directional selectivity. In: Touretzky DS (ed) Advances in neural information processing systems I. Morgan Kaufmann, New York, pp 477-484

Grzywacz NM, Amthor FR (1993) Facilitation in ON-OFF directionally selective ganglion cells of the rabbit retina. J Neurophysiol 69: 2188-2199

Grzywacz NM, Amthor FR, Merwine DK (1994) Directional hyperacuity in ganglion cells of the rabbit retina. Vis Neurosci 11: 1019-1025

Grzywacz NM, Tootle JS, Amthor FR (1997) Is the input to a GABAergic or cholinergic synapse the sole asymmetry in rabbit's retinal directional selectivity? Vis Neurosci 14: 39-54

Grzywacz NM, Amthor FR, Merwine DK (1998) Necessity of acetylcholine for retinal directionally selective responses to drifting gratings in rabbit. J Physiol 512: 575-581

Hassenstein B, Reichardt W (1956) Systemtheoretische Analyse der Zeit-, Reihenfolgen- und Vorzeichenauswertung bei der Bewegungsperzeption der Rüsselkafers, *Chlorophanus*. Z Naturforsch 11b: 513-524

He S (1994) Further investigations of direction-selective ganglion cells of the rabbit retina. PhD Thesis, Australian National University

He S, Masland RH (1997) Retinal direction selectivity after targeted laser ablation of starburst amacrine cells. Nature 389: 378-382

He S, Masland RH (1998) ON direction-selective ganglion cells in the rabbit retina: dendritic morphology and pattern of fasciculation. Vis Neurosci 15: 369-375

Ikeda H, Wright MJ (1972) Receptive field organization of 'sustained' and 'transient' retinal ganglion cells which subserve different functional roles. J Physiol (Lond) 227: 769-800

Jensen RJ (1995) Effects of Ca^{2+} channel blockers on directional selectivity of rabbit retinal ganglion cells. J Neurophysiol 74: 12-23

Kier CK, Buchsbaum G, Sterling P (1995) How retinal microcircuits scale for ganglion cells of different size. J Neurosci 15: 7673-7683

Kittila CA, Granda AM (1994) Functional morphologies of retinal ganglion cells in the turtle. J Comp Neurol 350: 623-645

Kittila CA, Massey SC (1995) Effect of ON pathway blockade on directional selectivity in the rabbit retina. J Neurophysiol 73: 703-712

Kittila CA, Massey SC (1997) Pharmacology of directionally selective ganglion cells in the rabbit retina. J Neurophysiol 77: 675-689

Koch C, Poggio T, Torre V (1982) Retinal ganglion cells: a functional interpretation of dendritic morphology. Philos Trans Roy Soc Lond B 298: 227-263

Koch C, Poggio T, Torre V (1983) Nonlinear interactions in a dendritic tree: localization, timing, and role in information processing. Proc Natl Acad Sci USA 80: 2799-2802

Koch C, Poggio T, Torre V (1986) Computations in the vertebrate retina: gain enhancement, differentiation and motion discrimination. Trends Neurosci 9: 204-211

Kogo N, Rubio DM, Ariel M (1998) Direction tuning of individual retinal inputs to the turtle accessory optic system. J Neurosci 18: 2673-2684

Kolb H, Nelson R (1984) Neural architecture of the cat retina. Prog Retinal Res 3: 21-60

Levick WR (1996) Receptive fields of cat retinal ganglion cells with special reference to the alpha cells. Prog Retinal Eye Res 15: 457-500

Levick WR, Thibos LN (1983) Receptive fields of cat ganglion cells: classification and construction. Prog Retinal Res 2: 267-319

Levick WR, Oyster CW, Takahashi E (1969) Rabbit lateral geniculate nucleus: sharpener of directional information. Science 165: 712-714

Linn DM, Massey SC (1992) GABA inhibits ACh release from the rabbit retina: a direct effect or feedback to bipolar cells? Vis Neurosci 8: 97-106

Linn DM, Blazynski C, Redburn DA, Massey SC (1991) Acetylcholine release from the rabbit retina mediated by kainate receptors. J Neurosci 11: 111-122

MacNeil MA, Masland RH (1998) Extreme diversity among amacrine cells: implications for function. Neuron 20: 971-982

MacNeil MA, Heussy JK, Dacheux RF, Raviola E, Masland RH (1999) The shapes and numbers of amacrine cells: matching of photofilled with Golgi-stained cells in the rabbit retina and comparison with other mammalian species. J Comp Neurol 413: 305-326

Marchiafava PL (1979) The responses of retinal ganglion cells to stationary and moving visual stimuli. Vision Res 19: 1203-1211

Mariani AP (1982) Association amacrine cells could mediate directional selectivity in pigeon retina. Nature 298: 654-655

Masland RH, Ames A (1976) Responses to acetylcholine of ganglion cells in an isolated mammalian retina. J Neurophysiol 39: 1220-1235

Masland RH, Mills JW, Hayden SA (1984) Acetylcholine-synthesizing amacrine cells: identification and selective staining by using radioautography and fluorescent markers. Proc Roy Soc Lond B 223: 79-100

Massey SC, Miller RF (1990) N-methyl-D-aspartate receptors of ganglion cells in rabbit retina. J Neurophysiol 63: 16-30

Massey SC, Mills SL (1996) A calbindin-immunoreactive cone bipolar cell type in the rabbit retina. J Comp Neurol 366: 15-33

Massey SC, Linn DM, Kittila CA, Mirza W (1997) Contributions of $GABA_A$ receptors and $GABA_C$ receptors to acetylcholine release and directional selectivity in the rabbit retina. Vis Neurosci 14: 939-948

Meister M, Lagnado L, Baylor DA (1995) Concerted signaling by retinal ganglion cells. Science 270: 1207-1210

Merwine DK, Amthor FR, Grzywacz NM (1995) Interaction between center and surround in rabbit retinal ganglion cells. J Neurophysiol 73: 1547-1567

Miles FA (1972) Centrifugal control of the avian retina. I. Receptive field properties of retinal ganglion cells. Brain Res 48: 65-92

Miles FA (1993) The sensing of rotational and translational optic flow by the primate optokinetic system. In: Miles FA, Wallman J (eds) Visual motion and its role in the stabilization of gaze. Elsevier, Amsterdam, pp 393-403

Millar TJ, Morgan IG (1987) Cholinergic amacrine cells in the rabbit retina synapse onto other cholinergic amacrine cells. Neurosci Lett 74: 281-285.

Miller RF (1979) The neuronal basis of ganglion-cell receptive-field organization and the physiology of amacrine cells. In: Schmitt FO, Worden FG (eds) The neurosciences: fourth study program. MIT Press, Cambridge, pp 227-245

Mills SL, Massey SC (1992) Morphology of bipolar cells labeled by DAPI in the rabbit retina. J Comp Neurol 321: 133-149

Nelson R (1977) Cat cones have rod input: a comparison of the response properties of cones and horizontal cell bodies in the retina of the cat. J Comp Neurol 172: 109-135

Oyster CW (1968) The analysis of image motion by the rabbit retina. J Physiol 199: 613-635

Oyster CW (1990) Neural interactions underlying direction-selectivity in the rabbit retina. In: Blakemore C (ed) Vision: coding and efficiency. Cambridge University Press, Cambridge, pp 92-102

Oyster CW, Barlow HB (1967) Direction-selective units in rabbit retina: distribution of preferred directions. Science 155: 841-842

Oyster CW, Takahashi E, Levick WR (1971) Information processing in the rabbit visual system. Doc Ophthalmol 30: 161-204

Oyster CW, Takahashi E, Collewijn H (1972) Direction-selective retinal ganglion cells and control of optokinetic nystagmus in the rabbit. Vision Res 12: 183-193

Oyster CW, Simpson JI, Takahashi ES, Soodak RE (1980) Retinal ganglion cells projecting to the rabbit accessory optic system. J Comp Neurol 190: 49-61

Oyster CW, Amthor FR, Takahashi ES (1993) Dendritic architecture of ON-OFF direction-selective ganglion cells in the rabbit retina. Vision Res 33: 579-608

Panico J, Sterling P (1995) Retinal neurons and vessels are not fractal but space-filling. J Comp Neurol 361: 479-490

Perry VH, Walker M (1980) Amacrine cells, displaced amacrine cells and interplexiform cells in the retina of the rat. Proc Roy Soc Lond B 208: 415-431

Peters BN, Masland RH (1996) Responses to light of starburst amacrine cells. J Neurophysiol 75: 469-480

Poggio T, Torre V (1981) A theory of synaptic interactions. In: Reichardt WE, Poggio T, (eds) Theoretical approaches in neurobiology. MIT Press, Cambridge, pp 28-38

Poznanski RR (1992) Modelling the electrotonic structure of starburst amacrine cells in the rabbit retina: a functional interpretation of dendritic morphology. Bull Math Biol 54: 905-928

Pu ML, Amthor FR (1990) Dendritic morphologies of retinal ganglion cells projecting to the nucleus of the optic tract in the rabbit. J Comp Neurol 302: 657-674

Rademaker GGJ, Ter Braak JWG (1948) On the central mechanism of some optic reactions. Brain 71: 48-76

Raviola G, Raviola E (1967) Light and electron microscopic observations on the inner plexiform layer of the rabbit retina. Am J Anat 120: 403-425

Reichardt W (1961) Autocorrelation, a principle for the evaluation of sensory information by the central nervous system. In: Rosenblith W (ed) Sensory communication. John Wiley, New York, pp 303-317

Rodieck RW (1988) The primate retina. In: Horst HD, Erwin J (eds) Comparative primate biology, Vol 4, neurosciences. Alan R Liss, New York, pp 203-278

Rodieck RW (1998) The first steps in seeing. Sinauer, Sunderland MA

Rosenberg AF, Ariel M (1991) Electrophysiological evidence for a direct projection of direction-sensitive retinal ganglion cells to the turtle's accessory optic system. J Neurophysiol 65: 1022-1033

Schiller PH, Malpeli JG (1977) Properties and tectal projections of monkey retinal ganglion cells. J Neurophysiol 40: 428-445

Simpson JI (1984) The accessory optic system. Ann Rev Neurosci 7: 13-41

Simpson JI, Soodak RE, Hess R (1979) The accessory optic system and its relation to the vestibulocerebellum. Prog Brain Res 50: 715-724

Simpson JI, Leonard CS, Soodak RE (1988) The accessory optic system of the rabbit. II. Spatial organization of direction selectivity. J Neurophysiol 60: 2055-2072

Smith RD, Grzywacz NM, Borg-Graham LJ (1996) Is the input to a GABAergic synapse the sole asymmetry in turtle's retinal directional selectivity? Vis Neurosci 13: 423-439

Smith RG, Freed MA, Sterling P (1986) Microcircuitry of the dark-adapted cat retina: functional architecture of the rod-cone network. J Neurosci 6: 3505-3517

Soodak RE, Simpson JI (1988) The accessory optic system of the rabbit. I. Basic visual response properties. J Neurophysiol 60: 2037-2054

Tauchi M, Masland RH (1984) The shape and arrangement of the cholinergic neurons in the rabbit retina. Proc Roy Soc Lond B 223: 101-119

Tauchi M, Masland RH (1985) Local order among the dendrites of an amacrine cell population. J Neurosci 5: 2494-2501

Taylor WR, Wässle H (1995) Receptive field properties of starburst cholinergic amacrine cells in the rabbit retina. Eur J Neurosci 7: 2308-2321

Taylor WR, Chen E, Copenhagen DR (1995) Characterization of spontaneous synaptic currents in salamander retinal ganglion cells. J Physiol 486: 207-221

Torre V, Poggio T (1978) A synaptic mechanism possibly underlying directional selectivity to motion. Proc Roy Soc Lond B 202: 409-416.

Vaney DI (1984) 'Coronate' amacrine cells in the rabbit retina have the 'starburst' dendritic morphology. Proc Roy Soc Lond B 220: 501-508

Vaney DI (1990) The mosaic of amacrine cells in the mammalian retina. Prog Retinal Res 9: 49-100

Vaney DI (1991) Many diverse types of retinal neurons show tracer coupling when injected with biocytin or Neurobiotin. Neurosci Lett 125: 187-190

Vaney DI (1994a) Patterns of neuronal coupling in the retina. Prog Retinal Eye Res 13: 301-355

Vaney DI (1994b) Territorial organization of direction-selective ganglion cells in rabbit retina. J Neurosci 14: 6301-6316

Vaney DI, Pow DV (2000) The dendritic architecture of the cholinergic plexus in the rabbit retina: selective labeling by glycine accumulation in the presence of sarcosine. J Comp Neurol 421: 1-13)

Vaney DI, Young HM (1988) GABA-like immunoreactivity in cholinergic amacrine cells of the rabbit retina. Brain Res 438: 369-373

Vaney DI, Levick WR, Thibos LN (1981a) Rabbit retinal ganglion cells. Receptive field classification and axonal conduction properties. Exp Brain Res 44: 27-33

Vaney DI, Peichl L, Wässle H, Illing RB (1981b) Almost all ganglion cells in the rabbit retina project to the superior colliculus. Brain Res 212: 447-453

Vaney DI, Peichl L, Boycott BB (1988) Neurofibrillar long-range amacrine cells in mammalian retinae. Proc Roy Soc Lond B 235: 203-219

Vaney DI, Collin SP, Young HM (1989) Dendritic relationships between cholinergic amacrine cells and direction-selective retinal ganglion cells. In: Weiler R, Osborne NN (eds) Neurobiology of the inner retina. Springer, Berlin, pp 157-168

Wallman J (1993) Subcortical optokinetic mechanisms. In: Miles FA, Wallman J (eds) Visual motion and its role in the stabilization of gaze. Elsevier, Amsterdam, pp 321-369

Wässle H, Peichl L, Boycott BB (1981) Dendritic territories of cat retinal ganglion cells. Nature 292: 344-345

Werblin F (1991) Synaptic connections, receptive fields, and patterns of activity in the tiger salamander retina. A simulation of patterns of activity formed at each cellular level from photoreceptors to ganglion cells. Invest Ophthalmol Vis Sci 32: 459-483

Wong RO (1990) Differential growth and remodelling of ganglion cell dendrites in the postnatal rabbit retina. J Comp Neurol 294: 109-132

Wyatt HJ, Daw NW (1975) Directionally sensitive ganglion cells in the rabbit retina: specificity for stimulus direction, size, and speed. J Neurophysiol 38: 613-626

Wyatt HJ, Daw NW (1976) Specific effects of neurotransmitter antagonists on ganglion cells in rabbit retina. Science 191: 204-205

Yang G, Masland RH (1992) Direct visualization of the dendritic and receptive fields of directionally selective retinal ganglion cells. Science 258: 1949-1952

Yang G, Masland RH (1994) Receptive fields and dendritic structure of directionally selective retinal ganglion cells. J Neurosci 14: 5267-5280

Young HM, Vaney DI (1991) Rod-signal interneurons in the rabbit retina: 1. Rod bipolar cells. J Comp Neurol 310: 139-153

Zhou ZJ, Fain GL (1995) Neurotransmitter receptors of starburst amacrine cells in rabbit retinal slices. J Neurosci 15: 5334-5345

Zhou ZJ, Fain GL, Dowling JE (1993) The excitatory and inhibitory amino acid receptors on horizontal cells isolated from the white perch retina. J Neurophysiol 70: 8-19

Identification of Mechanisms Underlying Motion Detection in Mammals

Michael Ibbotson

Developmental Neurobiology Group, Research School of Biological Sciences, Australian National University, Canberra, Australia

1. Introduction

Direction-selective visual neurones have been found in every mammalian visual system in which physiologists have searched for them. In the present article I will describe experiments that have revealed quite different motion processing mechanisms in two of the studied species, the rabbit and the wallaby, *Macropus eugenii*. Vaney et al. (this volume) review the current state of knowledge on direction-selective (DS) mechanisms in the rabbit retina. Here I will briefly recapitulate the proposed mechanisms that lead to DS responses in rabbit On-Off DS retinal ganglion cells, with specific reference to the non-linear interaction that is integral to the DS mechanism. The data from the rabbit retina will then be compared to results from horizontally tuned DS neurones in the nucleus of the optic tract (NOT) of the wallaby (e.g. Ibbotson et al. 1994, 1998, 1999). This comparison is interesting because it has been suggested that On-Off DS retinal ganglion cells form part of the input to the NOT in the rabbit (Oyster et al. 1972). If similar motion processing strategies are used in both the rabbit and wallaby retinas, it might be expected that evidence for the same DS mechanisms that are found in the rabbit On-Off DS neurones should be revealed in at least some of the wallaby NOT neurones (Ibbotson et al. 1994). As will be described, this is not the case, possibly indicating species-specific processing strategies.

Before discussing potential DS mechanisms, I address the reason why the non-linear mechanism is so important to the process of computing a directional signal. Several criteria must be met in order to calculate a directional response from a moving image (Borst and Egelhaaf 1989). First, two spatially displaced inputs are necessary because motion is a vector. Second, there must be a temporal asymmetry in the processing of the signals from the two inputs. Third, the interaction between the signals must be non-linear (Poggio and Reichardt 1973). The non-linearity is essential because in a purely linear system the mean time-averaged output would be the same for opposite directions of motion, even though the modulation strength for the two directions might be different (Ibbotson et al. 1999). Several types of non-linear interactions have been considered in biological motion detecting systems (Amthor and Grzywacz 1993a). The two that will be discussed here are division (Section 2) and multiplication (Section 3). These non-linearities are discussed because, respectively, they are thought to generate the non-linear operations in the motion detectors feeding into rabbit ON-OFF DS retinal ganglion cells and wallaby NOT neurones.

2. Non-linear mechanisms in rabbit DS retinal ganglion cells

The original study of direction-selectivity in On-Off DS cells in the rabbit retina described the directional response properties of the neurones and proposed two possible schemes for the DS mechanism (Barlow and Levick 1965), one based on inhibition and the other on excitation (e.g. Fig. 4 in Vaney et al., this volume). Both schemes have two spatially separated input channels and the signal in one channel is delayed to generate the temporal asymmetry required for a direction-selective detector. The signals from both channels interact in a nonlinear fashion after temporal filtering, and the nature of this interaction is the distinguishing feature between the schemes.

In the inhibitory scheme the non-linear stage can be modelled as a division-like process (Fig. 1a) while in the excitatory scheme it is modelled as a facilitatory interaction (e.g. Fig. 1b, see Section 3). The work of Barlow and Levick (1965) and subsequent experimenters (Amthor and Grzywacz 1993b) strongly favoured a division-like mechanism to produce the non-linear interaction in the motion detectors feeding the On-Off DS retinal ganglion cells of the rabbit. Figure 1a shows a version of the Barlow-Levick inhibitory scheme, as suggested by Torre and Poggio (1978), in which a division-like inhibitory mechanism forms the essential non-linear operation. A synaptic mechanism that can produce division-like interactions between incoming signals is shunting inhibition. In this type of synapse the reversal potential is close to the cell's resting potential. Inhibition is generated by increasing the membrane conductance and shunting incoming currents out from the cell. The division-like interaction occurs because shunting inhibition divides the excitatory currents by the membrane conductance.

To clarify the type of non-linear interaction in the rabbit On-Off DS neurones, Amthor and Grzywacz (1993b) recorded the responses of the cells to apparent motion in the null direction. Over a century ago, Exner (1875) showed that sequential flashing of two neighbouring light sources created an impression of apparent motion in humans. As with the work of Amthor and Grzywacz, modified forms of this stimulus are still commonly used today to reveal the response properties of direction-selective neurones (e.g. Franceschini et al. 1989; Egelhaaf and Borst 1992; Emerson et al. 1992; Ibbotson et al. 1998, 1999). For the rabbit studies, the apparent motion stimulus consisted of two slits in which the contrast could be increased or decreased in a stepwise fashion. First, Amthor and Grzywacz (1993b) measured the responses to stepped changes in stimulus brightness in each slit independently. Then they measured the responses to apparent motion in the null direction, i.e. the stimulus consisted of a contrast change in one slit followed a short time later by a contrast change in the second slit. A response-versus-contrast function was measured for single slit stimulation and for apparent motion. In the apparent motion experiments the contrast was held constant in the first slit while a range of contrasts were tested in the second slit. In this way, a response-versus-contrast function was measured for several different first slit contrasts. It was found that increasing the contrast of the first slit shifted the response-versus-contrast functions for apparent motion stimulation downwards. The observed shift could be modelled by dividing the response-versus-contrast function derived from single slit stimulation by a constant factor greater than one. Therefore, the null direction inhibition was concluded to be a non-linear, division-like process similar to that expected from Barlow and Levick's inhibitory scheme (see Torre and Poggio 1978).

Grzywacz and Amthor (1993) also studied the responses of the neurones to preferred direction motion using apparent motion stimulation. Again, they measured the responses to stepped changes in stimulus brightness in each slit independently. Then they measured the responses to simultaneous stimulation of both slits and noticed that the responses were more than twice the size of the sum of the responses to individual stimulation, suggesting a non-linear interaction between the two inputs that is not motion related. Finally, they measured the responses to apparent motion in the preferred direction. These responses were only slightly larger than the responses to simultaneous stimulation of both slits (Fig. 2 in Grzywacz and Amthor 1993). When measuring the response-versus-contrast functions produced by stimulating the second slit with several different first slit contrasts, they found that increasing the first slit contrast additively facilitated the responses elicited by the second slit. That is, the response-versus-contrast functions shifted upwards with higher first slit contrasts in parallel with the function derived during single slit stimulation. Therefore, Grzywacz and Amthor concluded that preferred direction facilitation is addition-like, i.e. linear. The work on the rabbit neurones suggests that motion is coded differently depending on the direction in which the stimulus moves relative to the motion detectors. For null direction motion, the mechanism is a division-like non-linearity but for preferred direction motion there is a linear facilitation.

3. Non-linear mechanisms in the nucleus of the optic tract

This section reviews experiments that show that a facilitatory mechanism similar to that expected in the Reichardt model (Reichardt 1961) is in operation in the motion detectors feeding the neurones of the nucleus of the optic tract (NOT) in the wallaby. In the subunits of the Reichardt model, signals from two spatially separated inputs are multiplied together after one signal is temporally filtered (Fig. 1b). As the magnitudes of the input signals are dependent on the contrast of the stimulus, it should be possible to show that the magnitudes of signals entering the inputs to the detector are related to the responses of the detector in a quadratic fashion. That is, if the contrast of the inputs is doubled, the motion detector response should be quadrupled. The complete Reichardt model (Fig. 1c) consists of two multiplicative subunits like those shown in figure 1b. The subunits have opposite preferred directions, one with an inhibitory output and the other with an excitatory output. The reason for the opponent arrangement of subunits is that individually the subunits respond quite strongly to non-moving stimuli (e.g. a flash of light) as well as moving images. As both subunits give the same response to direction-independent signals, that component of the signal can be removed by subtracting the output of one subunit from the response of the other. The direction-specific component of the response will remain because it is different for the two subunits. It is important to note that in the Reichardt detector a facilitatory non-linear interaction between input signals is responsible for both the excitatory response in the preferred direction and the inhibitory response in the anti-preferred direction.

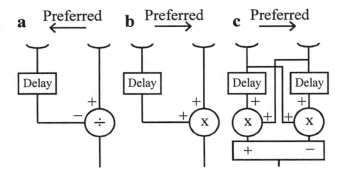

Fig. 1 Motion detector models. **a** A practical version of the Barlow-Levick (1965) model proposed by Torre and Poggio (1978). A division-like shunting inhibition forms the essential non-linear operation. **b** A facilitatory model where the nonlinear operation is a direct multiplication between the signals in the input channels of the detector. This model forms the subunits of the Reichardt model shown in **c** The Reichardt detector consists of two subunits with opposite preferred directions that feed into a subtraction stage.

Ibbotson et al. (1999) conducted a series of experiments on the wallaby NOT in which apparent motion was produced by presenting a step-like contrast change in one slit (A) followed by the same contrast step in an adjacent slit (B) a short time later. The increment in slit A produced a non-motion response and the subsequent contrast in slit B elicited a far larger motion response than occurred during stimulation of slit B alone. Stimulation of slits A and B simultaneously produced responses that were 30-50% larger than the response to stimulation of slit A or B alone. This non-motion facilitation is far less prominent than the facilitation obtained in rabbit On-Off DS ganglion cells by stimulating both slits simultaneously, which could be up to five times larger than the response to single slit stimulation (Grzywacz and Amthor 1993). In the wallaby NOT, the responses to apparent motion in the preferred direction were always larger than the responses obtained by taking the sum of the responses to single slit stimulation. This confirms that the motion-related interaction is non-linear. The apparent motion sequence in the opposite direction (B to A) elicited a small onset response to stimulation of slit B, then the cell's spontaneous firing rate was transiently reduced to zero with subsequent contrast increments in slit A. Motion opposite to the preferred direction produces a clear reduction in spontaneous activity in NOT neurones. This direction is therefore called the anti-preferred direction, rather than the null direction. The observation that motion in the anti-preferred direction leads to a large reduction of the spontaneous firing rate, is evidence for a powerful non-linearity. The anti-preferred response is clearly different to that observed in rabbit DS retinal ganglion cells, where null direction apparent motion can lead to excitatory responses (Barlow and Levick 1965; Amthor and Grzywacz 1993b).

Measuring responses to apparent motion stimuli with a range of contrasts generated response-versus-contrast functions for the neurones, which were used to determine the type of non-linear interaction that occurred between the input channels of the motion detectors. It is important to stress that in these experiments the contrasts in both slits (A and B) were the same for each of the contrasts tested. The curves in figure 2 show the response-versus-contrast functions of a typical NOT neurone for movement in the preferred (open circles) and anti-preferred directions (crosses) and for single slit responses (stars). For apparent motion in the preferred direction, the response-versus-contrast function increases in a manner similar to a quadratic function up to contrasts of approximately 25% (Fig. 2). At higher contrasts, the response size increases with increasing contrast at a slower rate and eventually saturates. The responses of the cells to single slit stimulation increased significantly above the spontaneous firing rate with increasing contrast but less rapidly than the apparent motion responses. Apparent motion in the anti-preferred direction had little effect at low contrasts but elicited a reduction in the spontaneous firing rate of the cells at contrasts >10% (Fig. 2). The results indicate that the responses to contrast changes in the second slit of the apparent motion sequence were strongly attenuated as compared to the single pulse responses, indicating a substantial inhibitory influence during apparent motion in the anti-preferred direction.

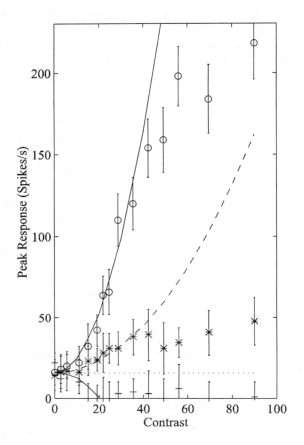

Fig. 2 Response-versus-contrast functions for an NOT neurone. Open circles show the peak response to preferred direction apparent motion, stars show the peak non-motion response in slits A and B and the crosses show the response to anti-preferred apparent motion. The peak response was the mean firing rate in the first 100 ms of the motion component of the response. The fitted lines are quadratic functions that were fit to the values for contrasts up to 25%. The fitted curves are shown at contrasts above 25% to demonstrate that the responses of the neurones have saturated. The horizontal dashed line is the spontaneous activity of the neurone. Error bars are standard deviations.

To compare data directly with results from rabbit On-Off DS ganglion cells, Ibbotson et al. (1999) employed a stimulus configuration similar to that of Grzywacz and Amthor (1993). In these experiments the contrast of one slit (A or B) was held constant and an apparent motion stimulus was generated by presenting a range of contrasts in the second slit a short time later (see Section 2). The response-versus-contrast functions of the responses to single slit contrast changes in each test area were also measured. It was found that multiplying the single flash curve by a constant factor for each first slit contrast provided a reasonable prediction for the size and shape of the motion-curves for preferred direction motion. This result would be expected if the interaction between input channels in the motion detectors is a multiplication. That is, the motion detectors resemble Reichardt-type motion detectors (Fig. 1c; Reichardt 1961; van Santen and Sperling 1985; Egelhaaf et al. 1989).

4. Motion opponency

In rabbit DS retinal ganglion cells, apparent motion in the preferred direction leads to a large increase in firing rate while motion in the null direction leads to a relatively small increase. These directional responses are ambiguous in that the same level of excitation may occur for motion in both directions. For example, a moderate increase in spike rate could be due to a low contrast pattern moving in the preferred direction or a high contrast pattern moving in the null direction. Unlike these retinal cells, NOT neurones in the wallaby and rabbit (Collewijn 1975) have high spontaneous activities. Motion in the preferred direction causes a large increase of firing rate in these cells, while motion in the anti-preferred direction reduces the spike rate below the spontaneous activity, i.e. the response is motion opponent. In terms of the direction of motion, the responses of NOT neurones are unambiguous, i.e. increased firing rate signals motion in the preferred direction while a firing rate below the spontaneous level signals motion in the anti-preferred direction. Motion opponent lateral geniculate neurones (LGN) in the rabbit have similar direction-selective properties to rabbit NOT neurones (Levick et al. 1969). It has been proposed that the responses of these LGN neurones come about by combining inputs from oppositely tuned On-Off DS retinal ganglion cells (see Vaney et al. this volume). That is, the input to an LGN neurone from one ganglion cell would be excitatory while the input from the neurone with opposite preferred direction would be inhibitory. This push-pull arrangement would generate a motion opponent response (Levick et al. 1969). The same logic can be applied to the neurones in the NOT. The motion opponent responses of wallaby NOT neurones suggest a local combination of directional precursor cells that are not necessarily themselves fully motion opponent (Ibbotson et al. 1994, 1999). That is, the precursor cells might have properties similar to the sub-units of the Reichardt detector (see Section 3).

5. Conclusions

The preferred direction responses of wallaby NOT neurones are generated by a powerful non-linear excitatory mechanism (Ibbotson et al. 1999). Due to an opponent combination of two units with opposite preferred directions (Fig. 1c), inhibitory responses in the NOT are understood as inverted outputs of the same mechanism. In rabbit On-Off DS retinal ganglion cells, the moderate preferred direction facilitation is generated by a linear mechanism while null direction inhibition is generated by a non-linear division-like mechanism (Fig. 1a; Amthor and Grzywacz 1993b; Grzywacz and Amthor 1993). This fundamental difference between the directional mechanisms in the rabbit retina and wallaby NOT can be interpreted in two ways. Firstly, the retinal input to the wallaby NOT originates from DS neurones with different properties to the DS On-Off ganglion cells in the rabbit. Studies of wallaby DS retinal ganglion cells are currently underway to assess this possibility. The input to NOT neurones in the rabbit appears to come primarily from On DS retinal ganglion cells rather than On-Off cells but functional input from the latter is present (Oyster et al. 1972; Pu and Amthor 1990). The directional mechanism present in the motion detectors feeding into the On DS cells in the rabbit is not known. Perhaps the mechanism in the On pathway of the rabbit has similarities to that found in the wallaby NOT pathway? Secondly, the input to the wallaby NOT may come mainly from another motion processing area, such as the cortex, and the DS mechanism in that region may operate differently to that in the On-Off DS cells of the rabbit. Certainly, evidence from the cat suggests that a motion detector mechanism with an excitatory facilitation for preferred direction motion exists in the visual cortex of that species (e.g., Emerson et al. 1992). Comparing the results from the wallaby NOT and rabbit retina demonstrate that care must be taken when deriving expectations about neuronal mechanisms in one species based on results from another. By comparing rabbit and wallaby data it was concluded that two quite different motion detector mechanisms are in operation in the two species, while it remains an open question whether each of these species can employ different types of fundamental mechanisms in different pathways. It must be concluded that no single motion detector mechanism can account for the directional response properties of all directional neurones in mammals.

References

Amthor FR, Grzywacz NM (1993a) Directional selectivity in vertebrate retinal ganglion cells. In: Miles FA, Wallman J (eds) Visual Motion and its role in the stabilization of Gaze. Elsevier, Amsterdam, pp 79-100

Amthor FR, Grzywacz NM (1993b) Inhibition in On-Off directionally selective ganglion cells in the rabbit retina. J Neurophysiol 69: 2174-2187

Barlow HB, Levick WR (1965) The mechanism of directionally selective units in rabbit's retina. J Physiol 178: 477-504

Borst A, Egelhaaf M (1989) Principles of visual motion detection. Trends Neurosci 12: 297-306

Collewijn H (1975) Direction-selective units in the rabbit's nucleus of the optic tract. Brain Res 100: 489-508

Egelhaaf M, Borst A, Reichardt W (1989) Computational structure of a biological motion detection system as revealed by local detector analysis. J Opt Soc Am A 6: 1070-1087

Egelhaaf M, Borst A (1992) Are there separate ON and OFF channels in fly motion vision? Visual Neurosci. 8: 151-164

Emerson RC, Bergen JR, Adelson EH (1992) Directionally selective complex cells and the computation of motion energy in cat visual cortex. Vision Res 32: 203-218

Exner S (1875) Experimentelle Untersuchung der einfachsten psychischen Processe. Pfluger's Arch Physiol 11: 403-432

Franceschini N, Riehle A, Le Nestour A (1989) Directionally selective motion detection by insect neurons. In: Stavenga DG, Hardie RC (eds) Facets of Vision. Springer Verlag, Berlin, Heidelberg, pp 360-390

Grzywacz NM, Amthor FR (1993) Facilitation in On-Off directionally selective ganglion cells in the rabbit retina. J Neurophysiol 69: 2188-2199

Ibbotson MR, Mark RF, Maddess TL (1994) Spatiotemporal response characteristics of direction-selective neurons in the nucleus of the optic tract and dorsal terminal nucleus of the wallaby, *Macropus eugenii*. J Neurophysiol 72: 2927-2943

Ibbotson MR, Clifford CWG, Mark RF (1998) Adaptation to visual motion in directional neurons of the nucleus of the optic tract. J Neurophysiol 79: 1481-1493

Ibbotson MR, Clifford CWG, Mark RF (1999) A quadratic nonlinearity underlies direction selectivity in the nucleus of the optic tract. Visual Neurosci 16: 991-1000

Levick WR, Oyster CW, Takahashi E (1969) Rabbit lateral geniculate nucleus: sharpener of directional information. Science 165: 712-714

Oyster CW, Takahashi E, Collewijn H (1972) Direction-selective retinal ganglion cells and control of optokinetic nystagmus in the rabbit. Vision Res 12: 183-193

Poggio T, Reichardt WE (1973) Considerations on models of movement detection. Kybernetik 13: 223-227

Pu ML, Amthor FR (1990) Dentritic morphologies of retinal ganglion cells projecting to the nucleus of the optic tract in the rabbit. J Comp Physiol 302: 657-674

Reichardt W (1961) Autocorrelation, a principle for the evaluation of sensory information by the central nervous system. In: Rosenblith WA (ed) Sensory communication. Wiley, New York, pp 303-317

Santen JPH, van Sperling G (1985) Elaborated Reichardt detectors. J Opt Soc Am A 2: 301-321

Torre V, Poggio T (1978) A synaptic mechanism possibly underlying directional selectivity to motion. Proc Roy Soc Lond 202: 501-508

Pathways in Dipteran Insects for Early Visual Motion Processing

John K. Douglass and Nicholas J. Strausfeld

Arizona Research Laboratories, Division of Neurobiology, University of Arizona, Tucson, USA

1. Introduction

In insects, as in vertebrates, neuroanatomical, electrophysiological, and modelling studies have provided insights regarding identities and connections among neurones that accomplish elementary motion detection. These studies include intracellular recordings from identified wide-field neurones that collate local information about motion, intracellular recordings from identified, mainly non-spiking small-field neurones that are candidates for a cardinal role in motion detection, and comparative anatomical studies of retinotopic neurones that are evolutionarily conserved across taxa. Nevertheless, many important features of motion processing in insects have yet to be revealed. This review concentrates on two questions: what are the identities and relationships among neurones that participate in elementary motion detection? And, are there distinct functional classes of elementary motion detectors (EMDs) in insects?

Previous reviews of insect motion processing have concentrated on recordings from wide-field motion-sensitive neurones in flies, on the possible roles of cholinergic and GABAergic systems in processing their inputs, and on modelling studies of hypothetical small-field motion detectors (Egelhaaf et al. 1988, 1989; Borst and Egelhaaf 1989; Egelhaaf and Borst 1993). Here, we focus on recent comparative anatomical and physiological studies of elements in a pathway that supports motion processing. Called the magnocellular pathway by analogy with comparable mammalian systems (Strausfeld and Lee 1991), the neurones of this pathway comprise relatively large-diameter and colour-insensitive elements that supply motion-sensitive tangential neurones in a tectum-like neuropil called the lobula plate. Intracellular recordings from identifiable small-field elements of this pathway provide a new level of understanding of motion detection in insects.

As in the vertebrate retina, motion detection in the insect optic lobe begins just two or at most three synapses removed from the photoreceptors. In flies, elementary motion detector circuits provide outputs to systems of collator neurones deep in the optic lobes. Certain of these elements provide information about global as well as local motion to premotor descending neurones. These descending neurones, in turn, integrate visual information with inputs from the halteres (organs of balance), and provide outputs to thoracic circuits that contribute to the stabilization of flight. Other collator neurones serve to relay ipsilateral information about motion to neuropils subserving the contralateral visual field. Such heterolateral connections provide inputs from both sides of the brain to nerve cells that integrate motion within the panoramic field of view of both eyes.

2. Spatial receptive fields of motion sensitive neurones adjust according to ambient light

One of these heterolateral collator neurones, called H1 (Hausen 1981), produces conventional action potentials and has a superficial axon near the front and dorsal brain surface, making it accessible for recording extracellularly the consequences of sequential stimulation of visual sampling units in the retina. A now classic experiment by Riehle and Franceschini (1982) demonstrated that the H1 neurone responds selectively to sequential illumination of adjacent visual sampling units, thereby establishing the minimum distance between inputs to an elementary motion detector circuit. Schuling et al. (1989), using a similar visual stimulus arrangement, showed that under conditions of very low ambient illumination ($5 \times 10^7 \text{q/cm}^2/\text{sec}$) or following dark-adaptation, more widely-separated visual sampling units contribute to these responses in H1. These conclusions confirmed previous inferences of spatial pooling, which had been based on behavioural (Pick and Buchner 1979) and electrophysiological (Srinivasan and Dvorak 1980) responses to wide-field motion and began at mean luminances 2-3 log units higher than those employed by Schuling et al. (1989). Together, these studies show that in flies, as in the rabbit, directional selectivity is maintained in dim ambient light by spatial pooling of inputs to the motion detectors. Whereas the rabbit retina recruits rods for this purpose (Vaney et al., this volume), in flies this is presumably accomplished by recruiting additional visual sampling points from the same achromatic pathway (involving photoreceptors R1-R6) that operates in bright light. An additional implication of these experiments is that peripheral mechanisms of light- and dark-adaptation (e.g. Dorlochter and Stieve 1997) permit the fly to maintain its highest spatial resolution for motion detection over an impressive range of photopic conditions: spatial pooling begins some 6-7 log units below the intensity of full sunlight.

3. What neurones underlie elementary motion detection circuits?

Even though the variety of cell morphologies in the retinas of birds and certain reptiles astonished Cajal (Cajal 1888), this was nothing compared to his reaction to the anatomy of the insect visual system. Cajal was profoundly moved by what he called the exquisite adjustment of neurones in the optic lobes of insects. Indeed, a characteristic feature of insect visual systems is the great number of identifiable nerve cells that are revealed by Golgi impregnations (Cajal and Sanchez 1915). For the housefly, *Musca*, more than 70 types of neurones contribute to a retinotopic column, many of which are repeated one per column (Strausfeld 1976). Likewise, in *Drosophila*, there are as many cell types, though fewer columns because there are fewer facets (Fischbach and Dittrich 1989). Different species of insects share many cell types in common, but others may be typical of the taxon. To the extent that motion detection is ubiquitous across taxa, it might be expected that neurones comprising elementary motion detection circuits are evolutionarily conserved. This hypothesis has been supported by neuroanatomical studies comparing the cellular organization of dipteran optic lobes across genera and demonstrating conserved neuronal morphologies and layer relationships in the magnocellular pathway to the lobula plate (Buschbeck and Strausfeld 1996). Similar morphological types of neurones have also been identified in Lepidoptera (Strausfeld and Blest 1970), and in Hymenoptera even though the latter lack a lobula plate. However, in honeybees, tangential neurones that are equivalent to those in dipteran lobula plate lie deep in the hymenopteran lobula. Neurones such as T5 cells, which in flies link the outer stratum of the lobula to the lobula plate (see Fig. 1), in Hymenoptera link the outer stratum of the lobula to layers of tangential neurones lying deep within the same neuropil (Strausfeld 1976).

A crucial finding has been that the morphologies and stratigraphic relationships within the lobula plate vary considerably among dipteran families, suggesting that the wide-field neurones in the lobula plate or its analogue are unlikely to be responsible for the process of elementary motion detection itself. For example, in robber flies, the lobula plate is equipped with horizontally oriented tangential neurones called HS cells, but not with vertically oriented VS cells (Buschbeck and Strausfeld 1997), both of which occur in calliphorids. Such differences contrast with the relationships among evolutionarily conserved and uniquely identifiable small-field retinotopic neurones, which supply or are peripheral to the lobula plate (Buschbeck and Strausfeld 1996). The conserved neurones that appear most likely to be involved in elementary motion detection include five morphological types of lamina interneurones (the monopolar cell types L1, L2, L4 and L5 and the type 1 basket T cell, T1), two types of transmedullary cells (called Tm1 and iTm), and two types of bushy T cells (called T4 and T5). Some of these are illustrated in figure 1. Others are detailed in Strausfeld and Nässel (1981) and Strausfeld and Lee (1991).

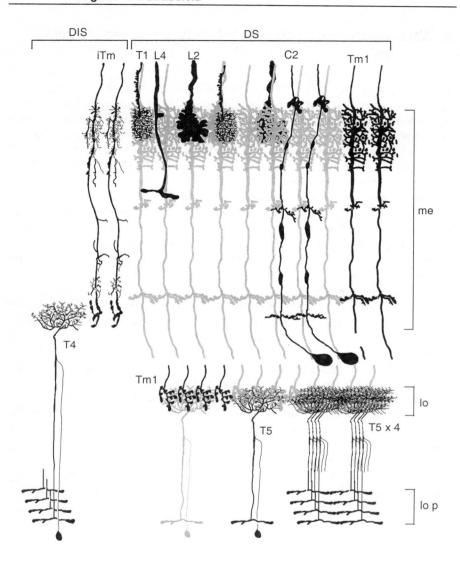

Fig. 1 Shapes of neurones in the medulla and lobula complex that are implicated as components of EMDs, shown as elements of 12 retinotopic columns. *T1*, *L4*, *L2*, and *C2* have peripheral inputs and/or outputs in the lamina (cf. Fig 2) and contribute, along with *Tm1*, to directionally sensitive channels (*DS*). The intrinsic transmedullary cell (*iTm*), which terminates in the medulla (*me*) at the level of the type 4 bushy T-cell (*T4*), is thought to be involved in directionally insensitive channels (*DIS*). The terminals of *Tm1* neurones in the lobula (*lo*) end on the dendrites of quartets of type 5 bushy T-cells (*T5*), the axons of which segregate to four levels in the lobula plate (*lop*) where they terminate onto layers of collator neurones (not shown).

4. The conserved pathway

Wide-field collator neurones of the lobula plate are exemplified by horizontal (HS) and vertical (VS) motion sensitive nerve cells found in calliphorid and muscid flies (Hausen 1982; Hengstenberg et al. 1982; reviewed by Hausen 1993). The dendrites of HS and VS cells receive retinotopically mapped synaptic inputs in parallel from two types of bushy T cells, the T4 and T5 neurones (Strausfeld and Lee 1991). T4 has its dendrites in the inner stratum of the medulla; those of T5 lie in the outer stratum of the lobula, which for anatomical reasons (Strausfeld and Lee 1991) and because it shows similar metabolic activity and immuno-cytological staining (Buchner and Buchner 1984; Schuster et al. 1993), can be viewed as a displaced layer of the inner medulla that is isolated from the remaining lobula neuropil beneath it. In general, T5 and T4 cells can be considered functional analogues of ganglion cells, with the inner layer of the medulla and outer stratum of the lobula corresponding to the inner plexiform layer of the vertebrate retina.

The dendritic fields of T4 and T5 appear identical, and their extent suggests that they receive direct inputs from about seven visual sampling units, compared to the dozen or so small-field channels that provide inputs to directionally selective ganglion cells in the rabbit (Vaney et al., this volume). T4 and T5 cells are unusual in that probably four of each type arise from each retinotopic column (see Section 5). This contrasts with all other types of retinotopic neurones in flies, which – with the possible exception of Tm1 – appear only as a single element in each column. The endings of T4 and T5 in the lobula plate also appear morphologically identical except that they terminate at four main levels corresponding to functionally identified strata that have distinct preferred motion directions. Based on these anatomical features alone, both T4 and T5 have long been considered prime candidates for participation in elementary motion detection circuits.

Intracellular recordings from identified T4 and T5 cells, however, suggest that both receive outputs from elementary motion detection circuits that already exhibit some degree of selectivity for motion direction. T5 shows a well-developed form of directional selectivity (Douglass and Strausfeld 1995) with excitatory and inhibitory responses defining preferred and null directions, respectively. These fully opponent responses suggest that T5 receives outputs subtracted from pairs of elementary motion detection circuits with opposite preferred directions. The existence of such a "subtraction" stage in the motion detection circuitry was predicted from modelling considerations, and supported by physiological recordings from H1 tangentials (Borst and Egelhaaf 1990). These responses also suggest that T5 neurones, or local interneurones onto which they may synapse in the lobula plate, could release either acetylcholine or GABA depending on whether the T5 neurone is depolarized or hyperpolarized. This suggestion is consistent with evidence for both GABAergic (Strausfeld et al. 1995) and cholinergic elements near the level of wide-field tangential neurones in the lobula plate (Brotz and Borst 1996).

Recordings also have been made from T4 neurones (Douglass and Strausfeld 1996). In contrast with T5, the T4 cells appear less concerned with encoding motion direction than with faithfully reporting local increases in light intensity, whether due to flicker or the passage of ON-edges across their receptive fields. Although the responses of T4 to motion are weakly direction-dependent, this could reflect a form of cross talk via indirect connections with direction-sensitive channels. Thus, the possible role of T4 in motion processing is uncertain.

If bushy T-cells (T4 and T5) are not themselves responsible for the first step in motion computation, what is the source of the well-developed directional sensitivity of T5 cells? Again, comparative studies have narrowed the candidates for the main inputs of bushy T-cells to two types of retinotopic neurones that penetrate through all the layers of the medulla. These are called transmedullary cells Tm1 (Figs. 1 and 2) and iTm (Fig. 1). Each occurs in every retinotopic column, and each possesses dendrites disposed at strata receiving the terminals of lamina monopolar cells. Each iTm neurone terminates in the innermost medulla stratum at the level of T4 dendrites. This is different from another small field transmedullary cell in the same column, Tm1, which sends its axon out of the inner face of the medulla, across the second optic chiasma and into the outer stratum of the lobula where it terminates among the dendrites of the type T5 bushy T-cells. Because of their consistent alignment with bushy T-cell dendrites, irrespective of the depth of the inner medulla in different taxa (Buschbeck and Strausfeld 1996), it is thought that the terminals of iTm neurones are presynaptic to the dendrites of T4 bushy-T cells. Likewise, the terminals of Tm1 neurones in the outer stratum of the lobula should be presynaptic to the dendrites of T5 bushy T-cells. T5 cells are visited by several Tm1 cells (Strausfeld and Lee 1991) from which they probably pool their inputs, as well as from three other types of transmedullary neurones that also terminate at the same level. These various types of transmedullary cells have been described from *Drosophila* (Tm1a, Tm9; Fischbach and Dittrich 1989) and *Phaenicia* (Tm1b, Tm9; Douglass and Strausfeld 1998).

The only recording to date from the Tm1 cell clearly revealed a form of directional selectivity (Douglass and Strausfeld 1995). Its response to one direction of motion was similar to a pure Off response, whereas the response to the opposite direction was an On-Off-like response. Thus, the anatomical evidence suggesting the involvement of Tm1 in motion processing is corroborated by direct physiological evidence for directional sensitivity. Tm1 may represent an initial, relatively undeveloped output of single elementary motion detection circuits, prior to the expected subtraction of outputs from elementary motion detection circuits with opposite preferred directions. One pressing issue is to determine whether there is only a single Tm1 cell in each retinotopic column. If so, the expected four preferred directions per column at the level of T5 cells (see below) would have to be generated from differential synaptic weights of pooled outputs from neighbouring Tm1 cells. The other small-field transmedullary cell types mentioned above that also terminate at T5 cell dendrites could in theory provide additional directionally selective inputs to T5 cells, but so far have not been shown to carry directional information.

Also unresolved is the location of the requisite lateral connections required for motion detection, particularly since the dendrites of each Tm1 cell in the outer medulla extend little, if at all, beyond its parent retinotopic column (Strausfeld and Lee 1991). Which neurones, then, are likely to provide these lateral connections? As we have noted previously (Douglass and Strausfeld 1995) one possibility is that the terminals of L2 monopolar cells provide inputs from adjacent visual sampling points (columns in the lamina called optic cartridges) to Tm1. This convergence could in principle be provided via the terminal processes of interneurones, called T1 basket cells, whose dendrites in the lamina pool inputs from adjacent optic cartridges via processes of lamina amacrine cells that are themselves postsynaptic to receptor axon terminals (Fig. 2a). In the medulla, T1 cell terminals are interposed between each L2 monopolar cell terminal and the dendrites of the corresponding Tm1 cell (Fig. 2b). Another interneurone, the midget monopolar cell (known as the L5 monopolar), also receives inputs from surrounding cartridges via tangential elements in the lamina, and in the medulla its terminal interposes between the ending of an L1 monopolar cell and the transmedullary cell dendrites in the same retinotopic column (Strausfeld, unpublished).

Other candidates for lateral connections exist in the lamina. One is the tripartite monopolar cell, known as the L4 neurone. Ensembles of L4 neurones provide a system of axon collaterals between neighbouring retinotopic columns that are presynaptic onto the axons of L1 and L2 monopolar cells and onto other L4 axon collaterals. At the level of the photoreceptor terminals, each L4 neurone is postsynaptic to amacrine cells (α fibers; Strausfeld and Campos-Ortega 1973, 1977) that are themselves postsynaptic to photoreceptor terminals (Fig. 2a). These amacrines are morphologically disposed to pool information from groups of 6-20 visual sampling units (Campos-Ortega and Strausfeld 1973). Thus, L4 neurones and the lamina amacrine cells together provide a system of connections between neighbouring retinotopic columns, as well as among more distant columns. This anatomical flexibility could help account for motion computation between neighbouring visual sampling units at higher light intensities, as well as motion computation between more distant visual sampling units at low light intensities, as suggested from the aforementioned studies of H1 motion sensitivity (Schuling et al. 1989). Also of great interest is that the axon collaterals of L4 neurones together provide a network of connections among lamina monopolar cells that is rectilinear, and which therefore could provide for four orthogonal preferred directions. In the medulla, L4 monopolar cell endings are also divided into three branches that connect three adjacent retinotopic columns (Fig. 2a; see also Strausfeld and Campos-Ortega 1972).

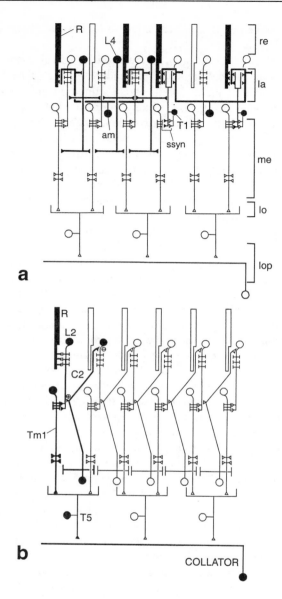

Fig. 2 Schematic of connections among neurones implicated in elementary motion detector circuits (see text). Described pathways are accentuated by heavy lines and filled triangles (presynaptic sites). **a** Connections between channels in the lamina (*la*; retina, *re*). Lamina amacrine cells (*am*) recruit inputs from photoreceptors (*R*) and synapse onto the tripartite monopolar cell (*L4*) and the basket T-cell (*T1*). L4 collaterals provide a lattice of connections onto monopolar cells (*L2*; see b) in adjacent retinotopic channels. L4 terminals provide a similar lattice between adjacent columns in the medulla (*me*). Dendrites of T1 cells are postsynaptic to amacrines. In the medulla, T1 endings are interposed between the endings of the L2 monopolar cells and the

Thus far, physiological evidence that L4 itself might be directionally sensitive is ambiguous. In a single recording from this neurone, responses to motion showed a direction-dependent phase shift (Douglass and Strausfeld 1995). However, because only one spatial frequency could be tested and the receptive field location was unknown, the results could be explained by a phase inequality between the receptive field and the position of the light meter that monitored the grating motion. As for the other lamina monopolar cell types, a recording from the midget monopolar cell (L5) demonstrated distinct responses to motion and flicker, but no directional sensitivity (Douglass and Strausfeld 1995). Other recordings from lamina monopolar cell types L1 and L2 during directional motion stimulation showed no evidence of directional sensitivity, and only a slight hint of directional selectivity has been noted in a single recording from L3 (Coombe et al. 1989; Gilbert et al. 1991). It should be remarked, however, that at the most peripheral levels of the elementary motion detection circuit, contributing neurones will not themselves exhibit directional responses to motion.

The type 2 centrifugal neurone (C2) provides a second and intriguing candidate for the required lateral connections in the motion detecting circuit. C2 neurones (Fig. 2b), along with the type 3 centrifugals (C3), occur in each retinotopic column (Strausfeld 1976). They provide feedback connections from their dendrites in the inner layer of the medulla back to the outer surface of the medulla and then to the lamina. Both the type 2 and 3 centrifugals (Strausfeld 1976) have GABAergic terminals in the lamina and their cell bodies are also strongly immunoreactive to antiGABA antisera (Datum et al. 1986; Strausfeld, unpublished). This feature is important when considering the lack of antiGABA immunoreactivity of these cells in the medulla. In the lamina, the terminals of type 2 centrifugal cells are presynaptic to L1 and L2 neurones distally, above the level at which these monopolar cells receive inputs from photoreceptor terminals (Meinertzhagen and O'Neil 1991, Strausfeld, unpublished). The terminals of type 3 centrifugal cells are presynaptic to the entire lengths of L1 and L2 monopolars. In the medulla, the type 2 centrifugal cells have dendrites at the level of T4 neurones, deep in the medulla, and a second set of dendrites at the level of the endings of L1 monopolar cells, which penetrate to about a third of the depth of the medulla. At the most distal level of the medulla type 2 centrifugal cells are presynaptic onto the terminals of L1 and L2. However, these synapses are not GABAergic, but occur in a thin stratum of neuropil that is immunoreactive to a

---

dendrites of the type 1 transmedullary cell (*Tm1*; see B), providing serial synapses (*ssyn*) and thus possible delay lines. **b** Connections between channels in the medulla. Receptors representing a single visual sampling point provide input to L2, which end at the level of type 1 transmedullary neurones, along with the terminal of T1. In the lobula (*lo*; in a) several Tm1s terminate on a single bushy T-cell (*T5*) that is presynaptic onto a collator neurone in the lobula plate (*lo p*; in a). Type 2 centrifugal neurones (*C2*) connect retinotopic channels. C2s receive inputs in the medulla (source unknown) and provide outputs at two distal levels: just above the ending of L2, and in an adjacent column in the lamina onto L2. The latter synapse is GABAergic, here designated by a negative (inhibitory) sign. The former is thought to be cholinergic and is designated by a positive (excitatory) sign.

subunit of the *Drosophila* acetylcholine receptor (Strausfeld, unpublished). Thus, type 2 centrifugal cells appear to contain two types of transmitter substances: possibly acetylcholine at the outer level of the medulla and GABA in the lamina.

Each C2 centrifugal cell appears to "jump" between retinotopic columns. That is to say, the C2 axon is displaced from its parent retinotopic column to an adjacent column at the level of its putatively cholinergic presynaptic site at the medulla's surface (Fig. 2b). This organization is intriguing, and made all the more so by the centrifugal cell's responses to grating motion (Douglass and Strausfeld 1995). These include a subtle orientation selectivity to vertical as opposed to horizontal gratings. Possibly the orientation selectivity reflects the vertical orientation of centrifugal neurone dendrites deep in the medulla at the level of T4 dendrites.

5. Are there distinct functional classes of elementary motion detectors?

It is often argued that insects are simpler than vertebrates, and therefore more tractable. However, this cannot be said of the lamina and medulla. Not only do these regions contain an astonishingly large variety of cell types, but the stratified organization of the medulla is at least as elaborate as the most stratified plexiform layers of vertebrates, such as those of birds (Cajal 1888). Thus, insect visual processing is not likely to be constrained by a shortage of neurones even though the raw visual inputs are at a lower spatial resolution than in most vertebrates.

It is possible, therefore, that comparative studies will uncover more systems of identified neurones that are shared across taxa having similar visual behaviours. Already, there is evidence that the Tm1-T5 pathway and its associated elements (the monopolar inputs and centrifugal neurones) are unlikely to be the only elementary motion detection circuit because other systems outside the lobula plate are directionally motion-sensitive. For example, the region of high acuity on the retina of male flies is relayed to systems of large collator neurones in corresponding regions of the lobula (Strausfeld 1991). These neurones, all motion sensitive and some directionally selective (Gilbert and Strausfeld 1991), supply systems of descending neurones (Gronenberg and Strausfeld 1991). The dendrites of male-specific motion sensitive neurones lie deep in the lobula and cannot be reached by any of the small retinotopic elements described so far. How, then, do they receive information about motion direction?

One possibility is that outputs from a single type of elementary motion detection circuit are distributed among several types of elements that project from the medulla to the lobula. Among these elements are Y-cells, so called because their bifurcating axons terminate in the lobula plate and deep in the lobula. Comparable neurones have been identified both in Diptera and in Lepidoptera (Strausfeld 1970; Strausfeld and Blest 1970). In addition, the medulla provides many other morphological types of transmedullary cells that terminate deep in the lobula. Although not yet investigated, it is possible that the inner medulla axon collaterals of these transmedullary neurones receive inputs from the small field Tm1 or iTm cells. If this occurs, it would allow distribution of motion information from a single set of elementary motion detecting circuits in the medulla to all levels through the lobula. Recordings from Tm and Y cells, which are difficult to penetrate due to their slender axons, show responses that are motion-modulated. Two identified Y-cells, Y 18 and CY 2, are directionally selective (Douglass and Strausfeld 1996, 1998). Thus information about directional movement across the retina can reach all levels of the lobula.

6. Segregated ON and OFF pathways and motion detection

Vertebrate retinae are characterized by an early segregation between ON and OFF channels, which begins at the photoreceptor outputs to bipolar cells and persists to some degree as separate populations of ON- and ON-OFF-directionally sensitive ganglion cells. Vaney et al. (this volume) make a further functional distinction between ON-OFF directionally selective ganglion cells as general purpose small-field motion detectors, and ON-directionally selective ganglion cells as early elements of segregated global motion processing pathways. So far, in flies, there is no evidence for an early segregation of ON and OFF channels, nor for a specialized small-field pathway devoted to global motion processing. All lamina monopolar cells (L1-L5) that receive direct or indirect inputs from photoreceptors (Strausfeld and Campos-Ortega 1977) and provide outputs to the medulla exhibit both ON and OFF responses to flicker (Laughlin 1981; Gilbert et al. 1991; Douglass and Strausfeld 1995). Correlation models of elementary motion detection successfully explain a variety of behavioural and electrophysiological observations by invoking a multiplicative interaction between input channels (Egelhaaf et al. 1989), and this interaction does not require separate ON and OFF pathways. If such pathways exist, supporting evidence would be provided by neurones that exhibit half-wave rectification, transmitting only ON or only OFF responses to combined ON and OFF inputs. In recordings from the wide-field motion-sensitive neurone H1 that were specifically designed to test for these properties, Egelhaaf and Borst (1992) found no evidence for separate ON and OFF mechanisms. Nevertheless, recordings from identified small-field neurones demonstrate half-wave rectification at levels peripheral to the lobula plate. The transmedullary neurone T1a (Gilbert et al. 1991) and the bushy T cell T4 (Douglass and Strausfeld 1996) have both shown predominantly ON responses, while an amacrine cell deep within the medulla responded only to light OFF (Douglass and Strausfeld 1996). ON and OFF segregation clearly exists in these neurones, but their relationship to motion processing is uncertain.

7. Distinct directional classes of small-field motion detectors: A feature shared by rabbit and fly?

Vaney et al. (this volume) discuss physiological evidence suggesting that the preferred directions of the ON-OFF directionally selective ganglion cells in the rabbit are clustered around four cardinal directions: downward, upward, temporal, and nasal. In flies, both anatomical and physiological evidence point to a similar organization. Anatomical studies indicate the presence of four identical small-field relay neurones (T4 cells) from each retinotopic column in the medulla, and at least two T5 cells per column in the outer lobula, all supplying collator neurones in the lobula plate (Fischbach and Dittrich 1989; Strausfeld and Lee 1991). The T4 and T5 cells each segregate their terminals to four major strata in the lobula plate, each stratum containing collators that are selective for one of four cardinal directions – downward, upward, progressive horizontal, and regressive horizontal – when stimulated with unidirectional wide-field motion (Hengstenberg 1982; Hausen 1993). Activity staining by ^{3}H 2-deoxyglucose during stimulation with panoramic motion confirms that these strata comprise a directional map in the lobula plate. Each layer accumulates radioactive deoxyglucose according to the motion direction to which the animal was exposed (Buchner and Buchner 1984; Buchner et al. 1984). Moreover, intracellular recordings from identified neurones, the dendrites of which are restricted to specific levels in the lobula plate, are consistent with this organization (Douglass and Strausfeld 1996, 1998). Finally, recordings from T5 neurones clearly demonstrate their directional selectivity, and in a manner that is consistent with an orthogonal pattern of preferred directions (Douglass and Strausfeld 1995).

8. Analysis of motion direction and speed

The physical properties of local motion are embodied in two parameters, direction and speed. Thus, in order for any motion processing system to make full use of motion information, information regarding both parameters should be acquired in some fashion. Early behavioural experiments involving optomotor responses, as well as models of classical correlation detectors and recordings from wide-field tangential neurones, suggested that motion detectors in insects are highly sensitive to the temporal frequency (cyc/s) of a pattern and thus do not provide an unambiguous measure of absolute speed (°/s). However, behavioural tests of the flight behaviour of fruit flies and bees in wind tunnels show that insects can use the angular speed of optic flow for range estimation (David 1982; Srinivasan et al. 1991).

Evidence is now accumulating that computational bases for separately analysing motion direction and speed exist even at the level of elementary motion detectors. There are several types of mechanisms that could participate in the analysis of speed information. First, because there are multiple candidate parallel pathways for delivering retinotopic motion information to the lobula plate (see above), it has been suggested that separate pathways may exist which are predominantly direction-sensitive or speed-sensitive (Srinivasan et al. 1993; Douglass and Strausfeld 1996). How could a speed-selective pathway arise? One obvious mechanism of speed discrimination involves comparing the responses of distinct motion detector subgroups with different sampling bases and temporal filter properties. This mechanism has indirect support from recordings at higher processing levels (Horridge and Marcelja 1992; O'Carroll et al. 1997) which suggest the existence of two classes of motion detector with distinct speed sensitivity ranges.

An unresolved difficulty is the often-cited preferential sensitivity to contrast frequency over absolute speed of the wide-field collators, which has conformed to predictions from theoretical correlation-type motion detectors. Must speed information then be extracted at higher levels that are postsynaptic to the lobula plate collators? A single elementary motion detection circuit that can reliably provide both direction and speed information would provide an elegant alternative. A recent re-examination of correlation-type motion detector models suggests an effective and biologically plausible design for such circuits. In a standard mirror-symmetrical elementary motion detection circuit with a subtraction stage, the gains at the outputs of the two "half-detectors" are usually defined to be approximately equal, and the circuit tends to be preferentially sensitive to contrast frequency. The same kind of circuit can be tuned to speed instead, simply by reducing the balance between the two half-detectors (Zanker et al. 1999).

Acknowledgements

Grant sponsors: National Institutes of Health, National Center for Research Resources [NCRR RO1 08688]; Office of Naval Research [ONR-N00014-97-1-0970].

References

Borst A, Egelhaaf M (1989) Principles of visual motion detection. Trends Neurosci 12: 297-306
Borst A, Egelhaaf M (1990) Direction selectivity of blowfly motion-sensitive neurons is computed in a two-stage process. Proc Natl Acad Sci USA 87: 9363-9367
Brotz TM, Borst A (1996) Cholinergic and GABAergic receptors on fly tangential cells and their role in visual motion detection. J Neurophysiol 76: 1786-1799
Buchner E, Buchner S (1984) Neuroanatomical mapping of visually induced nervous activity in insects by ^{3}H-deoxyglucose. In Ali MA (ed) Photoreception and vision in invertebrates. Plenum, New York, London, pp 623-634
Buchner E, Buchner S, Bülthoff I (1984) Deoxyglucose mapping of nervous activity induced in Drosophila brain by visual movement. J Comp Physiol A 155: 471-483

Buschbeck EK, Strausfeld NJ (1996) Visual motion detection circuits in flies: Small-field reti-notopic elements responding to motion are evolutionarily conserved across taxa. J Neurosci 16: 4563-4578

Buschbeck EK, Strausfeld NJ (1997) The relevance of neural architecture to visual performance: Phylogenetic conservation and variation in dipteran visual systems. J Comp Neurol 383: 282-304

Cajal SR (1888) Sur la morphologie et les connexions des éléments de la rétine des oiseaux. Anat Anz 4: 111-121

Cajal SR, Sánchez D (1915) Contribución al conocimiento de los centros nerviosos de los insec-tos. Parte I. Retina y centros opticos. Trab Lab Invest Biol Univ Madrid 13: 1-167

Campos-Ortega JA, Strausfeld NJ (1973) Synaptic connections of intrinsic cells and basket arborisations in the external plexiform layer of the fly's eye. Brain Res 59: 110-136

Coombe P, Srinivasan MV, Guy RG (1989). Are the large monopolar cells of the insect lamina on the optomotor pathway? J Comp Physiol A 166: 23-35

Datum K-H, Weiler R, Zettler F (1986) Immunocytochemical demonstration of γ-amino butyric acid and glutamic acid decarboxylase in R7 photoreceptors and C2 centrifugal fibers in the blowfly visual system. J Comp Physiol A 159: 241-249

David CT (1982) Compensation for height in the control of groundspeed by *Drosophila* in a new, "Barber's Pole" wind tunnel. J Comp Physiol 147: 485-493

Dorlochter M, Stieve H (1997) The *Limulus* ventral photoreceptor: Light response and the role of calcium in a classic preparation. Prog Neurobiol 53: 451-515

Douglass JK, Strausfeld NJ (1995) Visual motion detection circuits in flies: Peripheral motion computation by identified small-field retinotopic neurons. J Neurosci 15: 5596-5611

Douglass JK, Strausfeld NJ (1996) Visual motion-detection circuits in flies: Parallel direction- and non-direction-sensitive pathways between the medulla and lobula plate. J Neurosci 16: 4551-4562

Douglass JK, Strausfeld NJ (1998) Functionally and anatomically segregated visual pathways in the lobula complex of a calliphorid fly. J Comp Neurol 396: 84-104

Egelhaaf M, Borst A (1992) Are there separate ON and OFF channels in fly motion vision? Visual Neurosci 8: 151-164

Egelhaaf M, Borst A (1993) A look into the cockpit of the fly: Visual orientation, algorithms, and identified neurons. J Neurosci 13: 4573-4574

Egelhaaf M, Borst A, Reichardt W (1989) Computational structure of a biological motion-detec-tion system as revealed by local detector analysis in the fly's nervous system. J Opt Soc Am A 6: 1070-1087

Egelhaaf M, Hausen K, Reichardt W, Wehrhahn C (1988) Visual course control in flies relies on neuronal computation of object and background motion. Trends in Neurosci 8: 351-358

Fischbach K-F, Dittrich APM (1989) The optic lobe of *Drosophila melanogaster*. I. A Golgi analysis of wild-type structure. Cell Tissue Res 258: 441-475

Gilbert C, Penisten DK, DeVoe RD (1991) Discrimination of visual motion from flicker by identified neurons in the medulla of the fleshfly *Sarcophaga bullata*. J Comp Physiol A 168: 653-673

Gilbert C, Strausfeld NJ (1991) The functional organization of male-specific visual neurons in flies. J Comp Physiol A 169: 395-411

Gronenberg W, Strausfeld NJ (1991) Descending pathways connecting the male-specific visual system of flies to the neck and flight motor. J Comp Physiol A 169: 413-426

Hausen K (1981) Monocular and binocular computation of motion in the lobula plate of the fly. Verh Dtsch Zool Ges 1981: 49-70

Hausen K (1982) Motion sensitive interneurons in the optomotor system of the fly. I. The hori-zontal cells: Structure and signals. Biol Cybern 45: 143-156

Hausen, K. (1993) Decoding of retinal image flow in insects. In: Miles FA, Wallman J (eds) Visual motion and its role in the stabilization of gaze. Elsevier, Amsterdam, pp 203-235

Hengstenberg R (1982) Common visual response properties of giant vertical cells in the lobula plate of the blowfly *Calliphora*. J Comp Physiol 149: 179-193

Hengstenberg R, Hausen K, Hengstenberg B (1982) The number and structure of giant vertical cells (VS) in the lobula plate of the blowfly (*Calliphora erythrocephala*). J Comp Physiol 149: 163-177

Horridge GA, Marcelja L (1992) On the existence of 'fast' and 'slow' directionally sensitive motion detector neurons in insects. Proc Roy Soc Lond B 248: 47-54

Laughlin S (1981) Neural principles in the peripheral visual systems of invertebrates. In: Autrum H (ed) Comparative physiology and evolution of vision in invertebrates. Handbook of Sensory Physiology VII/6B. Springer, Berlin, pp 133-280

Meinertzhagen IA, O'Neil SD (1991) Synaptic organization of columnar elements in the lamina of the wild type in *Drosophila melanogaster*. J Comp Neurol 305: 232-263

O'Carroll DC, Laughlin SB, Bidwell NJ, Harris SJ (1997) Spatio-temporal properties of motion detectors matched to low image velocities in hovering insects. Vision Res 37: 3427-3439

Pick B, Buchner E (1979) Visual movement detection under light- an dark-adaptation in the fly *Musca domestica*. J Comp Physiol 134: 45-54

Riehle A, Franceschini N (1982) Response of a directionally selective movement detecting neuron under precise stimulation of two identified photoreceptor cells. Neurosci Lett Suppl 10: 5411-5412

Schuling FH, Mastebroek HAK, Bult R, Lenting BPM (1989) Properties of elementary movement detectors in the fly *Calliphora erythrocephala*. J Comp Physiol A 165: 179-192

Schuster R, Phannavong B, Schröder C, Gundelfinger ED (1993) Immunohistochemical localization of a ligand-binding and a structural subunit of nicotinic acetylcholine receptors in the central nervous system of *Drosophila melanogaster*. J Comp Neurol 335: 149-162

Srinivasan MV, Dvorak DR (1980) Spatial processing of visual information in the movement-detecting pathway of the fly. J Comp Physiol 140: 1-23

Srinivasan MV, Lehrer M, Kirchner WH, Zhang SW (1991) Range perception through apparent image speed in freely flying honeybees. Visual Neurosci 6: 519-535

Srinivasan MV, Zhang SW, Chandrashekara K (1993) Evidence for two distinct movement-detecting mechanisms in insect vision. Naturwiss 80: 38-41

Strausfeld NJ (1970) Golgi studies on insects. Part II. The optic lobes of Diptera. Phil Trans Roy Soc B 258: 175-223

Strausfeld NJ (1976) Atlas of an insect brain. Springer, Heidelberg

Strausfeld NJ (1991) Structural organization of male-specific visual neurons in calliphorid optic lobes. J Comp Physiol A 169: 379-393

Strausfeld NJ, Blest AD (1970) Golgi studies on insects. Part I. The optic lobes of Lepidoptera. Phil Trans Roy Soc Lond B 258: 81-134

Strausfeld NJ, Campos-Ortega JA (1972) Some interrelationships between the first and second synaptic regions of the fly's (*Musca domestica* L) visual system. In: Wehner R (ed) Information processing in the visual systems of arthropods. Springer, Heidelberg, Berlin, pp 23-30

Strausfeld NJ, Campos-Ortega JA (1973) The L4 monopolar neuron: a substrate for lateral interaction in the visual system of the fly *Musca domestica*. Brain Res 59: 97-117

Strausfeld NJ, Campos-Ortega JA (1977) Vision in insects: Pathways possibly underlying neural adaptation and lateral inhibition. Science 195: 894-897

Strausfeld NJ, Lee J-K (1991) Neuronal basis for parallel visual processing in the fly. Visual Neurosci 7: 13-33

Strausfeld NJ, Nässel DR (1980) Neuroarchitectures serving compound eyes of Crustacea and insects. In: Autrum H (ed) Comparative physiology and evolution of vision in invertebrates. VII/B. Springer, Berlin, pp 1-132

Strausfeld NJ, Kong A, Milde JJ, Gilbert C, Ramaiah L (1995) Oculomotor control in calliphorid flies: GABAergic organization in heterolateral inhibitory pathways. J Comp Neurol 361: 298-320

Zanker JM, Srinivasan MV, Egelhaaf M (1999) Speed tuning in elementary motion detectors of the correlation type. Biol Cybern 80: 109-116

Part II

Motion Signals for Global and Local Analysis

The Organization of Global Motion and Transparency
Oliver Braddick and Ning Qian

Combining Local Motion Signals: A Computational Study of Segmentation and Transparency
Johannes M. Zanker

Local and Global Motion Signals and their Interaction in Space and Time
Simon J. Cropper

The Organization of Global Motion and Transparency

Oliver Braddick[1] and Ning Qian[2]

[1]Department of Psychology, University College London, UK; [2]Center for Neurobiology and Behavior, Columbia University, New York, USA

Contents

1. Abstract

The visual system has the task of computing global motions associated with objects and surfaces. This task strongly involves extrastriate brain areas, particularly V5/MT. Motion transparency provides a particular challenge for understanding how global motions are computed and represented in the brain. Psychophysical experiments show that, for a single region, multiple motions can be quantitatively represented. However, at the most local scale, motion signals have a suppressive interaction so that only a single motion can be represented. Neurophysiological experiments show that this suppression is a property of MT, not of V1, reflecting a subunit structure within MT receptive fields and showing that transparency perception is related to MT rather than V1 activity. A full understanding of transparency perception and other global motion phenomena will require us to understand how perceived motions are related to the distribution of activity across a population of directionally selective neurones, and how the brain implements the representation of motions assigned to extended objects rather than to specific retinotopic locations.

2. Global representations of motion

Much research on visual motion processing has been concerned with the processes that measure image velocity in a local region of the image. The computational goal of motion vision is sometimes described as "computing the velocity field" i.e. obtaining valid measurements of this kind at each point in the field of view. However, it is not adequate to consider visual motion processing at a purely local level, for several reasons.

First, it is well known that purely local operations are not adequate to obtain a valid velocity field. The "aperture problem" means that, in regions of the image where variation is mostly along one dimension (i.e. contours) local operations can only measure the component of motion at right angles to the contour (Adelson and Movshon 1982). Directionally selective mammalian V1 neurones with small, oriented receptive fields will generally make ambiguous motion measurements of this kind. To derive the true velocity vector for such locations, information must be integrated from other locations on the same moving object. This process is believed to depend on "pattern-motion-selective neurones" of the kind found in the area known as V5 or MT in the primate brain (Movshon et al. 1986). Second, spatial integration is important to smooth noisy or sparse motion signals (Braddick 1993). Processes that disambiguate and smooth motion signals require operations over an extended area, but still, their goal can be regarded as achieving a representation of velocity for each point in the field.

However, the purpose of vision is not to compute fields but to create a representation that is useful for understanding, and acting on, the environment. Generally, the significant entities in the environment are large-scale structures, objects and surfaces. The visual organism needs to assign to these structures their motions, or properties of the structures which are derived from image motion, such as depth ordering or relation to the perceiver's trajectory. Thus the visual system has to compute global motions associated with objects and surfaces, not just the local motions associated with locations in the field.

The difference between local and global motion is most obvious when local motions are not uniform. For example, the turbulent flow of water in a stream, gravel being dumped from a truck, or a flock of birds taking to the air, each give the viewer a sense of overall or global direction despite containing a wide and disorderly range of local motions. Neural mechanisms must exist that can represent this global direction.

3. Experimental approaches to global motion

3.1 Motion coherence

Sensitivity to motion coherence has become an established measure of the performance of global motion mechanisms. It is tested with a random dot display in which some fraction of the dots are "signal" dots which share a common "coherent" motion, while the remaining "noise" dots are displaced at random. The noise can be defined in a number of different ways, but these do not appear to make much difference to the measurement of sensitivity (Scase et al. 1996). The direction of the signal dots can be detected when the coherence (proportion of signal dots) is as low as 5-7%. This direction cannot be derived by inspecting individual dots, but must depend on the integration of motion information over a large area of the display (Downing and Movshon 1989).

Several lines of evidence support the idea that this integration depends on processing in extrastriate cortical areas. First, lesions in MT of the macaque greatly impair coherence thresholds (Newsome and Pare 1988), and the presence of noise dots also has a very deleterious effect on motion discrimination in a patient with a lesion including the human homologue of this area (Baker et al. 1991). Secondly, the response of single neurones in macaque MT shows a systematic dependence on motion coherence (Britten et al. 1992). The estimated threshold coherence for such neurones agrees well with the animal's behavioural coherence threshold (Britten et al. 1993).

Third, a recent fMRI study (Braddick et al. 1998) has looked for cortical areas that are differentially activated by coherent motion compared to random noise (Fig. 1). Such activation is not found in V1, but is found in human V5 and some other extrastriate areas, particularly the area identified by Tootell et al. (1997) as V3A. In fact, V1 showed stronger responses to noise than to coherent motion. These findings confirm that in the human visual system, the integrative processes that underlie sensitivity to coherence occur beyond the initial stage of cortical processing in V1. They are also consistent with single unit findings in macaque, discussed below, that V1 cells respond much more strongly to dynamic noise than do cells in MT (Qian and Andersen 1994).

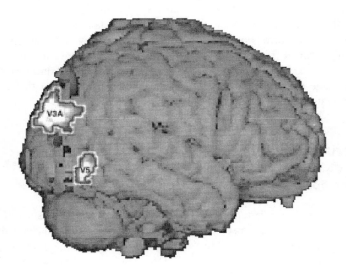

Fig. 1 Lateral view showing brain areas that are activated significantly more by coherent motion than by dynamic noise, in an individual subject. Highlighted areas are voxels within 3 cm of the cerebral surface with z > 3.09 (p<0.001) for this contrast in Statistical Parametric Mapping (Friston et al. 1995). The two major foci of activation marked correspond to areas identified as V5 and V3A. Other areas of differential activation were found on the ventral occipital surface, in the intraparietal sulcus and the superior temporal sulcus.

3.2 Directional distributions

An alternative approach to global motion is to use, rather than displays containing random noise, displays in which there is a broad distribution of elementary motions (Williams and Sekuler 1984). Human subjects' ability to extract global motion is exemplified by results from Watamaniuk et al. (1989), who showed that the mean direction of a random dot display could be judged with an accuracy of 2°, even when the motion vectors of the individual dots were distributed over a range of up to 120°. However, this performance should not be taken to imply that the motion is perceived as unitary. As in the real-world examples of the flock of birds or the tumbling gravel, subjects are aware that diverse local motions exist within the global flow.

In fact, global direction processing appears to be a flexible process that is capable of parsing the overall direction distribution differently for different purposes. This became apparent in experiments by Zohary et al. (1996). They tested subjects with skewed distributions of local motion vectors, in an attempt to determine whether global direction judgments depended on a mechanism using the peak or the mean of the velocity distribution. The result was that neither model alone gave an adequate account; rather subjects could freely switch between these two kinds of performance, as if they had access to the whole distribution of local motion vectors, and could perform different optional operations upon it to derive different global results. We return to this question below in discussion of bimodal and unimodal distributions of directional signals.

3.3 Multiple global motions: transparency

A particularly significant case where a distribution of directions does not automatically yield a single global outcome is the case of motion transparency. When there is a sufficiently large gap in the distribution of local directions (Smith et al. 1999), the distribution can be parsed into two separate global motions which are seen as spatially co-extensive and as superimposed in a transparent manner. This is not simply a special laboratory situation. It reflects the need to deal with visual situations where elements belonging to entities with different motions are interleaved or superimposed. Real-world examples are a moving object partially occluded by lacy foliage, or a moving shadow cast on a stationary, or differently moving, surface.

The phenomenon of transparency is an important test case for the nature of motion computation and representation. This is because it is incompatible with the simple goal of a single-valued velocity field. If perceived transparency is taken at its face value, it implies that the brain can represent two different motions at the same location. To make this point, it is important to test the nature of this multiple representation.

4. The analysis of motion transparency

4.1 Performance-based evaluation of motion transparency

Most earlier work on motion transparency has rested on simple subjective reports; does the display appear to contain two (or more) superimposed motions? Such reports cannot provide a very strong basis for understanding the underlying neural representation. In particular, they cannot clearly distinguish whether there is a true representation of two motions, or whether one motion is registered as segmented from its background. For instance, when signal dots are detected amid noise, they can be represented as a coherent entity that is distinct from the noise background of noise dots; this does not imply any necessary representation of the motion properties of the background, beyond its presence. Subjects given the choice between reporting one motion and two might well describe an asymmetrical representation of this kind as "two motions".

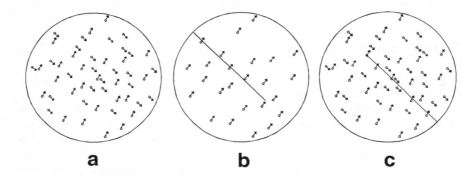

a   **b**   **c**

Fig. 2 a "Motion vs motion" display used in performance-based evaluation of transparency. The motions illustrated are 13° away from orthogonal, corresponding to the precision of subjects' judgments (s.d. of the psychometric function). **b** "Motion vs line" display serves as a control for the judgment illustrated in (a). The single motion illustrated is 5° away from being orthogonal to the line, corresponding to the precision of subjects' judgments (s.d. of the psychometric function). **c** "Transparent motions vs line" display. One motion is parallel to the line; the subject's task is to judge whether the other is orthogonal to the line.

If there are equivalent representations of the two velocities in a transparent display, subjects should have access to information about both of them. This can be tested by a perceptual task that necessarily depends jointly on both motions (Braddick 1997; Wishart and Braddick 1997a). Subjects are required to judge whether the angle between the directions of two superimposed streams of moving dots is greater or less than 90° (Fig. 2a). The orientation at which the pair of motions is presented is randomized from trial to trial, to ensure that there is no way to make the discrimination using one motion alone. An adaptive psychophysical method (Watt and Andrews 1981) is used to estimate the psychometric function for this discrimination. The standard deviation (s.d.) of this function is a measure of the precision with which the subject can use joint information about the two directions.

The results show that the angle between the motions can be judged with a precision of around 13°. Quantitative directional information is available from both dot streams jointly. By this criterion, the perception of transparency in this situation corresponds to a genuine multi-valued representation of velocity, not simply to a segregation of one motion from a background.

It is worth noting, however, that the precision, as reflected in the s.d. of these judgments, is somewhat impaired compared to that for a single motion direction relative to a line, which is around 5° (Fig. 2b). Some increase might be expected on a simple model of independent errors arising in the estimate of each of the pair of directions. This would predict a standard deviation about 7°, i.e. about half what is actually observed, so there must be some additional penalty associated with the transparent case. Such a penalty might arise from interference between the different local motions of dots in the two streams, or from an interaction at the level of the two global representations. In the latter case, it might be associated specifically with transparency, or it might be a more general penalty incurred by computing and comparing two directions at the same time.

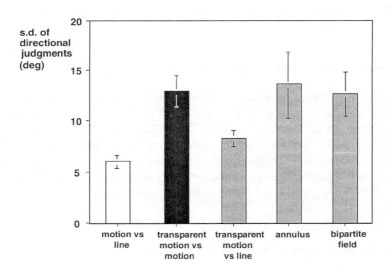

Fig. 3 Performance on directional judgments for the three conditions illustrated in figure 2a, b, c (columns 1-3) and for conditions where tow dot streams are separated into an annulus and central region (column 4) or a bipartite field (column 5). Mean and s.e. of results for 4 subjects are shown.

This issue has been tested by three further conditions. In one condition, the two motions are superimposed, but the direction judgment (relative to a line) has to be made for only one of them (Fig. 2c). This shows only a small loss of precision (s.d. around 8°) relative to a single motion, implying that local interference is not the major penalty in the dual motion judgment. In the second and third conditions, the required judgment is of the angle between two spatially separated motions. In one case these are a central region and a surrounding annulus; in the other, two halves of a bipartite field with a randomly oriented boundary. In these cases no interference will arise from superimposed local motions. Nonetheless, the loss of precision compared to judgment of a single stream is comparable to that for the transparent case; figure 3 shows the standard deviations obtained in each of the conditions. Thus the penalty in direction judgements is associated with the need to compute and compare two global representations of motions, whether or not these are spatially co-localized.

In summary, a measure based on visual performance shows that the visual system can represent two transparent motions in the same region of the field and operate with these represented motions as readily as when they are in adjoining regions. It should be noted, though, that this result does not tell us about the spatial properties of these representations. In some sense, two globally represented velocities occupy "the same place", but this leaves open whether at a local level, there is a multi-valued representation of velocity associated with each location. The study of locally paired displays, discussed below, throws some light on this question.

4.2 Transparent motions at the single-unit level

Along with the psychophysical investigation of transparency in random-dot patterns, there has also been an increasing number of physiological investigations of this topic in recent years (Snowden et al. 1991; Qian and Andersen 1994; Qian and Andersen 1995). Single unit studies have also tested plaid patterns (Movshon et al. 1986; Rodman and Albright 1989; Stoner and Albright 1992), where variations in stimulus parameters can change the perceived effect from that of two gratings sliding over each other transparently in different directions, to a single coherently moving plaid.

4.3 Directional suppression in V5/MT

Snowden et al. (1991) were among the first to record neurones from visual cortical areas V1 and MT of awake monkeys using transparent random dot patterns. MT (also known as V5) is a later stage than V1 along the well established motion processing pathway in primates (Andersen 1997). Intuitively, one might expect that neurones at the higher level would be able to represent the transparent motions as two independent global entities. However, the most prominent finding of the study was that for most MT cells, stimulated with a transparent display containing one set of dots moving in the cell's preferred direction and another in the antipreferred directions, responses were significantly reduced compared to a single set of dots moving in the preferred direction alone. This result indicates strong suppression in MT between the preferred and the antipreferred directions of motion, consistent with other physiological studies of this cortical area (Mikami et al. 1986a; Rodman and Albright 1987; Britten et al. 1992, 1993). In this review, we will use the terms "suppression" and "inhibition" interchangeably, both referring generically to any reduction in the preferred-direction response. We will not try to distinguish various specific mechanisms for the response reduction such as normalization, subtraction or averaging, although these might have quite different synaptic bases.

Snowden et al. (1991) found that this kind of suppression was much weaker in area V1. Since MT is usually considered to be a major site for motion analysis, the finding raises the question of how MT cells can represent transparent motions given the directional suppression in MT. Indeed, the subpopulation of directionally selective V1 cells whose responses to transparent displays are not much suppressed, would appear better able to support the perception of motion transparency. However, as Snowden et al. (1991) point out, motion perception is not unaffected by a superimposed transparent motion; psychophysical detection threshold is higher in the transparent condition than for unidirectional motion (Snowden 1989; Verstraten et al. 1996). They argued that this effect could be a correlate of the suppressive interaction seen in MT cells.

5. "Locally balanced" motions

5.1 The perceptual effects of locally balanced motions

Interactions between directions at different levels of the motion pathway can be examined by using a modification of the transparent random-dot display. In the psychophysical and physiological experiments discussed so far, dots belonging to the two streams are randomly distributed. Qian et al. (1994a, b) and Qian and Andersen (1994, 1995) noted that, in patterns of two components moving in opposite directions, the distribution of the dots had a critical effect upon the appearance of transparency. They designed "paired" dot patterns, composed of many (typically 100) randomly located pairs of dots. The two dots in each pair move across each other in opposite directions over a short distance (about 0.4°) and then jump to a new random location to repeat the process. This display can be compared with the "unpaired" case where the dots moving in opposite directions are positioned independently. Qian et al. observed that, although both types of stimuli contain two sets of dots moving in opposite directions, only the latter gives a perception of two transparent surfaces sliding across each other. The paired display appeared like flickering noise, with no percept of opposing motions.

Wishart and Braddick (1997b) and Braddick (1997) have shown that the effects of local pairing are reflected in directional judgments based on the two motions. They used the task of figure 2a with paired dots in short (0.14°) trajectories, varying the distance between centres of trajectories in a pair. Precision in judging orthogonal directions began to decline when the separation was reduced below 0.5°, and for separations of 0.2-0.1°, subjects found it impossible to make the judgments at all. A similar variation in performance occurred when paired trajectories had a common centre, and their length was manipulated. Testing at different speeds confirmed that the spatial separation of the dots was the critical variable (Fig. 4). It should be noted that in the case of orthogonal motions, the alternative to transparency may be coherent motion rather than flicker, an issue we discuss later.

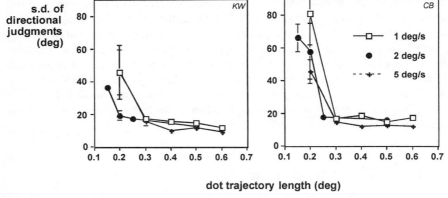

dot trajectory length (deg)

Fig. 4 Performance on directional judgements for locally paired dots moving in directions 90° apart, at 1-5°/s, with varying trajectory length. Panels show results of two subjects. At all speeds, performance radically deteriorates and variability (shown by s.e. bars) in-creases for trajectories shorter than 0.25°, indicating the abolition of motion transparency.

The effect of local pairing suggests that within any small area of the field, local motion detectors tuned to different directions have a strong interaction, such that only a single directional signal can survive at that location. In the case of opposed motions, the signals are locally balanced and the result is an abolition of any perceived motion. The co-representation of different motions in transparency can occur only at a scale coarser than this local interaction. In the case of unpaired dot patterns, local fluctuations of dot density can lead to locally unbalanced motion signals in opposite directions. For example, in one local area there might be three dots moving to the left and only one to the right; in a nearby location the opposite balance might occur. Transparency would then depend on the integration of two sets of spatially mixed local signals, into two global motions.

5.2 Neural responses to transparent and locally balanced motion stimuli

The local interaction which leads to the representation of motion being locally single-valued has been investigated neurophysiologically, through single unit recordings from behaving monkeys using paired and the unpaired dot patterns. Qian et al. (1994a,b) suggested that the interaction determining the presence or absence of transparency might correspond to the directional suppression observed in area MT. Two predictions can be made from this hypothesis. Firstly, if one set of dots move in the preferred direction and the other in the antipreferred direction for an MT cell, its response to both the paired and the unpaired dot patterns should be reduced compared to a single set of dots moving in the preferred direction alone. This is simply because the presence of motion in the antipreferred direction should always generate some suppression. Secondly, and more importantly, the amount of suppression should be stronger (i.e., the response smaller) for the paired dot patterns than for the unpaired ones, because the balanced opposing motion signals in the paired dot patterns should cancel each other more completely.

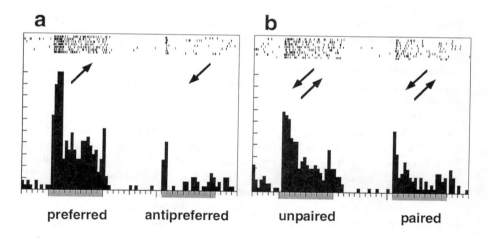

preferred   antipreferred   unpaired   paired

Fig. 5 An MT cell with significantly stronger response to the unpaired dot pattern than to the paired dot pattern. **a** Responses to single sets of random dots moving in its preferred and the antipreferred directions. **b** Responses to the paired and the unpaired dot patterns with one set of dots moving in the preferred direction and the other the antipreferred direction. The raster on the top of each diagram represent the spike records from several repeated trials. Each small dot in the raster represents the occurrence of a spike. The response histograms compiled from the rasters are shown at the bottom of the diagrams. The arrows below the rasters indicate the directions of motion. Each small division in the horizontal axis represent 10 ms. The one-second periods during which the stimuli are presented were marked by the thick black lines under the histograms. One small vertical division represents 7.7 spikes/s.

Qian and Andersen (1994) found that many MT cells did indeed behave as predicted. An example is shown in figure 5. Figure 5a shows the cell's responses to single sets of dots moving in its preferred and antipreferred directions respectively. It is evident that the cell is highly directionally selective, which is typical for MT cells. Figure 5b shows the cell's responses to the paired and the unpaired dot patterns. Both responses are weaker than the preferred direction response and the response to the paired dot pattern is significantly weaker than that to the corresponding unpaired dot patterns.

To see if the overall behaviour of MT cells is consistent with the predictions, the population results from a total of 91 recorded MT cells are summarized in figure 6. To characterize the cells quantitatively, two suppression indices for each cell were computed, one for the paired dot pattern and the other for the unpaired dot pattern along the preferred--antipreferred axis of motion. They are defined as:

$$SI_p = 1 - (paired\ response)/(preferred\ response),$$

$$SI_{up} = 1 - (unpaired\ response)/(preferred\ response).$$

The background firing rate was subtracted from all responses before calculation. These indices represent the percent reductions of a cell's responses to the paired and the unpaired dot patterns, respectively, in comparison with its preferred direction response. They therefore measure the degrees of suppression between preferred and antipreferred directions of motion. An index near zero indicates no suppression, a large value indicates strong suppression, and a negative value means that enhancement instead of suppression has occurred.

Figure 6 plots, for each cell, the unpaired suppression index, SI_{up}, against the paired suppression index, SI_p. First note that the suppression indices for almost all MT cells, whether measured with the paired or the unpaired dot patterns, are positive. This can be seen by projecting the dots in figure 6 along either coordinate axis. The result agrees with the previous finding by Snowden et al. (1991) that MT cells show significant suppression between different directions.

To examine how cells responded differently to the transparent and the non-transparent patterns, note that cells with similar responses (thus equal degrees of suppression) to the paired and the unpaired dot patterns lie near the diagonal line. The responses of these cells do not distinguish the transparent patterns from the non-transparent ones. Those falling well below the diagonal line show stronger suppression (or less response) for the paired dot patterns than for the unpaired ones. Finally, cells well above the diagonal line have the opposite behaviour: they show stronger suppression for the unpaired dot patterns than for the paired ones. There are significantly more cells below the diagonal line than above in figure 6 (Wilcoxon signed-rank test, $p < 0.0001$), consistent with the prediction.

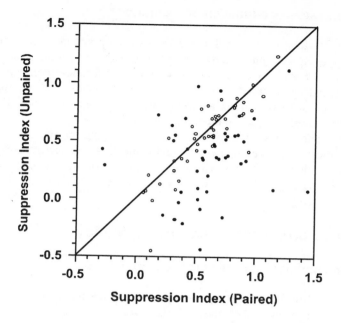

Fig. 6 MT population data. The suppression index for the paired dot pattern of each MT cell is plotted against its suppression index for the unpaired dot pattern. The cells with significantly different responses to the two types of patterns based on a two-tailed t-test are shown as solid dots. The rest are shown as open dots.

Each cell can also be considered individually. A two-tailed t-test was performed for each cell to see if its mean response to the unpaired dot pattern over repeated trials was significantly different from that to the paired dot pattern. The cells that did show a significant difference between the two patterns are plotted as solid dots in figure 6 while those did not are plotted as open dots. It was found that 40% of cells responded significantly more strongly to the unpaired dot patterns than to the paired dot patterns, while only 10% displayed the opposite behaviour. A closer examination indicated that the cells with the opposite behaviour are more responsive to flicker noise and therefore less selective to motion. The remaining 50% of cells did not show significant differences between displays of the two types. For these cells, the suppression was already very strong for the unpaired dot patterns, and there was perhaps no scope for further suppression when the paired dot patterns were used. When all recorded MT cells were averaged together, the whole population still showed significantly stronger responses to the transparent, non-paired, dot patterns than to the non-transparent paired ones. Therefore, measuring the residual responses of the directional suppression mechanism in the population of MT cells as a whole could reliably distinguish the two types of pattern.

5.3 MT suppression as a function of separation

If suppressive interaction within MT is the basis of the psychophysical effect of local pairing, the suppression should be reduced, like the psychophysical effect, when the local balance is reduced by spatial separation of differently moving dots. This effect was indeed observed by Snowden et al. (1991). In that study, displays were divided into either two or six adjacent but non-overlapping bands. Dots in alternate bands moved in opposite directions. Individual MT cells were recorded using these less balanced patterns as well as the overlapping transparent patterns. The results indicate that the suppression under the less balanced condition is indeed weaker than under the overlapping condition (shown by more points appearing below the diagonal lines in figure 9 of Snowden et al. (1991).

6. MT and V1 organization and transparency

6.1 Possible subunit structure of MT receptive fields

Since the physical difference between the paired and the unpaired dot patterns occurs at the scale of about 0.4°, these displays would best be distinguished by units operating at about this scale. The receptive fields of the MT cells recorded, however, were much larger (6-10° in diameter); a single receptive field could contain the whole of the paired or unpaired patterns. For MT cells to respond differentially to the paired and the unpaired dot patterns, it has to be assumed that the receptive field is composed of many small subunits of size about 0.4°, and that directional suppression occurs within each subunit. If it is further assumed that the response of an MT cell is determined by the sum of the thresholded outputs of all its subunits, the cell will then respond to the two types of dot patterns differently. The assumption of subunit structure in MT receptive fields is consistent with a study by Shadlen et al. (1993) who used stimuli containing local motion in one direction and global apparent motion in the opposite direction. They found that MT responses to such stimuli were largely determined by the local motion, suggesting that MT cells summate inputs to the subunits in their receptive fields.

However, there is evidence which makes it unlikely that inhibitory interactions occur only within subunits and not between them. First, the spatial range of directional interaction is three times as large in MT as in V1 (Mikami et al. 1986b). If V1 inputs provide the subunits, MT directionality must arise at least in part from mechanisms on a larger scale than the subunit fields. Second, Recanzone et al. (1997) recently recorded MT cells' responses to two small objects moving along different paths in the receptive field, and found suppressive interactions between the objects similar to those in the random dot case. Since the two objects were spatially close only briefly along their trajectories, the observed suppression must partially result from longer-range inhibitory interactions between different subunits in a receptive field. Clearly, more direct experimental evidence is needed to elucidate the structure and interactions of MT subunits. However, the difference between paired and unpaired patterns, both psychophysically and neurophysiologically, makes it clear that there is a large difference between the inhibitory interactions within a subunit field and any that go beyond it.

6.2 V1 responses to transparent and non-transparent motion stimuli

It is plausible that MT subunits should correspond to the inputs from individual V1 cells that build up the larger receptive fields of MT. If so, can the suppressive interactions that occur within subunit fields be found in V1 cells? To test this point, Qian and Andersen (1994) also recorded from V1 cells using the paired and the unpaired dot patterns. However, only 17% of V1 cells responded significantly more strongly to the unpaired dot patterns than to the paired ones, while 8% showed the opposite behaviour. The remaining 75% showed no significant difference of response to the two types of patterns. Moreover, unlike area MT, measurements of the average V1 responses did not reliably distinguish between the paired and the unpaired stimuli. V1 cells also showed overall much weaker suppression between different directions of motion, and stronger responses to the flicker noise whose effect in MT are discussed below.

The inhibitory interaction found within subunits of MT receptive fields cannot therefore be present in their inputs from V1. Rather, the interaction must be a feature of the connectivity that builds up the MT receptive field. For example, each subunit might receive mutually inhibitory inputs from a set of V1 cells tuned to the same spatial location but to different directions.

Because of the relatively weak suppressive interaction between different directions of motion in V1, cells in this area behave rather like unidirectional motion energy detectors (Emerson et al. 1992) that signal the presence of moving components in a pattern (Movshon et al. 1986; Snowden et al. 1991; Qian and Andersen 1994, 1995) regardless of the presence of other components. This means that the pattern of responses in MT cells correlates better than that in V1 cells with our perception of motion transparency.

6.3 Why directional suppression in MT?

It is natural to ask why there should be strong directional suppression in the organization of MT cells. The effect on transparency must be a side-effect, since it is hard to see any functional reason why paired dot patterns should not be perceived as transparent. One possible explanation is that suppression may be useful for enhancing the directional selectivity of neurones (Barlow and Levick 1965; Mikami et al. 1986a; Rodman and Albright 1987). Indeed, a positive correlation was found between the cells' directional index and their suppression index. Another, and perhaps more important, functional role of the observed directional suppression could be motion noise reduction. Noise, for the motion system, is any change in the stimulus light intensity distribution that is not generated by coherent motion. For example, motion noise is generated when the overall intensity of the scene changes as when the light is turned on or off, or when an object or part of an object, appears or disappears. Such situations are quite common in a natural environment. When leaves and branches move in a forest, they also generate flicker noise due to the changes of reflectance with the orientation of their surfaces, and randomly occlusions or removals of occlusions. Such spatially and temporally uncorrelated noise has a relatively uniform spatiotemporal Fourier spectrum and, from the point of view of motion energy detectors (Adelson and Bergen 1985; Watson and Ahumada 1985), it contains equal amounts of "motion" signals in all directions. Mutual suppression among detectors tuned to different directions of motion can greatly reduce the responses of these detectors to noise.

Qian and Andersen (1994) tested this hypothesis by recording from MT cells using two types of noise pattern. These were identical to the paired and unpaired dot patterns, except that instead of moving, each dot stayed in a fixed random location during its lifetime (7 frames or 117 ms) and then was replotted at a new location. Thus, in the "paired" noise, the two dots of a pair are plotted stationary but transient in nearby locations. If directional suppression is important for noise reduction, then cells with stronger suppression should respond less to the noise patterns. An example of an MT cell's responses to the noise patterns is shown in figure 7, together with its responses to preferred, antipreferred, paired and unpaired motions. This cell shows very strong directional inhibition, and also the small noise responses that would be expected if the inhibition serves to suppress noise.

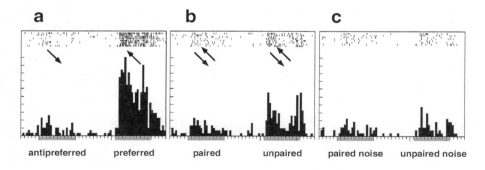

Fig. 7 Noise response of an MT cell. **a** Responses to single sets of random dots moving in its preferred and the antipreferred directions. **b** Responses to the paired and the unpaired dot patterns with one set of dots moving in the preferred direction and the other the antipreferred direction. **c** Responses to the paired and the unpaired noise patterns of flickering dots. One small vertical division represents 7.8 spikes/s. See figure 5 for an explanation of other details.

This relationship can be quantified over population in terms of two noise indices:

$$NI_p = \text{(paired noise response)/(preferred response)},$$

$$NI_{up} = \text{(unpaired noise response)/(preferred response)}.$$

The background rate was again subtracted before calculation. Figure 8a shows that cells with a large index (i.e. strong noise response) tend to show a low suppression index (Spearman rank correlation = -0.75, p < 0.0001) in the case of the paired noise and motion patterns. The cells' responses to the paired dots and the paired noise patterns were similar. The correlation is weaker but still highly significant (Spearman rank correlation = -0.52, p < 0.0001) for the unpaired case (Fig. 8b). The stronger relationship in the former case is presumably because the paired dot patterns provide a better probe for the cells' local suppressive mechanisms.

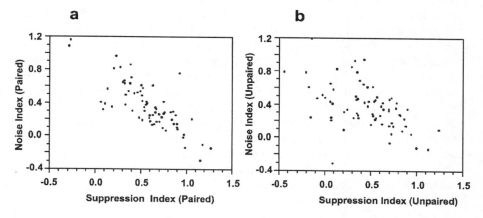

Fig. 8 Noise index verses suppression index for MT cells. **a** The paired case; **b** the unpaired case.

This discussion suggests that there is a conflict between noise reduction and the representation of multiple motions; local directional suppression in MT reduces responses to noise but makes it impossible for MT cells to represent more than one motion vector in a small area. Fortunately, there are other cues present in the real world that helps to minimize the conflict. For example, different objects in the natural environment will often be separated in depth and hence binocular disparity. Psychophysical experiments and computer simulations (Qian et al. 1994a, b) have demonstrated the effect of the disparity cue; paired dot patterns can become perceptually transparent when binocular disparity introduced between the two dots in each pair. These findings suggest that the directional suppression in MT should be strongest when the two different directions of motion are in the same disparity plane, and decrease with increasing separation in disparity (Qian and Andersen 1994). This prediction has been subsequently verified to be consistent with MT physiology (Bradley et al. 1995).

7. MT responses and perception

7.1 Attentional modulation

Subjectively, we can shift our attention between the two moving components in a transparent motion stimulus. Lankheet and Verstraten (1995) have shown that such switches of attention can affect the induction of motion aftereffects (MAEs); attention to one component significantly increased the MAE in the opposite direction. If area MT is involved in transparent motion perception, can MT activity be modulated by attention in an analogous way? Such modulation has indeed been demonstrated recently by Treue and Anderson (1996). They placed two dots, moving in opposite directions back and forth, in MT cells' receptive fields. At any time, one dot moved along the preferred direction of a cell and the other along the anti-preferred direction. They found that MT cells' activities were enhanced when the monkeys paid attention to the dot in the preferred direction, and were reduced when they attended the anti-preferred dot, even though under both conditions, the sensory inputs in the receptive fields were identical. The median attentional modulation in MT was found to be as large as 86%. Although these stimuli were quite different from transparent random dot patterns, it is plausible that the enhanced activity for the attended stimulus could be the basis for the increased adaptation found in Lankheet and Verstraten's study.

These effects raise the intriguing possibility that attention might act to reduce the suppression between different directions of motion, and consequently facilitate transparent motion perception, in the same way as does the visual cue of binocular disparity. In locally paired displays, it does not seem to be possible to perceive one of the superimposed motions by directing attention to it. Nonetheless, the hypothesis should be tested for MT cells with transparent random dot patterns in Treue and Maunsell's experimental paradigm.

7.2 Directional interactions at different angles

The physiological data discussed above was mostly concerned with random dot components moving in directions 180° apart. Snowden et al. (1991) also investigated interactions at smaller relative angles. They measured tuning curves of the suppressive effect on responses of MT cells to dots moving in the preferred direction, from another set of dots moving at different directions on different trials. These suppression tuning curves had a width of about 90°, similar to the width of ordinary directional tuning curves. A simple interpretation of the finding is that

MT cells show strong suppression for any two directions differing by more than 90°, and less suppression when the angular difference is smaller. Alternatively, Simoncelli and Heeger (1998) demonstrated that the result can be explained by a directionally isotropic mechanism of divisive inhibition (normalization). According to this view, with smaller angular differences the second set of dots excites the preferred direction more strongly, thus cancelling out a larger proportion of the uniform directional suppression and showing a reduced net suppressive effect.

These physiological data can be compared with psychophysical observations on locally paired stimuli with various directional differences. Qian et al. (1994a) noted informally that paired dot patterns with directional differences of 90° or 135° between the two dots in each pair were appeared neither transparent nor coherent, just as in the 180°-case. However, with a direction difference of only 45°, they observed a single coherent motion along the average direction of the two moving components. In all these cases, motion transparency was observed when the dots were not paired.

More recently, Curran and Braddick (1999, 2000) have tested subjects' ability to extract directional information from locally paired dot displays using a method similar to that illustrated in figure 1c. With direction differences of 60, 90, and 120°, they found that subjects could reliably judge a global direction of motion, although performance (s.d. in the range 10-18°) was somewhat impaired relative to directional judgment of a single set of dots. The judged direction followed closely the average of the two component directions. Measurements when the paired sets of dots had different speeds, and judgments of speed as well as direction, confirmed that the perceived velocity corresponded to the vector average of the components under a range of direction/speed conditions. These results imply that at no angle up to at least 120° is there a wholly suppressive interaction between locally paired directions. Of course, with equal speeds differing in direction by 180°, the vector average of the two velocities would be zero, a result indistinguishable from suppression of directional motion signals. It remains to be explored whether any effective combination of directions occurs in the range between 120-180°, although it should be realized that in this range the vector average velocity would become progressively smaller. The results are not incompatible with physiological models involving an interplay between summation and inhibitory effects, but suggest that a more detailed quantitative exploration of these models will be necessary to test them against psychophysical data.

In an earlier section on motion coherence, we mentioned the response of MT cells to coherently moving "signal" dots embedded in randomly moving noise (Newsome et al. 1989; Britten et al. 1992, 1993). The detection of the coherent signal in such displays must be influenced, and possibly may be aided, by the suppressive interactions between different directions of motion in MT. Simoncelli and Heeger (1998) found that the experimental dependence of MT responses on signal-to-noise ratio can be modelled by their directionally uniform, divisive inhibitory mechanism.

7.3 Bimodal versus unimodal population activity distribution

When a transparent random dot pattern containing two independent sets of dots moving in opposite directions is presented, the population activity of MT cells as a function of their preferred directions would be expected to form a bimodal distribution, with the two peaks approximately centred around the two directions in the stimulus. Although it is appealing to suppose that perception of transparent motion might correspond to a bimodal neuronal activity distribution in MT, such a relation cannot be expected to hold in general. Since the typical direction tuning width of MT cells is about 90°, transparent patterns containing two directions at an acute angle would most likely produce a merged unimodal distribution. This is indeed what Treue et al. (2000) found physiologically. They also found that the merging of two distributions in the transparent motion condition produced a wider distribution of activity in MT than in the unidirectional condition, and suggested that the brain could use the width of the population distribution as an indicator of the presence of transparent motion. Treue et al. further noted that a population distribution of a certain width can be generated either by two motion components, or by an appropriate mixture of more than two components. They then demonstrated through psychophysical experiments that observers could not tell the stimuli with different mixtures apart, and reported the minimum number of directions consistent with the population activity.

The experiment of Treue et al. illustrates the problem of parsing a distribution of activation into the components arising from different moving entities in the visual environment. This parsing operation appears to be flexible and may be influenced by the visual task the subject is performing. So, for example, in the experiments of Watamanuik et al. (1989) subjects could use broad directional distributions to define a single global direction with high precision. However, Zohary et al. (1996) showed that with asymmetrical distributions, the same distribution could either yield a global direction judgement or alternatively be parsed into a dominant direction corresponding to the peak plus a separate "background". Another aspect of the parsing operation is illustrated by the results of Smith et al. (1999). They tested judgements based on motion transparency for pairs of broad directional distributions with a gap between them. The size of gap that yielded effective transparency was not fixed, but depended on the overall range of directions represented. They concluded that parsing the internal representation of directions into two motions did not depend only on local features such as peaks and gaps, in the directional distribution. Rather, the parsing operation took account of the distribution as a whole. The overall range of directions of distributions shown in Smith et al. (1999) was wider than that in the experiment of Treue et al. (2000). Given this difference, it is not yet clear whether the parsing rules proposed by Treue et al., deriving transparency from a unimodal distribution, are compatible with those suggested by Smith et al., which imply bimodal distributions. It is also possible that the way the distributions were parsed in each case depend on the specific visual tasks which subjects were performing.

8. Transparency in plaid patterns

8.1 MT responses to transparent and coherent plaids

A plaid pattern consists of two superimposed gratings with different orientations and moving in different directions. The perceptual problem posed by such displays was first analysed by Hans Wallach (Wuerger et al. 1996), and following the work of Adelson and Movshon (1982) they have been widely used for psychophysical and physiological investigations of how the brain combines local motion measurements. Depending on a variety of factors (such as the differences in contrasts, spatial frequencies, and binocular disparities, and the relative angle between the two gratings), the two component gratings in a plaid can either appear to slide across each other transparently, or move together coherently in a single direction. An early physiological study of plaid coherence was provided by Movshon et al. (1986) who demonstrated that V1 cells responded to the individual motion components, but that a subpopulation of MT cells appeared to combine these components and respond to the single coherent direction, as perceived.

More recently, Stoner et al. (1990) introduced a clever manipulation of plaid coherence versus transparency, which they subsequently employed to study the neural correlates of both coherence and transparency in area MT (Stoner and Albright 1992). Instead of adding two moving sinusoidal gratings as done by Adelson and Movshon (1982), Stoner et al. constructed plaids by superimposing two drifting square wave gratings, and discovered that the perception depended critically on the luminance of the diamond-shaped intersections between the two gratings. Specifically, they found that when the intersection luminance relative to that of the gratings satisfied the rule of static transparency, a surface segmentation cue, the pattern tended to appear as two independent gratings sliding across each other transparently. Otherwise, a single coherently moving plaid was more likely to be perceived. An alternative explanation of these observations could come from the Fourier energy in the coherent pattern motion direction provided by the intersections (Simoncelli and Heeger 1998; Wilson and Kim 1994). However, Stoner and Albright (1996) later provided evidence against the Fourier interpretation by introducing a novel plaid stimuli in which surface segmentation cues could be varied independently of changes in the Fourier energy distribution.

Stoner and Albright (1992) recorded responses of MT cells to plaids whose intersection luminance was manipulated so as to generate either transparent or coherent percepts. They compared the cells' directional tuning curves under the two conditions, and found that neurones were more tuned to the component directions in the condition yielding transparency, and to the single combined pattern direction under the coherent condition. These results support those with random dot patterns, in suggesting a neural substrate for perceptual transparency and coherence in area MT.

8.2 Random dot stimuli compared with plaid patterns

Although both random dot stimuli and plaid patterns have been used in studies of motion transparency, they have some quite different properties. They can be viewed, respectively, as highly simplified versions of two different real world situations of motion transparency: (1) partial and patchy occlusions between moving objects (e.g., a tiger moving behind waving bushes), and (2) overlapping motions involving semi-transparent surfaces (e.g., an object seen through the window of a moving car). They pose different problems for the visual system; the ambiguities which arise in plaids from the coupling of motion and orientation are absent in a dot pattern, where the modulation of luminance is fully two-dimensional.

There are different reasons for using each kind of stimulus in electrophysiological and psychophysical research. Random dot patterns have a broad spatiotemporal frequency spectrum and therefore excite most V1 and MT cells effectively. Also the small sizes of dots make them suitable for investigating the spatial ranges of interaction through the pairing manipulation. On the other hand, the narrow spatiotemporal spectra of plaids render them less convenient in physiological experiments because most cortical cells are frequency tuned, and cells do not respond well unless the grating frequency matches their preferred frequencies. In addition, cells in different areas often have different tuning characteristics, which makes comparison across different areas, for example V1 and MT, somewhat difficult. However, there are important questions that cannot be investigated with random dot patterns. Since the dots moving in different directions rarely overlap, the luminance of intersections has negligible effect. In contrast, plaid patterns have large intersection regions whose luminance can be readily manipulated to study the influence of static transparency cues. In addition, the directional ambiguity of each component grating in plaids makes them suitable for investigating how the brain solves the "aperture problem". However, when these issues are not the focus of investigation, the presence of intersection regions and corners in plaids may complicate the interpretation of results. Thus the two types of patterns appear to complement each well other in research on multiple motion directions and transparency.

Despite the differences between random dot stimuli and plaid patterns, experimental investigations with both types of patterns point to some common properties and mechanisms for motion transparency. Psychophysical experiments indicate that surface segmentation cues such as binocular disparity, spatial frequency, and static transparency enhance transparent motion perception. Physiological recordings on both types of patterns agree that MT activities are well correlated with our perception of motion transparency, or the lack of it (Movshon et al. 1986; Snowden et al. 1991; Stoner and Albright 1992; Qian and Andersen 1994), while V1 cells, in contrast, appear to be more concerned with representing individual components in a motion pattern through motion energy computations, regardless of the pattern's perceptual appearance (Movshon et al. 1986; Snowden et al. 1991; Qian and Andersen 1994, 1995).

9. Computational considerations

9.1 The role of suppressive interactions

Qian et al. (1994b) demonstrated through computer simulations that a motion energy computation, followed by disparity and spatial frequency specific suppression among different directions of motion can indeed explain the perceptual difference of the transparent and non-transparent displays used in their psychophysical experiments (Qian et al. 1994a). According to the discussion above, these two stages approximately correspond to physiological responses seen in areas V1 and MT respectively. Qian et al. (1994b) found that the non-transparent displays, such as paired dot patterns, generate relatively weak responses in the simulations (opponent or normalized energies) at the suppression stage due to strong cancellations of locally balanced motion signals. In fact, these responses are not significantly higher than those generated by flicker patterns. On the other hand, the perceptually transparent displays, such as unpaired dot patterns, or paired dots with disparity, generate much stronger responses along more than one direction of motion at the suppression stage due to the presence of unbalanced motion signals. These responses in different directions are located either in different but mixed small areas, as in the unpaired dot pattern, or in different disparity or spatial frequency channels over the same spatial regions, as in the paired dot pattern with binocular disparity. A later stage could integrate these responses in different directions separately to form two overlapping transparent surfaces. Simulation examples for the unpaired, paired, and flicker patterns, and for the paired dots pattern with disparity cue are shown in figure 9.

The idea that transparent motion perception can be viewed as detection of locally unbalanced motion signals in different spatial locations, or in different disparity (or other) channels, is also consistent with a recent selection model by Nowlan and Sejnowski (1995) who proposed that motion computation should be based on those regions in the visual field where the velocity estimates are most reliable. For a transparent motion display containing two directions of motion, the most reliable regions for motion estimation are clearly those with strong unbalanced motion signals in one direction or the other. For the non-transparent paired dot patterns (without disparity cues), such reliable regions do not exist.

The suppression among different directions of motion makes it impossible for a stimulus to generate strong responses along more than one direction of motion in each small spatial area at the opponent stage when there are no other cues in the stimulus, such as disparity or spatial frequency. The minimum size of the small area is determined by the size of the front-end filters. In this regard, the suppression stage is rather like the pooling or regularization step commonly used in machine vision systems (Horn and Schunk 1981; Lucas and Kanade 1981; Hildreth 1984; Heeger 1987; Poggio et al. 1988; Grzywacz and Yuille 1990). Such a step provides a means to solve the aperture problem and to average out noise in the initial measurements. At the same time, it prevents those models from having more than one velocity estimate over each area covered by the pooling operator. In this connection, it is interesting to note that some versions of the pooling procedures for combining local gradient constraints are equivalent to a mixture of subtractive and divisive types of suppression (Simoncelli 1993). The agreement between our simulations and the psychophysical observations implies that machine vision systems can be made more consistent with transparent motion perception if the pooling operation in these systems is restricted to small areas and to each frequency and disparity channel. We suggest that the difficulty most machine vision systems have with motion transparency can be partly attributed to the fact that these systems typically apply pooling operations over a relatively large region, and that they usually do not explore other cues such as disparity and spatial frequency to restrict the scope of pooling.

We identified the suppression stage with the pooling operation in machine vision systems above. Since the function of the pooling operation is to solve the aperture problem and to average out noise, we suggest that directional suppression in MT has similar functions. Indeed unidirectional energy detectors like V1 cells suffer the aperture problem (Movshon et al. 1986). Also, V1 cells are very responsive to dynamic noise patterns made of flickering dots (Qian and Andersen 1994). Therefore, directional V1 cells are not adequate for representing transparent motion, and in fact, as a population, they could not reliably distinguish transparent patterns from non-transparent ones. The suppression in MT could help to solve these problems, just as the pooling operations do in machine vision systems. The noise response of MT cells is indeed much lower than that of V1 cells (Qian and Andersen 1994). There is also evidence that the human visual system may solve the aperture problem by averaging local motion measurements (Ferrera and Wilson 1990, 1991; Rubin and Hochstein 1993). We have discussed above how the outcome of the suppressive interaction appears to be a vector averaging, at least for a certain range of stimuli. A negative effect of the suppression is the reduced sensitivity to transparent motion. This problem is minimized, however, by applying suppression locally and by restricting it within each disparity and spatial frequency channel, since multiple motions in the real world are usually not precisely balanced in each local area, and different objects tend to have different disparity and spatial frequency distributions. While the inhibition among the cells within each disparity and spatial frequency channel could help to combine V1 outputs into a single motion signal at each location in order to solve the aperture problem and to reduce noise, different disparity and spatial frequency channels make it possible to represent multiple motions at the same spatial location.

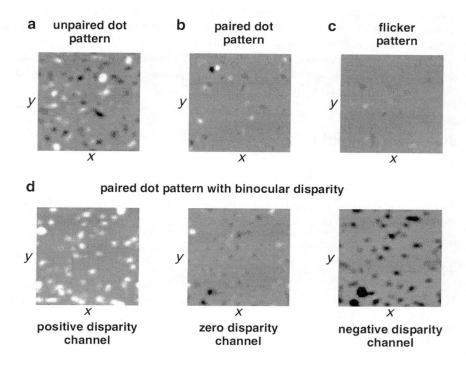

Fig. 9 Computer simulations of opponent motion energies at the motion suppression stage for **a** unpaired dot pattern, **b** paired dot pattern, **c** flicker noise pattern, and **d** paired dot pattern with binocular disparity. White and black colors code for rightward and leftward opponent energies, respectively, and grey color indicates little motion energy. In (d), opponent energies in three different disparity channels are shown.

9.2 Representation of motion, objects, and scenes

Since local suppressive mechanisms mean that the representation of motion is locally single valued, the perception of transparent motions must depend on linking spatially interdigitated signals into global entities. This can be considered as a special case of the wider process of perceptual grouping. The global entities which results from a grouping process arises from signals that have their own locations specified in a retinotopic matrix of signals. However, the global entities themselves do not have a location that is readily specified in this way, especially in the case of transparency which implies that two objects occupy overlapping locations. Therefore, the representation of transparent objects, and indeed, objects in general, cannot easily be understood as based on the kind of retinotopic array that is familiar from the early stages of cortical processing. The representation, rather than being indexed by spatial locations, must be indexed by objects – perhaps the kind of representation that has been referred to as an "object file" (Triesman 1988). Nonetheless, the information in such a representation must allow it to be associated with (distributed) spatial locations. The way in which such an representation can be implemented in a neural system is a challenge for future accounts of neural computation. Motion transparency may provide a useful route to attack this important general problem.

Acknowledgments

The work from OJB's laboratory described here was supported by project grant G9509331 and programme grant G7908507 from the Medical Research Council. NQ was supported by a Sloan Research Fellowship and an NIH grant (MH54125) during the preparation of this chapter.

References

Adelson EH, Bergen JR (1985) Spatiotemporal energy models for the perception of motion. J Opt Soc Am A 2: 284-299

Adelson EH, Movshon JA (1982) Phenomenal coherence of moving visual patterns. Nature 300: 523-525

Andersen RA (1997) Neural mechanisms of visual motion perception in primates. Neuron 18: 865-872

Baker CL, Hess RF, Zihl J (1991) Residual motion perception in a 'motion-blind' patient, assessed with limited-lifetime random dot stimuli. J Neurosci, 11: 454-481

Barlow HB, Levick WR, (1965) The mechanism of directionally selective units in the rabbit's retina. J Physiol 178: 477-504

Braddick O (1993) Segmentation versus integration in visual motion processing. Trends in Neurosci 16: 263-268

Braddick O (1997) Local and global representations of velocity: transparency, opponency, and global direction perception. Perception 26: 995-1010

Braddick OJ, Hartley T, O'Brien J, Atkinson J, Wattam-Bell J, Turner R (1998) Brain areas differentially activated by coherent visual motion and dynamic noise. NeuroImage 7: S322

Bradley DC, Qian N, Andersen RA (1995) Integration of motion and stereopsis in middle temporal cortical area of macaques. Nature 373: 609-611

Britten KH, Shadlen MN, Newsome WT, Movshon JA (1992) The analysis of visual motion: a comparison of neuronal and psychophysical performance. J Neurosci 12: 4745-4765

Britten KH, Shadlen MN, Newsome WT, Movshon JA (1993) Responses of neurons in macaque MT to stochastic motion signals. Visual Neurosci 10: 1157-1169

Curran W, Braddick OJ (1999) Perceived motion direction and speed of locally balanced stimuli. Perception 28 (suppl): 49

Curran W, Braddick OJ (2000) Speed and direction of locally-paired dot patterns. Vision Res: in press

Downing CJ, Movshon AJ (1989) Spatial and temporal summation in the detection of motion in stochastic random dot displays. Invest Ophthalmol Vis Sci (Suppl) 30: 72

Emerson RC, Bergen JR, Adelson EH (1992) Directionally selective complex cells and the computation of motion energy in cat visual cortex. Vision Res 32: 203-218

Ferrera VP, Wilson HR (1990) Perceived direction of moving two-dimensional patterns. Vision Res 30: 273-287

Ferrera VP, Wilson HR (1991) Perceived speed of moving two-dimensional patterns. Vision Research 31: 877-894

Friston KJ, Holmes AP, Worsley KJ, Poline JB, Frith CD, Frackowiak RSJ (1995) Statistical parametric maps in functional imaging: a general approach. Human Brain Mapp 2: 189-210

Grzywacz NM, Yuille NL (1990) A model for the estimate of local image velocity by cells in the visual cortex. Proc R Soc Lond A 239: 129-161

Heeger DJ (1987) Model for the extraction of image flow. J Opt Soc Am A 4: 1455-1471

Hildreth EC (1984) Computations underlying the measurement of visual motion. Art Intell 23: 309-355

Horn BKP, Schunck B (1981) Determining optical flow. Art Intell 17: 185-203

Lankheet MJM, Verstraten FJ (1995) Attentional modulation of adaptation to two-component transparent motion. Vision Res 35: 1401-1412

Lucas BD, Kanade T (1981) An iterative image registration technique with an application to stereo vision. Proc 7th Internat Joint Conf Art Intell, Vancouver: 674-679

Mikami A, Newsome WT, Wurtz RH (1986a) Motion selectivity in macaque visual cortex. I. Mechanisms of direction and speed selectivity in extrastriate area MT. J Neurophysiol 55: 1308-1327

Mikami A, Newsome WT, Wurtz RH (1986b) Motion selectivity in macaque visual cortex. II. Spatiotemporal range of directional interactions in MT and V1. J Neurophysiol 55: 1328-1339

Movshon JA, Adelson EH, Gizzi MS, Newsome WT (1986) The analysis of moving visual patterns. In: Chagas C, Gattass R, Gross C (eds) Experimental Brain Research Supplementum II: Pattern recognition mechanisms. Springer, New York, pp 117-151

Newsome WT, Britten KH, Movshon JA (1989) Neuronal correlates of a perceptual decision. Nature 341: 52-54

Newsome WT, Paré EB (1988) A selective impairment of motion processing following lesions of the middle temporal area (MT). J Neurosci 8: 2201-2211

Nowlan SJ, Sejnowski TJ (1995) A selection model for motion processing in area MT of primates. J Neurosci 15: 1195-1214

Poggio T, Torre V, Koch C (1988) Computational vision and regularization theory. Nature 317: 314-319

Qian N, Andersen RA (1994) Transparent motion perception as detection of unbalanced motion signals II: Physiology. J Neurosci 14: 7367-7380

Qian N, Andersen RA (1995) V1 responses to transparent and non-transparent motions. Exp Brain Res 103: 41-50

Qian N, Andersen RA, Adelson EH (1994a) Transparent motion perception as detection of unbalanced motion signals I: Psychophysics. J Neurosci 14: 7357-7366

Qian N, Andersen RA, Adelson EH (1994b) Transparent motion perception as detection of unbalanced motion signals III: Modeling. J Neurosci 14: 7381-7392

Recanzone GH, Wurtz RH, Schwarz UC (1997) Responses of MT and MST neurons to one and two moving objects in the receptive field. J Neurophysiol 78: 2904-2915

Rodman HR, Albright TD (1987) Coding of visual stimulus velocity in area MT of the macaque. Vision Res 27: 2035-2048

Rodman HR, Albright TD (1989) Single-unit analysis of pattern-motion selective properties in the middle temporal visual area (MT). Exp Brain Res 75: 53-64

Rubin N, Hochstein S (1993) Isolating the effect of one-dimensional motion signals on the perceived direction of moving two-dimensional objects. Vision Res 33: 1385-1396

Scase MO, Braddick OJ, Raymond JE (1996) What is noise for the motion system? Vision Res 16: 2579-2586

Shadlen MN, Newsome WT, Zohary E, Britten KH (1993) Integration of local motion signals in area MT. Soc Neurosci Abstract 19: 1282

Simoncelli EP (1993) Distributed analysis and representation of visual motion. Ph.D. Thesis, MIT, Cambridge

Simoncelli EP, Heeger DJ (1998) A model of neuronal responses in visual area MT. Vision Res 38: 743-761

Smith AT, Curran W, Braddick OJ (1999) What motion distributions yield global transparency and spatial segmentation? Vision Res 39: 1121-1132

Snowden RJ (1989) Motions in orthogonal directions are mutually suppressive. J Opt Soc Am A 7: 1096-1101

Snowden RJ, Treue S, Erickson RE, Andersen RA (1991) The response of area MT and V1 neurons to transparent motion. J Neurosci 11: 2768-2785

Stoner GR, Albright TD (1992) Neural correlates of perceptual motion coherence. Nature 358: 412-414

Stoner GR, Albright TD (1996) The interpretation of visual motion: Evidence for surface segmentation mechanisms. Vision Res 36: 1291-1310

Stoner GR, Albright TD, Ramachandran S (1990) Transparency and coherence in human motion perception. Nature 344: 153-155

Tootell RBH, Mendola JD, Hadjikhani NK, Ledden PJ, Liu AK, Reppas JB, Sereno MI, Dale AM (1997) Functional analysis of V3A and related areas in human visual cortex. J Neurosci 17: 7060-7078

Treue S, Maunsell JHR (1996) Attentional modulation of visual motion processing in cortical areas MT and MST. Nature 382: 539-541

Treue S, Hol K, Rauber HJ (2000) Seeing multiple directions of motion – physiology and psychophysics. Nature Neurosci 3: 270-276

Triesman, A (1988) Features and objects. Quart J Exp Psychol A 40: 201-237

Verstraten FA, Fredericksen RE, van Wesel RJ, Boulton JC, van de Grind WA (1996) Directional motion sensitivity under transparent motion conditions. Vision Res 36: 2333-2336

Watamaniuk SNJ, Sekuler R, Williams DW (1989) Direction perception in complex dynamic displays - the integration of direction information. Vision Res 29: 47-59

Watson AB, Ahumada AJ (1985) Model of human visual-motion sensing. J Opt Soc Am A 2: 322-342

Watt RJ, Andrews DP (1981) APE: Adaptive probit estimation of psychometric functions. Curr Psychol Rev 1: 205-214

Williams DW, Sekuler R (1984) Coherent global motion percepts from stochastic local motions. Vision Res 24: 55-62

Wilson HR, Kim J (1994) A model for motion coherence and transparency. Visual Neurosci 11: 1205-1220

Wishart KA, Braddick O (1997a) Performance-based measures of motion transparency. Invest Opthalmol Vis Sci 38: S75

Wishart KA, Braddick OJ (1997b) Performance based measures of transparency in locally-balanced motions. Perception 26 (Suppl): 86

Wuerger S, Shapley R, Rubin N (1996) "On the visually perceived direction of motion" by Hans Wallach: 60 years later. Perception. 25: 1317-1368

Zohary E, Scase MO, Braddick OJ (1996) Integration across directions in dynamic random-dot displays: vector summation or winner-take-all? Vision Res 16: 2321-2331

Combining Local Motion Signals: A Computational Study of Segmentation and Transparency

Johannes M. Zanker

Visual Sciences Group, Research School of Biological Sciences, Australian National University, Canberra, Australia

1. Introduction

Behaviourally significant information has often to be extracted from complex distributions of motion signals, which the visual system has to segment into meaningful components. Under special conditions two different motion directions are present simultaneously in the same region of the visual field: a situation called motion transparency. A closer look at the output of motion detector arrays will help us to understand what strategies may be employed in higher motion processing stages to segment the image and how complex distributions of motion directions can be represented. I used motion-defined gratings to investigate what kind of information is represented in the output of a simple two-dimensional motion detector model (2DMD). When motion-defined stripes are wide, the 2DMD output shows clearly separable regions which can easily be detected by spatial filters operating on such motion signals. When the stripes are too narrow to be resolved by such filters, the 2DMD output still reflects the presence of two motion directions, which nevertheless can be discriminated from pure noise. Only if the grain of the moving dots is below the receptive field size of the local motion detectors does the stimulus become indiscriminable from pure noise. These simulation results correspond to the psychophysical observations on segmentation and transparency, and relate well to a processing structure suggested by psychophysical and electrophysiological experiments (see Braddick and Qian, this volume). To understand motion processing we have to consider a variety of mechanisms that may serve to analyse the distributions of local motion signals, from simple spatio-temporal filters to high-level pattern recognition strategies.

2. Motion-defined gratings

In a variety of psychophysical experiments the basic properties of human motion perception have been analysed using isolated moving objects or extended homogenous motion patterns (reviewed by Sekuler et al. 1990). However, the optic flow generated by moving observers in a three-dimensional world (Gibson 1979; Nakayama 1985) produces complex distributions of motion signals that might be regarded as a critical test to assess strategies of extracting the behaviourally significant motion information. Motion-defined boundaries, for instance, inform an observer about the relative distance of objects. In practice the visual system is faced with dynamic noise superimposed on the pattern of motion signals. Noise can be induced by perturbations through optical media, occluding foliage swirling in the wind, or the limited transmission fidelity of the neural system itself. Furthermore, multiple motion signals can appear in the same region of the visual field when the environment contains transparent surfaces. In order to understand how the visual system copes with such situations, we need well defined psychophysical tasks and stimuli beyond the detection (Braddick 1980) or recognition (Regan 1991) of motion-defined shapes. An appropriate stimulus should allow for systematic manipulations of its spatial and temporal layout, and at the same time set a behaviourally relevant task such as segmenting regions of distinct motion signals. In this paper I discuss a class of motion stimuli, namely motion-defined gratings, which have been increasingly used in psychophysics during the last 15 years (Nakayama and Tyler 1981; Van Doorn and Koenderink 1982; Watson and Eckert 1994; Zanker 1996). In a typical configuration these stimuli consist of "imaginary" stripes in which randomly distributed dots are

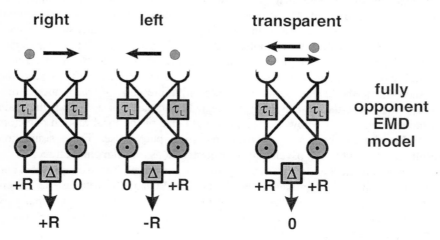

Fig. 1 Schematic sketch of correlation model. Directional responses of half detectors and full detectors to dot motion to the right, to the left, and simultaneous motion in both directions.

moving in alternating directions parallel to the stripe boundaries. Whereas such periodic motion stimuli can be easily segmented into stripes of coherent motion, as long as the patches are large enough, they can also give rise to the sensation of transparent motion, i.e. both motion directions in the whole stimulus field, when the stripes are very narrow. Finally, with very narrow stripes the stimulus may no longer be discriminated from pure dynamic noise, i.e. dot patterns changing randomly in space and time. There are thus two transitions, from segmentation to transparency, and from transparency to noise, which illustrate different aspects of motion processing. How does such a stimulus paradigm help us to analyse the computational demands for analysing spatial and spatiotemporal aspects of motion distributions?

The present discussion focuses not so much on the actual performance of human observers, but rather on the question of what kind of information can, in principle, be extracted by a simple motion detector network, and by higher-level filters operating on its output. Such a computational study not only illustrates the processing requirements for a biological visual system but may also help to design machine vision systems capable of operating under realistic conditions. In this study I used an elementary motion detector (EMD) model of the correlation type (for review, see Reichardt 1987), which can account for a wide range of experimental data in insect and human motion perception (van Santen and Sperling 1985; Borst and Egelhaaf 1989). It is a representative of luminance based motion detectors and could be replaced by other models without affecting the major results (Adelson and Bergen 1985; Watson and Ahumada 1985). The crucial aspect of the local mechanism of motion detection is that it combines signals from at least two points in space in an asymmetrical nonlinear operation after sending one of them through a temporal filter, thus monitoring spatiotemporally correlated displacements. With regard to these fundamental operations, the problem of transparency reduces to the problem of local averaging.

The basic situation is illustrated with highly schematic sketches in figure 1. An isolated dot passing the motion detector from left to right elicits a positive response in the left multiplication unit (due to the temporal coincidence of the delayed signal from the left input element and the unfiltered signal from the right input element) and no response in the right multiplication unit (due to the temporal desynchronization of the delayed signal from the right input element and the unfiltered signal from the left input element). This leads to a positive final output of the fully opponent model which combines two antisymmetrical subunits by subtracting the two multiplication outputs from each other. An isolated dot passing the detector from right to left leads to the inverse situation of a negative output of the fully opponent model, with no response from the left and a positive response from the right multiplication unit. When two dots are moving within the receptive field of the detector simultaneously, the two isolated responses superimpose and cancel each other in the opponent model output. This condition is identical to a situation in which motion is absent. Based on this stage of processing, a perceptual system would not detect transparent motion. The response at the processing level before the subtraction stage, however, does reflect the presence of moving dots. The presence of two simultaneous motion directions could therefore, in principle, be detected by analysing motion signals represented at this stage of "half-detectors".

Whereas at first sight this appears to be a possible processing strategy for perceiving transparency, there are two problems associated with it. First, it is well known that the directional selectivity of half-detectors is weak (Mikami et al. 1986; Borst and Egelhaaf 1989), and a because of this lack in selectivity a flicker stimulus and a transparent motion stimulus would generate similar responses. Second, the experiments described in the keynote article (see Qian et al. 1994a) indicate that close proximity of oppositely moving dots destroys the percept of transparency, contrary to what simple model considerations predict (see above). I will in the following therefore analyse what motion information is represented by the fully opponent model. The hypothesis is that a two-dimensional array of correlation type motion detectors (Zanker et al. 1997) may already provide an adequate representation of motion transparency. This approach complements the simulations summarized in the keynote paper (see Qian et al. 1994b) by applying similar mechanisms (half-detectors and opponency corresponding to motion energy extraction and a directional suppression stage) on a different class of stimuli, namely dot pairs moving in close proximity in opposite directions.

3. Qualitative simulation results

The basic building blocks of the two-dimensional motion detector (2DMD) model are EMDs which, in a simple implementation (default circuit diagram indicated in figure 1), each receive input from two neighbouring locations of the stimulus patterns, interacting in a nonlinear way after temporal filtering to provide a directionally selective signal. DOGs (Difference of Gaussians), with excitatory centre and inhibitory surround balanced so as to exclude any DC components from the input (cf. Srinivasan and Dvorak 1980; Marr and Hildreth 1980), were used as bandpass filters in the input lines. To prevent aliasing, the diameter of the receptive field (as measured between zero-crossings from excitatory to inhibitory regions) was set to about twice the sampling distance between the two inputs (Götz 1965). This sampling distance was used as a fundamental spatial model parameter, and was varied between 1 and 8 pixels. The time constant of the first-order lowpass filter was used as a fundamental temporal model parameter, and was usually set to the duration of one frame of the stimulus sequence. The signal from one input line was multiplied with the temporally filtered signal from the other line, and two antisymmetric units of this kind were subtracted from each other, with equal weights leading to a fully opponent EMD with high directional selectivity.

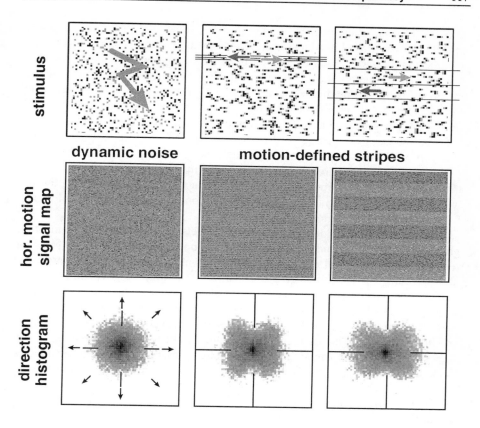

Fig. 2 Motion signal maps and direction histograms generated by the 2DMD model for noise (left column), fine (middle column), and coarse (right column) motion-defined horizontal gratings (stimuli sketched in top row). The dark and bright regions in the horizontal motion signal maps (middle row) correspond to the dominating motion components to the left and right, respectively. These are also reflected by the two-lobed direction histograms (bottom row).

Sequences of 8 frames were used as motion stimuli and each frame consisted of 256 × 256 pixels. Dark and bright dots, each covering 2 x 2 pixels, were distributed randomly in space with equal probability; the contrast was 100%. In a control stimulus all pixels were replaced between frames at random, thus leading to dynamic noise. In the test stimuli, gratings were defined by alternating stripes of leftward and rightward motion (displacement of 1 dot per frame) of all dots within the stripes leading to motion-defined square-wave gratings. The sequences of stimulus frames were processed by two orthogonal 2D-arrays of correlators (two sets of 256 x 256 EMDs), one oriented along the horizontal, the other along the vertical axis of the computer images. This led to a two-dimensional motion signal distribution, the 2DMD output, with pairs of horizontal and vertical components for each image point. In figures 2 and 3, some examples of these motion signal maps are plotted in a greyscale code in which direction of horizontal motion is indicated by the intensity of a given pixel above (rightwards) and below (left-wards) average grey. The 2D-maps of motion responses were converted into direction histograms by digitizing the local EMD responses at 8 bit resolution and then counting the number of occurrences for a given pair of horizontal and vertical responses. The greylevel at which a given two-dimensional histogram bin is plot-ted here indicates the number of occurrences for each corresponding horizontal and vertical motion component (horizontal and vertical positions in the histo-gram). The histograms are scaled relative to the maximum signal strength so that the majority of response magnitudes are accommodated. The zero histogram bin was excluded from analysis because it usually contains a huge number of counts from the static image regions.

If the stripes are wide enough, human subjects have no problems segmenting gratings made from random dots moving in opposite direction into separate regions defined by local motion signals. When motion-defined stripes are very narrow, human subjects perceive transparency, i.e. they see two directions of motion simultaneously in the same region of the visual field. The same stimuli were used as input for the 2DMD model. For wide stripes the output gives rise to a clear modulation of the motion signal in space, indicated by the alternating dark and bright regions in the right column of figure 2. The direction histogram, which shows the average distribution of motion directions across the whole stimulus field, has a typical two-lobed shape that reflects the presence of two (horizontal) motion directions in the stimulus. The middle column of figure 2 shows the 2DMD response for the case of fine motion-defined gratings. It is obvious that it would be difficult, or impossible, to segment the signal map into clear horizontal stripes of alternating greylevel because local signals blend into each other. Never-theless, the overall pattern seems to be more homogenous than a response pattern generated by stimulating the 2DMD model with a pure dynamic noise stimulus (shown as control condition in the left column). The signal maps for fine motion-defined gratings lack of vertical directions as compared to the maps produced by the dynamic noise control condition. This difference between the two conditions is reflected by the shape of the corresponding direction histograms, which however do no longer inform about the spatial distribution of signals. The histogram for the noise stimulus is rotationally symmetric, indicating equal contributions from all motion directions, while fine motion-defined stripes generate two-lobed output histograms, indicating a dominance of horizontal over vertical motion directions across the stimulus field. The same basic results are achieved with different spatial or temporal parameters of the model, or with local averaging of the 2DMD output. The predominance of the two opposite horizontal directions corresponds to the perception of transparency, and suggests perceptual mechanisms which are able to retrieve different aspects of motion signal maps to generate different sensations like transparent motion or segmentation. This higher-level processing resembles the parsing operations, discussed by Braddick and Qian (this volume) in the context of retrieving different aspects of motion direction distributions in psycho-physical experiments, which separate compound stimuli into individual components. The crucial question then is how such information can be extracted from the signal distribution at the lower processing layers.

4. Quantitative simulation results

So far the simulation results have been treated qualitatively by inspection of typical examples of motion signal maps and direction histograms in search of any obvious patterns. To understand how higher-level perceptual mechanisms may analyse different aspects of the motion signal maps, a quantitative measure is needed for how well segmented and transparent motion signals can be detected. A somewhat artificial but informative approach to such a quantitative analysis is illustrated in figure 3. A specialized operation to recover the aspect of *transparency* needs to detect the simultaneous presence of different motion directions or, in other words, needs to measure the modulation of the direction histogram. To this end, the direction distribution (computed for the complete stimulus field) is re-plotted in a different format. The integrated motion vector length, $\Sigma_{x,y}R(\rho=\alpha)$,

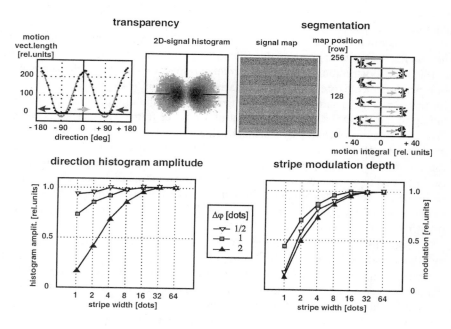

Fig. 3 Quantitative analysis of motion signal maps and histograms resulting from motion-defined gratings with variable stripe width. Two operations, estimating the amplitude of the direction histogram (left) and the modulation depth of the signal maps (right), provide measures for segmentation and transparency, respectively. These two estimates are plotted as a function of stripe-width in the bottom diagram, with the EMD sampling base $\Delta\varphi$ as parameter.

i.e. the sum of all local 2DMD response magnitudes R where the direction ρ of the motion signal takes a particular value α, is plotted as function of motion direction α. In this data format the two peaks in the left top panel of the figure correspond to the presence of two opposite motion directions. In order to use the direction histogram amplitude as a transparency measure, a $\cos2\alpha$ function was fitted to this direction distribution with a least square method (grey line in panel). To measure *segmentation* performance, the horizontally integrated horizontal motion vector length, $\Sigma_x R_x(y)$, i.e. the sum of horizontal response components R_x across image rows x, is calculated as function of vertical position in the map, y, (black dots in right top panel of figure 3). The modulation depth of the spatial motion signal distribution is estimated by a least square fit of a square-wave (grey line in panel) with variable amplitude to this spatial profile. This modulation depth indicates how well the stripes could be segmented in the signal maps derived by the 2DMD model from motion-defined gratings. The horizontally integrated vertical motion vector length, $\Sigma_x R_y(y)$, i.e. the vertical response components R_y summed across image rows x, serves as a baseline measurement which does not carry any segmentation signal (data not shown).

The results of estimating direction histogram amplitude and stripe modulation depth from the 2DMD output are plotted in the two lower panels of figure 3 as a function of motion stripe width (inversely proportional to the spatial frequency of the motion-defined grating), and with the EMD sampling base $\Delta\varphi$ as parameter (indicated by different symbols in figure 3). *Stripe modulation depth* shows a saturation-like maximum for stripe widths above 16 dots and decreases for smaller stripes to approach values close to zero for stripes that are only a single dot wide. The variation of the model parameter only weakly affects the segmentation performance expressed by this measure, apart from a slightly enhanced performance when the sampling base $\Delta\varphi$ is identical to the dot size (i.e. an optimal tuning of the EMD for the displacement of 1 dot per frame). It is clear, however, that segmentation is limited by a critical stripe width, below which motion-defined gratings are hard to resolve. On the other hand, at least for EMDs with small $\Delta\varphi$, the *direction histogram amplitude* depends much less on stripe width. When $\Delta\varphi$ is chosen to be smaller than the dot size, the histogram amplitude stays close to its maximum for all stripe widths. Only the EMD with large sampling base, tuned to low spatial frequencies, cannot detect two simultaneous motion directions when the stripes get very narrow.

Comparing the two measures extracted from the 2DMD output, histogram amplitude and modulation depth, the important conclusion is that if a local motion detector of sufficient spatial resolution is used to generate the basic motion information then two motion directions remain visible when stripes can no longer be resolved. This boundary between a segmentation domain and a transparency domain is related to the structure of the motion stimulus. Two to four rows of horizontal dot motion are required to pool across spatial inhomogeneities, in order to segment the motion-defined gratings reliably. A transition from the transparency domain to the noise domain is observed when the receptive field size of the input elements, which is locked to the sampling base, is large compared to the grating so that two oppositely moving dots are allowed to stimulate the EMD synchronously. This result for motion-defined gratings presented here resembles the modelling and physiological data, presented by Braddick and Qian in this keynote paper, for stimuli in which sparse dots moving in opposite directions are paired with each other at various distances leading to a transition from transparency to noise. Taken together, these data support the view that the motion signals from the opponency stage of the EMD limit the perception of transparency, and that signals from half-detectors (which are less directionally selective, as mentioned in Section 2) are not used to extend the possible range for detecting transparency.

5. Conclusions

This computational study of a two-dimensional array of fully-opponent motion detectors indicates that, despite its simple architecture, such a model reflects characteristic properties of human perception, namely the perception of transparency and the segmentation of a stimulus into motion-defined patterns. In order to predict the experimental results, specific aspects of motion signal distribution need to be extracted by operations designed for the particular purpose, namely detecting peaks in the direction distribution of the whole stimulus field and detecting spatial modulations of directional responses, respectively. The quantitative evaluation of the 2DMD model output leads to a different critical stripe width for the ability to segment motion-defined gratings and the ability to detect transparency. The ability to detect transparency furthermore depends on the spatial tuning of the local EMD model. This basic result parallels a number of observations using stimuli with locally balanced and unbalanced dot motion, which are reviewed in the keynote paper of this chapter.

Suggesting particular algorithms to parse one or the other percept from the motion signal distributions, and thus analysing segmentation and transparency quantitatively, goes beyond the simulations reviewed by Braddick and Quian in this volume. However, some basic aspects of my simulations nicely correspond to the results reported by Braddick and Qian. In both approaches, (a) high directional selectivity of the local motion sensing mechanism is crucial, in order to discriminate transparent motion from noise, and (b) the transition from the detection of transparency to the prevalence of noise is linked to the relation between the receptive field geometry of local motion detectors and the proximity of motion signals in opposite directions. Differences in the detailed model structure seem not to be really critical and are anyway difficult to distinguish experimentally. Under certain conditions the models can be formally equivalent (see Hildreth and Koch 1987): Braddick and Qian describe a motion-energy detector with low directional selectivity that provides motion signals which are subsequently improved by a local "directional suppression" mechanism. In the EMD model that I used, these two processing steps are implemented in a single, and simple, structure, a correlation-type "half-detector" and an opponency stage that leads to cancellation of non-directional response components. The local nature of the interaction between opposite directions, and the similarity of the directional tuning curves for the interacting components, prompted Braddick and Qian to assume an explicit suppression mechanism between co-localized V1 outputs of opposite polarity, which is acting within small sub-regions of the receptive fields of integrating V5/MT units. This hypothetical intermediate processing stage fits very well with the mirror-symmetrical structure of the correlation model that inherently generates the local direction opponency. It furthermore resembles the actual circuit diagrams of insect neurones that are believed to implement a similar spatial integration as the V5/MT neurones (see also Newsome et al., this volume, for a comparison of coding strategies in these neurones). The simulations presented here demonstrate in particular that the lower limit to perception of transparency should depend on the relation between stimulus geometry and the size (sampling base and receptive field size) of the local motion detector. Such a prediction can be tested experimentally by isolating motion detectors of a particular size, for instance with spatially filtered dot patterns (Zanker 1999).

Of course, the simple processing structure put forward here can at best be regarded as one possible component of the actual cortical processing. In particular, for reasons of parsimony, it disregards other kinds of information that are important for the segmentation process such as stereoscopically defined depth or spatial frequency composition. Indeed, in their contribution Braddick and Qian report various pieces of experimental evidence that such information influences the response properties of V5/MT neurones. The emphasis here was on possible strategies to process motion signal distributions, which surely could be improved by including signal selection criteria, be it stereo information, be it attentional focus on particular image features. One of the challenging questions for future research is to identify particular neurones involved in such higher-level processing. Are there individual neurones which not only code for a single direction, or for two motion directions, but which code for the state of transparency as such? Do they indicate the transition between spatial segmentation and transparency? Specific responses of neurones to motion-contrast have been reported (Egelhaaf 1985; Allman et al. 1985; Albright and Chaudhuri 1989) supporting the view that the segmentation of the visual field into regions of different directions or speeds is highly relevant in a variety of biological contexts. Motion-defined (second-order) motion stimuli were originally designed to test the limits of human motion vision with respect to the complexity of processing mechanisms (Lu and Sperling 1995; Zanker 1996). The fact that human observers can utilize such elaborated motion extraction mechanisms, however, suggests that the visual system may be adapted to the analysis of motion distributions in a functional context (such as segmenting complex image flow), which might not be obvious at first sight.

In summary, it was shown here by means of model simulations how spatial and spatiotemporal aspects of motion signal distributions can be extracted by applying specific computational strategies. Proceeding from these fundamental observations, we now need to analyse a variety of filter operations which use local motion signals as inputs, ranging from simple spatiotemporal operations as suggested here for a idealized segmentation problem, to high-level pattern recognition mechanisms, which are necessary to interpret biological motion or motion patterns used in animal communication (Zeil and Zanker 1997). It will be challenging to relate such mechanisms for dealing with motion signal distributions to the computational demands imposed by the signal distributions in the real world (cf. Eckert and Zeil on natural image statistics in this volume). Such an approach will raise a number of questions about the implementation of generic and purpose-built operations in biological systems and mechanisms of task-driven input selection in a modular processing architecture. Finally, it will give us clues to understand the functional specialization within the visual stream, such as specific processing roles of V1, V5/MT and MST in the primate cortex.

References

Adelson EH, Bergen JR (1985) Spatiotemporal energy models for the perception of motion. J Opt Soc Am A 2: 284-299

Albright TD, Chaudhuri A (1989) Orientation selective responses to motion contrast boundaries in Macaque V1. Soc Neurosci Abstr 15: 232

Allman JM, Miezin FM, McGuinness EL (1985) Direction- and velocity-specific responses from beyond the classical receptive field in the middle temporal visual area (MT). Perception 14: 105-126

Borst A, Egelhaaf M (1989) Principles of visual motion detection. Trends Neurosci 12: 297-306

Braddick OJ (1980) Low-level and high-level processes in apparent motion. Phil Trans Roy Soc Lond B 290: 137-151

Egelhaaf M (1985) On the neuronal basis of figure-ground discrimination by relative motion in the visual system of the fly. II. Figure-detection cells, a new class of visual interneurones. Biol Cybern 52: 195-209

Gibson JJ (1979) The ecological approach to visual perception. Lawrence Erlbaum Associates, Hillsdale, New Jersey

Götz KG (1965) Die optischen Übertragungseigenschaften der Komplexaugen von Drosophila. Kybernetik 2: 215-221

Hildreth E-C, Koch C (1987) The analysis of visual motion: From computational theory to neuronal mechanisms. Ann Rev Neurosci 10: 477-533

Lu Z-L, Sperling G (1995) Attention-generation apparent motion. Nature 377: 237-239

Marr D, Hildreth E-C (1980) Theory of edge detection. Proc Roy Soc Lond B 207: 187-217

Mikami A, Newsome WT, Wurtz RH (1986) Motion selectivity in Macaque visual Cortex. I. Mechanisms of direction and speed selectivity in extrastriate area MT. J Neurophys 55: 1308-1327

Nakayama K (1985) Biological image motion processing: a review. Vision Res 25: 625-660

Nakayama K, Tyler CW (1981) Psychophysical isolation of movement sensitivity by removal of familiar position cues. Vision Res 21: 427-433

Qian N, Andersen RA, Adelson EH (1994a) Transparent motion perception as detection of unbalanced motion signals. I. Psychophysics. J Neurosci 14: 7357-7366

Qian N, Andersen RA, Adelson EH (1994b) Transparent motion perception as detection of unbalanced motion signals. III. Modeling. J Neurosci 14: 7381-7392

Regan D (1991) Spatial vision for objects defined by colour, contrast, binocular disparity and motion parallax. In: Regan D (ed) Vision and visual dysfunction 10 Spatial vision. Macmillan Press, Houndmills, pp 135-178

Reichardt W (1987) Evaluation of optical motion information by movement detectors. J Comp Physiol A 161: 533-547

Sekuler R, Anstis SM, Braddick OJ, Brandt T, Movshon JA, Orban GA (1990) The perception of motion. In: Spillmann L, Werner JS (eds) Visual perception. The neurophysiological foundations. Academic Press, San Diego, pp 205-230

Srinivasan MV, Dvorak RD (1980) Spatial processing of visual information in the movement-detecting pathway of the fly. J Comp Physiol 140: 1-23

Van Doorn AJ, Koenderink JJ (1982) Spatial properties of the visual detectability of moving spatial white noise. Exp Brain Res 45: 189-195

Van Santen JPH, Sperling G (1985) Elaborated Reichardt detectors. J Opt Soc Am A 2: 300-321

Watson AB, Ahumada AJ (1985) Model of human visual-motion sensing. J Opt Soc Am A2: 322-342

Watson AB, Eckert MP (1994) Motion-contrast sensitivity: visibility of motion gradients of various spatial frequencies. J Opt Soc Am A 11: 496-505

Zanker JM (1996) On the elementary mechanism underlying secondary motion processing. Proc Roy Soc Lond B 351: 1725-1736

Zanker JM (1999) Generating Motion-Defined Gratings from Spatially Filtered Random Dot Patterns. Invest Ophth Vis Science 40: S 422

Zanker JM, Hofmann MI, Zeil J (1997) A two-dimensional motion detector model (2DMD) responding to artificial and natural image sequences. Invest Ophthalmol Vis Sci 38: S 936

Zeil J, Zanker JM (1997) A glimpse into crabworld. Vision Res 37: 3417-3426

Local and Global Motion Signals and their Interaction in Space and Time

Simon J. Cropper

Department of Physiology, University of Melbourne, Victoria, Australia

1. Introduction

The general issue of how we integrate local (isolated) signals into what is often termed a global percept is fundamental within vision research. The contribution by Braddick and Qian (this volume) deals specifically with the integration of local motion signals into a global motion percept but the combination of dispersed signals is a general problem with which the visual system must deal continuously. To address both the issues of general signal integration and the specific case of motion-signal integration, this commentary takes three parts. Firstly, an experimental stimulus is usually designed to simulate some component of a natural scene, so the definition of "global motion" will be discussed in the context of these artificial stimuli and also in relation to a natural image. Secondly, the parameters of the elementary local inputs to the global percept will be outlined in order to link the current commentary to the fundamental motion-detecting operations discussed in the first chapter of this book. Thirdly, I will discuss a recent discovery regarding interconnections in V1, their influence upon spatial-orientation coding and their potential relationship to the resultant percept of transparency or coherence between orthogonal motion signals in the same restricted area of space and time. I will outline a hypothesis that relates to the interaction of different fundamental image features, such as form and motion, in the process of image segmentation. Specifically, it is suggested that co-operative interactions between V5 and V1 may be instrumental in detecting borders and discontinuities in the input, facilitating basic perceptual processes such as the ability to discriminate an object from the background in an image.

2. What is global motion?

"Global motion", as opposed to "local motion", is the perceived direction of a dynamic input when that direction is the result of a combination of many individual motion signals within the stimulus. Although individual local motion vectors remain visible, if not easily independently discriminable, a single overall direction of the pattern motion – the global motion – is also perceived. It is a matter of definition whether there is really a global percept when all local signals are identical to each other and to the overall percept of motion. As a general rule one assumes that not all contributing local signals imply movement in that same global direction when discussing motion in these terms. Furthermore, the distinction of global motion as opposed to local motion strongly suggests that disparate motion signals are integrated from several locations in order to extract the overall direction. In the experimental stimulus designed to examine such a process, the assumption is that it is this integration process that is being quantified. In a hierarchy of motion-processing stages within the visual system the extraction of a global motion-vector can be thought of as the stage following the initial local-motion detection. It is important to remember that this integration process is different from simple local operations when drawing conclusions from experimental data collected using sparse-dot stimuli (cf. Edwards et al. 1996). In terms of stimulus properties, the process of global motion extraction can be described as corresponding to an extraction of the second-order temporal structure of the stimulus (Julesz 1971), the first-order temporal structure being described by the local motion-vectors which must be somehow related to one another to give the global (second-order) motion vector.

The degree to which the perception of global direction relies upon the integration of disparate local signals, or is simply signalled by a few elemental vectors, may simply depend upon the precise mathematical structure of the stimulus rather than on the properties of the underlying detection mechanism (Braddick and Qian, this volume; Newsome and Pare 1988; Scase et al. 1996). However, when human performance is quantified in terms of a simple binary direction-discrimination task (using motion-coherence as independent variable as is the norm (Braddick and Qian, this volume), this precise structural definition is not critical to the psychophysical response (Scase et al. 1996). This result could be due to the relatively coarse ($\pm 180°$) nature of the psychophysical measurement (Cox and Derrington 1994), or to the insensitivity of the visual system to the precise structure of the stimulus. The relationship of the local motion vectors to the global structure is, however, an important parameter of the stimulus as experimental examination of the proposed integration of local motion vectors requires a pattern free of low spatial-frequency cues that may provide a direct global motion solution (Smith et al.1994; Scase et al. 1996; Fredericksen and Richman 1999).

An early report of the effects of local vectors on the perceived (global) direction of a whole stimulus was provided by Wallach (1935; reviewed by

Wuerger et al. 1996). He observed that the perceived direction of motion of a line passing through an aperture was dependent upon the aperture shape relative to the moving line and that as the line passed through the aperture the perceived direction of motion would change as the aperture borders relative to the line-ends changed. As the motion vector normal to the line does not vary, the perceived direction change must be due to an interaction between the local motion-vectors and the aperture border. This observation is the precursor to much of the current work designed to reveal how such local vectors may be combined.

The later work using sparse random-dot kinematograms as stimuli provided a versatile method for examining the integration of local motion vectors by allowing researchers to isolate particular local vectors and to examine how they interact with one another; a task much harder to solve with line segments such as those used by Wallach. It is also worth noting that the utility of the random dot stimulus works with the assumption, directly or indirectly, that the artificial global motion stimulus relates well to natural scenes and may therefore provide informative data on how the task is done in the natural environment. For instance, natural motion signals are often fragmented and it is a principal goal of the visual system to extract some overall velocity field by integration of dispersed local signals (Hildreth 1984; Koenderink 1986) whether they arise from a fragmented source or simply become effectively fragmented because of the finite nature of the receptive fields that detect the local motion vectors. A caveat to this is that the integration of disparate signals only takes place over a relatively restricted area of space and time since the greater the integration period, the more coarse the resolution of the image characterized as a result of the process. An additional constraint is placed upon the integration process because not only similar motion signals are of interest, but also local-motion contrast that is relevant to segmentation. It is essential to identify moving objects within the image and to segregate these from other (moving and non-moving) areas. Somehow the system must not only integrate over a region of common motion-signals but delineate where in space and time this commonality ends: i.e., where there is a motion discontinuity.

A simple approach to this segmentation/integration problem may be to analyse motion simultaneously at multiple spatial scales and to let the coincidence of the location of motion continuity and discontinuity across spatial scale govern the recombination of local signals. In other words, if a given discontinuity is consistent in its location across several spatial scales, then it may be considered to be a border rather than a noisy motion stimulus. In the usual global-motion examples of a flock of birds or a truck depositing a load of gravel, salient motion vectors of the natural object are commonly regarded to be sparse and uninformative when processed independently. However, in these situations it is likely that there is a strong global motion vector at a coarse spatial scale compared to the many conflicting local vectors elicited by the individual elements at a fine spatial scale. Thus, the lowest spatial frequency components of the image would provide the global motion direction perceived even though higher spatial frequency components may have different directions of motion and constitute the local dynamic structure. The integration of local vectors, rather than providing the global-direction cue as such, may augment the coarse spatial-scale signal through a consistency in integrated direction-signal, despite the fact that there is still salient coding of all the individual noise vectors. This suggestion is supported by the finding that the local motions remain visible and their direction discriminable in a relatively simple composite motion plaid stimulus where a clear "pattern" or global motion is perceived (Cropper et al. 1996) It is also the case that many neurones in V5 are sensitive to all aspects of the input, that is both the global pattern motion and the local noise components (Braddick and Qian, this volume). Recent work has suggested that the resultant perception of coherence or transparency in a given dynamic field of local motion vectors may well rely on the fact that the whole population of local vectors remains explicit at the level of V5 and it is the shape of the response of the population of sensitive neurones as whole which underlies the behavioural response rather than some specific peak or average signal elicited by a single "integrating" neurone (Treue et al. 2000). In addition to this there is clear experimental evidence from psychophysics that motion analysis proceeds in parallel at multiple spatial scales in the a similar way to that which spatial analysis is thought to operate (Movshon et al. 1986; Morgan 1992; Cropper et al. 1996; Nishida et al. 1997). This is complemented by models of motion processing adopting a similar strategy (Johnston et al. 1992; Wilson et al. 1992; Wilson and Kim 1994; Zanker 1996). It is perhaps surprising that little work has been carried out concerning the integration of local motion vectors across spatial scale.

It is interesting to note that the suggestion above resembles the apparent importance of spatial phase-coherence or consistency across scale in the identification of edges and borders in space. The Harmon-Julesz blocked image of Abraham Lincoln is the classic example for this inference (Harmon and Julesz 1973; Morrone and Burr 1997) as improved upon by Dali ("Lincoln in Dalivision", 1977). It has recently been shown that the recognition of these blocked images is dependent upon the consistency of the position of edges across spatial scale and that shifts of the high spatial frequency information by the blocking process makes the image harder to recognize. However if the location of these high spatial frequency borders is obscured, by the phase-scrambling induced by blurring for instance, the image becomes recognisable (Hayes 1990, 1994, 1998). This more recent interpretation differs from the original explanation which suggested that the low spatial frequency edge information is masked by the high spatial frequencies. (Harmon and Julesz 1973). If spatiotemporal phase coherence were equally important, then multiple motion-analysis across scale may provide a tool for identification of a potential border and facilitate subsequent discrimination of the appropriate scale of integration. The underlying assumption of this argument is that some degree of rigidity or consistency is important in identifying a moving object in the scene.

In some situations, the visual system is able to extract meaningful information from what may seem particularly complex patterns of motion-vectors for which quite different rules of integration are needed rather than simply the extraction of a single global direction. An example of such complex motion patterns is a so-called "biological motion" stimulus where a moving animate body is perceived from the movements of a few isolated dots (Johansson 1973). In addition to ensuring that the scale of the integration does not obscure an important border in the image, the visual system needs to connect related areas of motion such that the profile of a moving "body" may be identified. Experimentally, it is the norm to identify particular points on the "body" upon which to locate the point light sources which comprise the stimulus components. Within such an animate stimulus, there are clear areas with non-rigid and inconsistent relationships between them yet the system combines them into some meaningful form. It is pertinent to note that the disruption of the biologically relevant coherence between the chosen points of motion (usually the joints) is achieved by changing the relative phase of movement such that the individual vectors remain constant but their relationship to each other changes (Blake 1993). Such complex processing underlying the perception of biological motion requires quite different rules of integration to be implemented in order to reach the desired response whether that be psychophysical or physiological (Johansson 1973; Hoffman and Flinchbaugh 1982; Vaina et al. 1990; Mather et al. 1992; Blake 1993; Oram and Perrett 1993; Neri et al. 1998).

Whether this kind of biological motion stimulus constitutes rigid or non-rigid motion depends upon the extent of the moving border. For instance, if the region between the knee and the foot on a moving "body" were to deform the interpretation would be quite different from a realistically simulated moving body. Since usually only the points of motion discontinuity are marked in the stimulus, this remains a speculation. It also means, however, that the perception of biological motion and experiments dealing with such do not impinge particularly upon the issue of rigidity and its environmental importance. More closely linked to the relevance of rigidity is the extraction of structure from motion (Marr 1982; Wuerger and Landy 1993a; Wuerger and Landy 1993b) where the perception of the 3D structure is dependent upon defining the components using rules consistent with the motion of a rigid or smoothly changing surface.

In summary, the rules of signal integration have not been fully explored to date, but need to be related to the utility of the global motion stimulus as a psychophysical tool. It seems possible that the scale of integration may provide some framework for these rules but it is also clear that this process will not be simple.

3. Elementary inputs to global motion: dots and plaids

Although it was originally thought that only luminance signals provided an input to the motion system, it is now more likely that this is not the case. Evidence suggests that chromatic, contrast and motion-contrast signals provide a compelling and consistent motion signal which is utilized by the system (Badcock and

Derrington 1985; Cavanagh and Favreau 1985; Derrington and Badcock 1985; Mullen and Baker 1985; Nakayama 1985; Badcock and Derrington 1987; Chubb and Sperling 1988; Badcock and Derrington 1989; Derrington et al. 1993; Werkhoven et al. 1993; Cropper 1994; Cropper and Derrington 1994; Holliday and Anderson 1994; Ledgeway 1994; Ledgeway and Smith 1994a; Ledgeway and Smith 1994b; Smith 1994; Ledgeway and Smith 1995; Lu and Sperling 1995; Cropper and Derrington 1996a; Ledgeway and Smith 1997; Cropper 1998) although the underlying mechanism remains controversial. The environmental constraints placed upon the system and its development means that it remains most sensitive to the motion of a luminance signal when considered in terms of a spatiotemporal orientation discrimination (Adelson and Bergen 1985) and this seems to hold for both the discrimination of local motion signals and for the integration of those signals into a percept of global motion.

Generally the interaction of two disparate motion signals depends upon the spatiotemporal properties of the component coding the signal. The more similar the two signals are, the more likely they are to be integrated although, as usual, there are exceptions to this. It has been claimed that local motion signals coded by both colour and by luminance-contrast contribute to a global motion percept but that their contributions are generally weaker than the contribution from luminance signals (Croner and Albright 1994; Edwards and Badcock 1995; Edwards et al 1996; Edwards and Badcock 1996a; Edwards and Badcock 1996b; Bilodeau and Faubert 1999).

Although at least one report claims there is no percept of global motion elicited by *purely* chromatic local motion signals (Bilodeau and Faubert 1999) it is clear that adding chromatic information to the elementary luminance-coded motion signals can affect discrimination performance. This principally seems to be either through the addition of further identifying features to the signal dots (Croner and Albright 1994), or by solving an ambiguity when the luminance polarity changes between frames (Edwards and Badcock 1996a). Luminance and chromatic signals do not seem to combine in a straightforward manner and nor do signals of different colours. However, when there are two populations of dots making, in a sense, two different signal-groups, a luminance population will bias the perceived motion of a purely chromatic (coherent) population when their respective directions are less than 30 degrees apart (Heidenreich and Zimmerman 1995). It is worth noting that this is more of a motion-capture phenomenon than global motion *per se* and indicates the possible interaction between two global-motion vectors.

A contrast-coded local motion vector is affected by, and combines with, a luminance-coded local vector but a luminance coded signal population is not affected by contrast-coded noise (Edwards and Badcock 1995). This peculiar result may be explained by the suggestion that the mechanism integrating contrast-coded local motion input is also sensitive to the luminance coded local motion, but not vice versa (Edwards and Badcock 1995). This suggestion may be partially consistent with a limited amount of physiological data (Zhou and Baker 1993) and may even be argued to be a necessary consequence of some non-linear pre-processing (Zanker 1995) but really remains highly speculative given the current evidence.

However, as if to ensure there are no clear rules for integration, if a luminance signal-dot (or presumably noise-dot) inverts its polarity between frames of the motion sequence then there is no global percept unless a consistent chromatic signal is also present to resolve the ambiguity (Edwards and Badcock 1996a). If however the signal population is made up of both light and dark dots then they seem to combine linearly such that the polarity of the local motion vectors is not important (Edwards and Badcock 1996b).

An issue that has not really been experimentally examined to date, although it is mentioned by Braddick and Qian (this volume), is the relationship between the global sparse-dot motion and plaid stimuli, and indeed the relationship of both of these to the original bar stimuli used by Wallach (1935). The three patterns are similar in the sense that they require integration of local motion signals in order to produce the global percept. The lack of any extensive spatial interaction between the components of a sparse dot motion pattern is significantly different from the situation in plaid stimuli where the two-dimensional structure relies on the spatial interaction of the one-dimensional components as do models of their detection (Wilson et al. 1992; Wilson and Kim 1994) and the bar stimuli where the vectors are all derived from a single moving "object". The general similarity rule that initially was a defining feature of plaid coherence seems to hold in plaid and dot patterns (Movshon et al. 1986).

Coherent two-dimensional plaid patterns are generally only perceived when the one-dimensional components are similar in the same critical properties of spatial frequency, statistical order and colour (Movshon et al. 1986; Krauskopf and Farell 1990; Smith and Edgar 1991; Kim and Wilson 1993; Krauskopf and Farell 1993; Cropper et al. 1996). In addition, a clear and predictable integration seems to occur only between local motion vectors defined by similar dots (but see Edwards and Badcock (1996b) as explained above). An exception to this rule of similarity in the case of plaid stimuli becomes apparent when an interaction between two components of different spatial frequency or colour is sufficient in area to constitute a discrete feature of the stimulus in itself and the direction is perceived independently (Kim and Wilson 1993; Cropper et al. 1996). This "discrete feature" aspect of the stimulus can in itself be a problem when attempting to examine the nature of the underlying neural processes (Braun 2000). The distinction between the situations created by dot and grating components can be a subtle one but the general rule of similarity observed for the integration of motion vectors within both types of patterns is both experimentally and ecologically pertinent. Experimentally the result is relevant for our attempt to find some common motion integration strategy. Ecologically it is likely that a single moving body is made up of principally similar basic signals within the neural image and initially integrating across these common vectors will likely give the most parsimonious response given that other constraints such as rigidity and common depth are satisfied.

It seems to me that it would be useful to address experimentally the issue of the relationship between single and multiscale components (i.e., gratings and dots) and their spatial inter-relationships in a stimulus. To date the closest analogue to this aspect is the work concerning motion-capture (Ramachandran and Inada 1985; Ramachandran and Cavanagh 1987; Yo and Wilson 1992; Bressan and Vallortigara 1993; Scase and Braddick 1994; Morrone and Burr 1997) and that dealing with coherence in plaids made from different types of components (Krauskopf and Farell 1990; Smith and Edgar 1991; Kim and Wilson 1993; Wilson and Kim 1994; Kim and Wilson 1996). It is clear that the component properties of plaids and contrast gratings are critical in the ultimate motion percept (Cropper and Badcock 1995; Cropper and Johnston 1999), an inference that needs to be directly related to the global motion percept and extended to sparse random dot kinematograms.

In summary, the strongest elemental input to the global percept is undoubtedly from luminance-coded local motion signals, but it is odd that there seems to be such a poor input – if any (Bilodeau and Faubert 1999) – from purely chromatic signals. Not only do chromatic gratings support a perfectly adequate motion percept independent of a luminance signal (Cavanagh et al. 1984; Cavanagh and Anstis 1991; Cropper and Derrington 1994; Cropper and Derrington 1996b) but receptive fields in V5 are also sensitive to this colour-coded local motion (Saito et al. 1989; Gegenfurtner et al. 1994). If neurones in V5 are responsible for integrating the local signals into a global signal, as reviewed by Braddick and Qian in this volume, why then are dispersed chromatic signals not integrated into a global pattern response? If it is just an issue of stimulus strength, and V5 cells are less vigorous in their response to chromatic stimuli, then one would have expected this contrast dependency to have been evident in the data collected so far (Bilodeau and Faubert 1999). The chromatic contrast of a luminance-defined dot also clearly contributes to the global motion percept (Edwards and Badcock 1996a), raising the question as to whether this is simply the symptom of a chromatic "label" added to the signal in question (Croner and Albright 1994) or due to a distinct chromatic motion signal that actually is integrated. The latter seems unlikely because although the chromatic signal may aid in solving an ambiguity (Edwards and Badcock 1996a) it does not seem to contribute to the global integration process as an independent motion signal (Bilodeau and Faubert 1999).

4. The modularity of processing and the global motion signal

Using sparse dot kinematograms to investigate motion integration makes sense as the dots can provide an unambiguous motion signal because, unlike gratings, their luminance is modulated in two spatial dimensions within the bounds of the receptive field. However, one problem with drawing broad conclusions from these symmetrical dot stimuli may be that the data are collected in the absence of any specific orientation cues, which would both be present in the natural input and which play an important role in the initial stages of cortical processing. Under natural conditions it is more than likely that spatial orientation and motion (i.e. coding in x, y, t) are associated in the processing stream, as they are in the image, and motion information could be preferentially collated along collinear contour elements. Many of the receptive fields in V1 are orientation (x y) selective as well as direction (x t) selective and it is these units that provide both direct and indirect input to the integrating neurones in area V5 (see Braddick and Qian, this volume). Recent work has introduced a stimulus which allows one to control the stimulus energy contained within particular orientation bands (Braun 2000; Schrater et al. 2000). Such an approach may facilitate the characterization of aspects of integration of motion signals which so far were untapped by sparse dot stimuli. In these orientation-filtered random-dot patterns motion energy seems to be summed across all orientation bands to contribute to the final perceived direction whether a given band contains useful information or not (Schrater et al. 2000). Using sparse-dot stimuli, we essentially assume that each isolated local motion-vector is extracted prior to the stage of global integration and as such has passed the stage of summation across orientation bands. If the rules of motion-vector integration hold regardless of energy distribution in orientation bands, then conclusions based on sparse dot kinematograms may be unaffected. However, as is implied by recent work and comment (Cropper and Badcock 1995; Braun 2000; Schrater et al. 2000) and suggested below, this may not be the whole story. A useful manipulation to link the more traditional sparse-dot stimulus with the filtered random-dot fields (Schrater et al. 2000) may be to give the elements of the sparse dot kinematograms some orientation in space. This may provide an alternative experimental approach, which examines more directly the integration of isolated motion vectors but also addresses the spatial orientation issues discussed by Schrater et al.. The question remains as to what framework may guide the experimentation to link these two potential levels of motion extraction, if indeed this is what they are.

Recent work has considerably enhanced the degree to which we think spatial orientation coding may proceed at the level of V1 (Das and Gilbert 1997; Das and Gilbert 1999; Eysel 1999). New emphasize is given to the functional significance of interconnections between the individual cells constituting the orientation columns in V1 each structured with the "pinwheel" arrangement of orientation sensitivity (Braitenberg and Braitenberg 1979; Bonhoeffer and Grinvald 1991; Sejnowski et al. 1988), which is an extension of the originally proposed columnar structure (Hubel and Wiesel 1977) describing how each hypercolumn (containing all orientation sensitivities in its constituent neurone sensitivities mapped to a given spatial location) is arranged radially around a central axis (Braitenberg and Braitenberg, 1979). Both excitatory and inhibitory synaptic connections between individual neurones vary in their range (i.e. they are remote and local), and connect both separate and overlapping classical receptive fields. Excitatory interactions between remote receptive fields with similar orientation selectivity could be very useful in "binding" a similar contour across an extended region of space in a situation where the organization of V1 would lead to a discontinuous representation of such extended contours relative to a given receptive field size (Eysel 1999). On the other hand, an inhibitory connection between neighbouring receptive fields (abutting in visual space and therefore lying between adjacent hypercolumns) with preferred orientations perpendicular to each other could code for a discontinuity in the contour, a corner or a T-junction (Das and Gilbert 1999). The strength of the inhibitory connection decreases with increasing distance between the receptive fields localizing the effect to nearby regions of visual space. However, if the receptive fields are actually overlapping rather than abutting then the effect is to sharpen the orientation tuning by suppressing the responses to orientations away from the peak orientation sensitivity of the receptive field rather than code for a discontinuity (Sejnowski et al. 1988; Crook et al. 1998; Das and Gilbert 1999).

Since it seems there is some commonality in structure exhibited throughout the sensory cortex (Mountcastle 1957; Hubel and Wiesel 1962; Hubel and Wiesel 1974), it is not unlikely that coding of spatiotemporal orientation at either the level of V1 or V5 may be similar. If motion discontinuities were coded in a similar way to spatial contour discontinuities, the orthogonal motion suppression noted by several studies (Snowden 1989; Snowden et al. 1991; Qian and Andersen 1994) may be explained on exactly the same basis as outlined by Das and Gilbert (1999). That is, the suppression of output observed when two orthogonal motions are presented together is the result of the inhibitory interaction between two spatially abutting receptive fields of orthogonal direction-selectivity. The data collected from V5 receptive fields are also consistent with the spatial proximity of the motion signals in the stimulus and whether suppression is observed or not: suppression of the receptive field output is only observed at a critical (adjacent) proximity of the orthogonal motion vectors. If the opposing motions are either too close or too far apart in spatial location then the suppressive effect disappears (Snowden et al. 1991). It is possible that the effect of spatially coinciding orthogonal motion signals (to give the presumed single-unit analogy to perceptual transparency) actually sharpens the directional tuning of the cell. If such a mechanism existed, V5 cells, when taken as a population, may have the additional capacity to take motion integration one stage further and code motion boundaries and discontinuities in the same way in which V1 may code orientation discontinuities. The percept of transparency may be the result of two overlapping receptive fields signalling, equally vigorously, two different directions of motion essentially existing in the same region of space and time (see Zanker, this volume). In this scheme, the spatial proximity or coding properties of the input stimuli prevent any specific inhibitory interaction which causes only one resultant motion to be gleaned and this is consistent with the shape of the response of a population of neurones being critical to the resultant percept (Treue et al. 2000). Whatever the case, the situation of varying degrees of coherence or transparency perceived in structurally very similar displays obviously requires quite specific rules about which signals to integrate and which signals to keep separate. The importance being placed on the shape of a given response-surface may overcome some of the problems associated with having to implement such rules so rigidly and also explains some of the apparent behavioural anomalies with the underlying physiology (Groh 2000; Treue et al. 2000). Despite this recent work and as implied earlier, this is a fundamental yet relatively poorly understood issue when considering the image as a whole.

In support of an intrinsic and continuing link between space and space-time orientation, recent psychophysical evidence has illustrated the utility of a potentially informative "streak" in the neural image due to the rapid motion of a spatially localized stimulus such as Gaussian blobs (Geisler 1999). This work showed that masking in the spatial-orientation (xy) domain had a specific effect upon detection in the spatiotemporal-orientation (xt) domain, suggesting that the neural streak can support detection of the moving stimulus by interaction between xy and xt tuning at the level of V1. The greater implication of this result is that it shows the importance of linking orientation across space and time which is potentially inherent in every stimulus, however spatially localized we consider it to be on the input level. In a similar sense, the increasing amount of work illustrating the degree and importance of reciprocal connections between different cortical areas, such as feedback running back from V5 to V1 (as well as V2 & V3) can have the effect of modifying the response of these areas to motion (Hupe et al. 1998). The utility of the response of V5 in localizing the motion boundary in order to modify the overall V1 response could potentially be linked to an orientation code (Das and Gilbert 1999; Geisler 1999) to aid in the localization of edges and boundaries. In fact, it makes good sense to retain as much orientation information as possible when analysing and integrating local motions in order to facilitate the reconstruction of moving borders from the disparate input signals.

5. Conclusion

One of the major goals of the visual system is to integrate a wide variety of seemingly disparate signals back into some semblance of order such that an image that as been ultimately fragmented on the retina and in its cortical representation is reassembled in order to allow us to effectively navigate in a complex environment. The book in general, and in particular this Part, have attempted to give some background and structure to the way in which this may be accomplished for motion signals. One of the more critical rules that seems to be adopted, which has a direct relationship to the structure of the visual environment, is that motion signals coded by similar stimulus properties seem to be preferentially integrated. Furthermore, this spatial integration seems to draw a principal input from luminance-coded local-motion signals, although contributions from other signals cannot be ruled out. In particular an additional "labelling" cue from colour seems particularly efficacious in augmenting or destroying a global motion luminance signal, which again can be understood as tool to operate in a patchy natural environment.

As an extension to the empirical data available so far, the current commentary seeks to suggest that the spatial and spatiotemporal signals constituting the environment may be more intrinsically linked than previously thought, despite our inclination toward the theory of modularity in the system. A general idea about interactions between local spatial and temporal orientation selective receptive fields is put forward which is consistent with a recent finding in V1 which implies that a reciprocal interaction between V1 and V5 may well account for many of the basic tasks of image segmentation in space and time.

Acknowledgements

The author was supported by a QE2 Fellowship and project grants from the Australian Research Council and The Wellcome Trust. I thank Johannes Zanker, David Badcock, Sophie Wuerger, Michael Johnston, Jochen Zeil and Umberto Castiello who provided critical and insightful feedback on this chapter.

References

Adelson EH, Bergen JR (1985) Spatiotemporal energy models for the perception of motion. J Opt Soc Am A 2: 284-299

Badcock DR, Derrington AM (1985) Detecting the displacement of periodic patterns. Vision Res 25: 1253-1258

Badcock DR, Derrington AM (1987) Detecting the displacements of spatial beats: a monocular capability. Vision Res 27: 793-797

Badcock DR, Derrington AM (1989) Detecting the displacements of spatial beats: no role for distortion products. Vision Res 29: 731-739

Bilodeau L, Faubert J (1999) Global motion cues and the chromatic motion system. J Opt Soc Am A 16: 1-5

Blake R (1993) Cats perceive biological motion. Psychol Sci 4: 54

Bonhoeffer T, Grinvald A (1991) Iso-orientation domians in cat visual cortex are arranged in pinwheel-like patterns. Nature 353: 429-431

Braitenberg V, Braitenberg C (1979) Geometry of orientation columns in the visual cortex. Biol Cybern 33: 179-186

Braun J (2000) Targeting visual motion. Nature Neurosci 3: 9-11

Bressan P, Vallortigara G (1993) What induces capture in motion capture? Vision Res 33: 2109-2112

Cavanagh P, Anstis S (1991) The contribution of color to motion in normal and color-deficient observers. Vision Res 31: 2109-2148

Cavanagh P, Favreau OE (1985) Colour and luminance share a common motion pathway. Vision Res 25: 1592-1601

Cavanagh P, Tyler CW, Favreau OE (1984) Perceived velocity of moving chromatic gratings. J Opt Soc Am A 1: 893-899

Chubb C, Sperling G (1988) Drift-balanced random stimuli: A general basis for studying non-Fourier motion perception. J Opt Soc Am A 5: 1986-2006

Cox MJ, Derrington AM (1994) The analysis of motion of two-dimensional patterns: do fourier components provide the first stage? Vision Res 34: 59-72

Croner LJ, Albright TD (1994) Segmentation by color improves performance on a visual motion task. Invest Ophthalmol Vis Sci 35: S 1643

Crook JM, Kisvárday ZF, Eysel UT (1998) Evidence for a contribution of lateral inhibition to orientation tuning and direction selectivity in cat visual cortex: reversible inactivation of functionally characterized sites combined with neuroanatomical tracing techniques. Eur J Neurosci 10: 2056-2075

Cropper SJ (1994) Velocity discrimination in chromatic gratings and beats. Vision Res 34: 41-48

Cropper SJ (1998) The detection of luminance and chromatic contrast modulation by the visual system. J Opt Soc Am A 14: 1969-1986

Cropper SJ, Badcock DR (1995) Perceived direction of motion: It takes all orientations. Perception 24: 106a

Cropper SJ, Derrington AM (1994) Motion of chromatic stimuli: First-order or second-order? Vision Res 34: 49-58

Cropper SJ, Derrington AM (1996a) Detection and motion detection in chromatic and luminance beats. J Opt Soc Am A 13: 401-407

Cropper SJ, Derrington AM (1996b) Rapid colour-specific detection of motion in human vision. Nature 379: 72-74

Cropper SJ, Johnston A (1999) The motion of contrast envelopes: peace and noise. Invest Ophthalmol Vis Sci 40: S 119

Cropper SJ, Mullen KT, Badcock DR (1996) Motion coherence across cardinal axes. Vision Res 36: 2475-2488

Das A, Gilbert CD (1997) Distortions of visuotopic map matches orientation singularities in primary visual cortex. Nature 387: 594-598

Das A, Gilbert CD (1999) Topography of contextual modualtions mediated by short-range interactions in primary visual cortex. Nature 399: 655-661

Derrington AM, Badcock DR (1985) The low level motion system has both chromatic and luminance inputs. Vision Res 25: 1874-1884

Derrington AM, Badcock DR Henning GB (1993) Discriminating the direction of second-order motion at short stimulus durations. Vision Res 33 1785-1794

Edwards M, Badcock DR (1995) Global motion perception: no interaction between the first- and second-order motion pathways. Vision Res 35: 2589-2602

Edwards M, Badcock DR (1996a) Global-motion perception: Interaction of chromatic and luminance signals. Vision Res 36: 2423-2432

Edwards M, Badcock DR (1996b) Global-motion perception: Interaction of the ON and OFF pathways. Vision Res 36: 2849-2858

Edwards M, Badcock DR, Nishida S (1996) Contrast sensitivity of the motion system. Vision Res 36: 2411-2421

Eysel U (1998) Turning a corner in vision research. Nature 399: 641-643

Fredericksen RE, Richman S (1998) Calculating fourier spectra of random dot motion stimuli: Interpretations of some experimental results. Invest Ophthalmol Vis Sci 39: S 1075

Gegenfurtner KR, Kiper DC, Beusmans JMH, Carandini M, Zaidi Q, Movshon JA (1994) Chromatic properties of neurones in MT. Visual Neurosci 11: 455-466

Geisler WS (1999) Motion streaks provide a spatial code for motion direction. Nature 400: 65-69

Groh JM (2000) Predicting perception from population codes. Nature Neurosci 3: 201-202

Harmon LD, Julesz B (1973) Masking in visual recognition: Effects of two-dimensional filtered noise. Science 180: 1194-1197

Hayes A (1990) 'Blocked' images: why less is more for recognition. Perception 19: 267

Hayes A (1994) Multi-resolution local-phase coding of image structure; an explanation of the 'blocked Lincoln' effect. Australian J Psychol 46: 13

Hayes A (1998) 'Phase filtering' disrupts continuity through spatial scale of local image-structure. Perception 27: 81

Heidenreich SM, Zimmerman GL (1995) Evidence that luminant and equiluminant motion signals are integrated by directionally selective mechanisms. Perception 24: 879-890

Hildreth EC (1984) The computation of the velocity field. Pro Roy Soc Lond B 221: 189-220

Hoffman DD, Flinchbaugh BE (1982) The interpretation of biological motion. Biol Cybern 42: 195-204

Holliday IE, Anderson SJ (1994) Different processes underlie the detection of second-order motion at low and high temporal frequencies. Proc Roy Soc Lond B 257: 165-173

Hubel DH, Wiesel TN (1962) Receptive fields, binocular interactions, and functional architecture in cat's visual cortex. J Physiol 160: 106-154

Hubel DH, Wiesel TN (1974) Sequence regularity and geometry of orientation columns in the monkey striate cortex. J Comp Neurol 158: 267-294

Hubel DH, Wiesel TN (1977) Functional architecture of macaque monkey visual cortex. Proc Roy Soc Lond 198: 1-59

Hupe JM, James AC, Payne BR, Lomber SG, Girard P, Bullier J (1998) Cortical feedback improves discrimination between figure and background by V1, V2 and V3 neurones. Nature 394: 784-787

Johansson G (1973) Visual perception of biological motion and a model for its analysis. Perc and Psychophys 14: 201-211

Johnston A, McOwan PW, Buxton H (1992) A computational model for the analysis of some first-order and second-order motion patterns by simple and complex cells. Proc Roy Soc Lond B 250: 297-306.

Julesz B (1971) Foundations of cyclopean perception. University of Chicago Press, Chicago

Kim J, Wilson HR (1993) Dependence of plaid motion coherence on component grating directions. Vision Res 33: 2479-2489

Kim J, Wilson HR (1996) Direction repulsion between components in motion transparency. Vision Res 36: 1177-1187

Koenderink JJ (1986) Optic flow. Vision Res 26: 161-180

Krauskopf J, Farell B (1990) Influence of colour on the perception of coherent motion. Nature 348: 328-331

Krauskopf J, Farell B (1993) Precise definition of the cardinal directions based on the coherence of plaids. Invest Ophthalmol Vis Sci 34: 1046

Ledgeway T (1994) Adaptation to second-order motion results in a motion aftereffect for directionally ambiguous test stimuli. Vision Res 34: 2879-2889

Ledgeway T, Smith AT (1994a) The duration of the motion aftereffect following adaptation to first-order and second-order motion. Perception 23: 1211-1219

Ledgeway T, Smith AT (1994b) Evidence for seperate motion-detecting mechanisms for first and second-order motion in human vision. Vision Res 34: 2727-2740

Ledgeway T, Smith AT (1995) The perceived speed of second-order motion and its dependence on stimulus contrast. Vision Res 35: 1421-1434

Ledgeway T, Smith AT (1997) Changes in perceived speed following adaptation to first-order and second-order motion. Vision Res 37: 215

Lu Z-L, Sperling G (1995) The functional architecture of human visual motion perception. Vision Res 35: 2697-2722

Marr D (1982) Vision: A computational investigation into the human representation and processing of visual information. WH Freeman and Company, San Francisco

Mather G, Radford K, West S (1992) Low-level visual processing of biological motion. Proc Roy Soc Lond B 249: 149-155

Morgan M (1992) Spatial filtering precedes motion detection. Nature 355: 344-346

Morrone MC, Burr DC (1997) Capture and transparency in coarse quantized images. Vision Res 37: 2609-2629

Mountcastle VB (1957) Modality and topographic properties of single neurons of cat's somatic sensory cortex. J Neurophysiol 20: 408-434

Movshon JA, Adelson EH, Gizzi MS, Newsome WT (1986) The analysis of moving visual patterns. Exp Brain Res Suppl 11: 117-152

Mullen KT, Baker CL (1985) A motion aftereffect from an isoluminant stimulus. Vision Res 25: 685-688

Nakayama K (1985) Biological image motion processing: a review. Vision Res 25: 625-660

Neri P, Morrone MC, Burr DC (1998). Seeing biological motion. Nature 395: 894-896

Newsome WT, Pare EB (1988) A selective impairment of motion perception following lesions of the middle temporal visual area (MT). J Neurosci 8: 2201-2211

Nishida S, Ledgeway T, Edwards M (1997) Dual multiple-scale processing for motion in the human visual system. Vision Res 37: 2685-2698

Oram MW, Perrett DI (1993) Responses of anterior superior temporal polysensory (STPa) neurons to 'biological' motion. J Cogn Neurosci 6: 99-116

Qian N, Andersen RA (1994) Transparent motion perception as detection of unbalanced motion signals. II. Physiology. J Neurosci 14: 7367-7380

Ramachandran VS, Cavanagh P (1987) Motion capture anisotropy. Vision Res 27: 97-106

Ramachandran VS, Inada V (1985) Spatial phase and frequency in motion capture of random dot patterns. Spatial Vis 1: 57-67

Saito H, Tanaka K, Isono H, Yasuda M, Mikami A (1989) Directionally selective response of cells in the middle temporal area (MT) of the macaque monkey to the movement of equi-luminous opponent colour stimuli. Exp Brain Res 75: 1-14

Scase MO, Braddick OJ (1994) Motion contrast and capture between gratings and dots. Invest Ophthalmol Vis Sci 35: S 2076

Scase MO, Braddick OJ, Raymond J (1996) What is noise for the motion system? Vision Res 16: 2579-2586

Schrater PR, Knill DC, Simoncelli EP (2000) Mechanisms of visual motion detection. Nature Neurosci 3: 64-68

Sejnowski TJ, Koch C, Churchland PS (1988) Computational Neuroscience. Science 241: 1299-1306

Smith AT (1994) The detection of second-order motion. In: Smith AT, Snowden RJ (eds) Visual Detection of Motion. Academic Press Limited, London, pp 145-176

Smith AT, Edgar GK (1991) Perceived speed and direction of complex gratings and plaids. J Opt Soc Am 8: 1161-1171

Smith AT, Snowden RJ, Milne AB (1994) Is global motion really based on spatial integration of local motion signals? Vision Res 34: 2425-2430

Snowden RJ (1989) Motions in orthogonal directions are mutually suppressive. J Opt Soc Am A 6: 1096-1101

Snowden RJ, Treue S, Erickson RG, Andersen RA (1991) The response of area MT and V1 neurons to transparent motion. J Neurosci 11: 2768-2785

Treue S, Hol K, Rauber H-J (2000) Seeing multiple directions of motion - physiology and psychophysics. Nature Neurosci 3: 270-276

Vaina LM, Lemay M, Bienfang DC, Choi AY, Nakayama K (1990) Intact 'biological motion' and 'structure from motion' perception in a patient with impaired motion mechanisms: A case study. Visual Neurosci 5: 353-369

Wallach H (1935) Über visuell wahrgenommene Bewegungsrichtung. Psychologische Forschung 20: 325-380

Werkhoven P, Sperling G, Chubb C (1993) The dimensionality of texture-defined motion: a single channel theory. Vision Res 33: 463-485

Wilson HR, Kim J (1994) A model for motion coherence and transparency. Visual Neurosci 11: 1205-1220

Wilson HR, Ferrera VP, Yo C (1992) A psychophysically motivated model for two-dimensional motion perception. Visual Neurosci 9: 79-97

Wuerger SM, Landy MS (1993a) Role of chromatic and luminance contrast in inferring structure from motion. J Opt Soc Am A 10: 1363-1372

Wuerger SM, Landy MS (1993b) Structure from motion for chromatic and luminance stimuli. J Opt Soc Am A 10: 1363-1372

Wuerger SM, Shapley R, Rubin N (1996) "On the visually perceived direction of motion" by Hans Wallach: 60 years later. Perception 25: 1317-1367

Yo C, Wilson HR (1992) Moving two-dimensional patterns can capture the perceived directions of lower or higher spatial frequency gratings. Vision Res 32: 1263-1269

Zanker JM (1995) Of models and men: mechanisms of human motion perception. In: Papathomas TV, Chubb C, Gorea A, Kowler E (eds) Early vision and beyond. MIT press, Boston, pp 156-165

Zanker JM (1996) On the elementary mechanism underlying secondary motion processing. Proc Roy Soc Lond B 351: 1725-1736

Zhou YX, Baker CLJ (1993) A processing stream in mammalian visual cortex for non-Fourier responses. Science 261: 98-101

Part III

Optical Flow Patterns

Extracting Egomotion from Optic Flow: Limits of Accuracy and Neural Matched Filters

Hans-Jürgen Dahmen[1], Mattihas O. Franz[2] and Holger G. Krapp[3]

[1]Lehrstuhl für Biokybernetik, Biologisches Institut, Universität Tübingen,
Tübingen; [2]DaimlerChrysler AG, Research and Technology, Ulm; [3]Lehrstuhl für
Neurobiologie, Universität Bielefeld, Bielefeld, Germany

Contents

1. Abstract

In this chapter we review two pieces of work aimed at understanding the principal limits of extracting egomotion parameters from optic flow fields (Dahmen et al. 1997) and the functional significance of the receptive field organization of motion sensitive neurones in the fly's visual system (Franz and Krapp 1999). In the first study, we simulated noisy image flow as it is experienced by an observer moving through an environment of randomly distributed objects for different magnitudes and directions of simultaneous rotation R and translation T. Estimates R', of the magnitude and direction of R, and t', of the direction of T, were derived from samples of this perturbed image flow and were compared with the original vectors using an iterative procedure proposed by Koenderink and van Doorn (1987). The sampling was restricted to one or two cone-shaped subregions of the visual field, which had variable angular size and viewing directions oriented either parallel or orthogonal with respect to the egomotion vectors R and T. We also investigated the influence of environmental structure, such as various depth distributions of objects and the role of planar or spherical surfaces. From our results we derive two general rules how to optimize egomotion estimates: (i) Errors are minimized by expanding the field of view. (ii) Sampling image motion from opposite directions improves the accuracy, particularly for small fields of view.

From the iterative algorithm we derived a fast, non-iterative "matched filter" to extract R' and t', which under many conditions yields results very similar to those obtained by iteration. Its structure shows striking similarities to the receptive field organization of wide-field motion sensitive neurones in the visual system of the fly (Krapp and Hengstenberg 1996), but there are characteristic differences. To explain these differences, we developed a more elaborate version of this approach in which the statistical properties of the fly's environment and behaviour, i.e. the distribution of object distances and flight directions, are taken into account. A matched filter was directly derived from an optimization principle that minimizes the variance of the filter output caused by noise and distance variabilities. The optimized filters were then compared to the detailed organization of the receptive fields of the fly's wide-field neurones. Our analysis suggests that these neurones are not optimal for estimating the magnitude of R and t', but rather for consistently encoding the presence and the sign of rotatory or translatory flow fields along a particular set of axes.

2. Introduction

Many visually controlled behaviours require a fast and reliable determination of egomotion. One source of egomotion information is the characteristic pattern of retinal image shifts which are induced when an observer moves relative to the surroundings. These so-called optic flow fields (Gibson 1950) are thought to be analysed by the visual system of many spices – including human – to gain information on different aspects of egomotion (reviews: Miles and Wallman 1993; Lappe 1999; Lappe et al. 1999). In particular, flying and swimming animals have to rely on optic flow to monitor their true movement in space because they operate in a drifting and turbulent medium the movement of which cannot be detected directly.

In this chapter two questions regarding egomotion estimates from optic flow fields are addressed. First: How reliable can egomotion parameters be determined in principle from visual cues? And second: Can these considerations shed some light on the functional role of certain motion-sensitive neurones in the fly brain?

At any given moment egomotion can be uniquely decomposed into a rotation R and a linear displacement T (Koenderink 1986). This fact has lead to a large body of theoretical and experimental work addressing the problem of how accurately R and T can be extracted from optic flow (for a review see Heeger and Jepson 1992). Because all points in the environment along one viewing direction are projected to the same point in the image, the monocular 2D image of the 3D environment is ambiguous. From the image flow alone an observer can thus only estimate the direction and the size of ego-rotation R', and the direction t' (unit vector) of translation but not its absolute speed. As far as the information about the surrounding world is concerned, image flow only reveals the relative distances of contours. Whenever absolute distances or the true speed of locomotion need to be known, additional information (often from a non-visual source) is required either about the true speed of the observer, or about the absolute distance of at least one (sufficiently close) contour.

The limits of accuracy of the egomotion estimates, R' and t', depend on a number of parameters, including the spatial structure of the environment, the density and distribution of contours, and the design of the visual system in terms of the orientation and extent of its visual field. The accuracy with which local motion vectors can be measured imposes additional constraints. In a first step, we therefore analyse the effects the environmental topography and the visual field have on the reliability of R' and t' by simulating spherical and planar surroundings, and by restricting the solid angle of measuring optic flow. In contrast to a camera-like coordinate system with a planar image, we prefer algorithms formulated in a spherical coordinate system because many animals, especially arthropods possess nearly panoramic visual fields. The global analysis of the principal limits for estimating egomotion parameters from optic flow leads to the development of a fast "matched filter". The structure of this filter shows striking similarities with the receptive field organization of wide field motion sensitive neurones in flies (Krapp and Hengstenberg 1996; Dahmen et al. 1997; Krapp et al. 1998).

The third optic neuropil of the fly visual system, the lobula plate, contains about 60 wide-field, directionally selective interneurones (Hausen 1984; Hausen and Egelhaaf 1989; cf. also Warzecha and Egelhaaf, this volume) which integrate the outputs of retinotopically organized arrays of many local motion detectors (EMDs; review: Reichardt 1987). At a given location, retinal image shifts are analysed by sets of at least 6 EMDs which differ in their preferred directions (Buchner 1976; Götz et al. 1979). Tangential neurones that are thought to be involved in visual course stabilization and gaze control have been divided into two distinct functional and anatomical groups: the horizontal system (HS; Hausen 1982a, b) and the vertical system (VS; Hengstenberg et al. 1982; Hengstenberg 1982). The HS comprises three neurones which are named according to the orientation of their receptive fields and their dominant sensitivity to the direction of image motion. In a crude approximation, all three HS neurones respond to horizontal front-to-back motion. HSN (N = north) is highly sensitive to wide-field motion within the dorsal visual field, HSE (E = equatorial) responds maximally to respective motion in the equatorial region, and the receptive field of HSS (S = south) covers the ventral visual space. The VS neurones, in contrast, were thought to preferentially receive input from EMDs that are tuned to vertical downward motion. In early studies, however, it had already been noted that the local preferred directions of the tangential neurones are not always confined to either horizontal or vertical orientations alone (Hausen 1981; Hengstenberg 1981). It has now become clear that the distribution of local preferred directions and motion sensitivities within the receptive field of individual tangential neurones show striking similarities with the distribution of velocity vectors in optic flow fields (Krapp and Hengstenberg 1996; Krapp et al. 1998). These findings suggest that each tangential neurone is adapted to process image flow generated by a specific movement of the insect.

Although the receptive field organization of the tangential neurones shows a good qualitative correspondence to the above mentioned matched filter model (Dahmen et al. 1997), there are systematic differences. With respect to the sensitivity distribution within the receptive fields of the VS neurones, a marked dorso-ventral asymmetry was obtained which is not reproduced by the model of Dahmen et al. (1997). We therefore review a further study that tries to explain these differences from the statistical properties of the fly's environment and the velocity characteristic of its motion detectors (Franz and Krapp 2000). In this approach, matched filters are derived from an optimality criterion that minimizes the variance of the filter output caused by noise and distance variability between different scenes. To obtain quantitative predictions of the receptive fields of tangential neurones, it is necessary to model (1) the regional differences in distance distribution in the visual field, (2) the average distribution of flight directions, and (3) the velocity characteristic of the fly's motion detector. The resulting matched filters accurately reproduce the receptive fields of several tangential neurones.

3. Limiting factors for the extraction of egomotion parameters

3.1 An iterative procedure for extracting egomotion parameters from optic flow

The limits of accuracy of $\mathbf{R}'$ and $\mathbf{t}'$ were studied in "numerical experiments" (for details, see Dahmen et al. 1997), much in the same way as described by Koenderink and van Doorn (1987). We used their model of a rigid world consisting of N fiducial points at fixed positions in space that are given by the vectors $\mathbf{D_i}$ (i = 1...N). The visual system is represented by a unit sphere onto which each fiducial point is imaged. The vantage point in its centre is the origin of the coordinate system. The distances of the fiducial points from the vantage point are D_i. The viewing directions (unit vectors) towards the fiducial points $\mathbf{d_i} = \mathbf{D_i}/D_i$ are called markers. When the system moves on a curved path, described by rotation $\mathbf{R}$ and translation $\mathbf{T}$, image velocities $\mathbf{p_i}$ at the markers $\mathbf{d_i}$ are generated according to:

$$\mathbf{p_i} = \partial \mathbf{d_i}/\partial t = -(\mathbf{T} - (\mathbf{T} \bullet \mathbf{d_i})\mathbf{d_i})/D_i - [\mathbf{R} \times \mathbf{d_i}] \tag{1}$$

where $\bullet$ means the scalar product and [$\times$] the vector product. Introducing the "relative nearness" $\mu_i = |\mathbf{T}|/D_i$ (for details, see Koenderink and van Doorn 1987, Dahmen et al. 1997) equation (1) can be rewritten as:

$$\mathbf{p_i} = -\mu_i (\mathbf{t} - (\mathbf{t} \bullet \mathbf{d_i})\mathbf{d_i}) - [\mathbf{R} \times \mathbf{d_i}] \tag{1a}$$

Image velocities for at least five markers are necessary in order to solve equation (1). Usually local image velocity estimates $\mathbf{p_i}$ at more than five markers are available. In real life these estimates are subject to noise and errors, so that generally there will be no solution to the simultaneous equations (1). The best we can do is to find a "best estimate" $\mathbf{R}'$, $\mathbf{t}'$, μ_i' in the sense that the average error $E = (1/N) \sum|\mathbf{p_i}'-\mathbf{p_i}|^2$ is minimized, where $\mathbf{p_i}'$ are derived from $\mathbf{R}'$, $\mathbf{t}'$, μ_i' via equation (1a). Minimizing E with respect to variations $\delta\mathbf{R}'$, $\delta\mathbf{t}'$ and $\delta\mu_i'$ under the constraint $|\mathbf{t}'| = 1$ leads to the following three equations:

$$\mathbf{t}' = -k\{av(\mu_i'\mathbf{p_i}) + [\mathbf{R}' \times av(\mu_i'\mathbf{d_i})] - av(\mu_i'^2 (\mathbf{t}' \bullet \mathbf{d_i})\mathbf{d_i}) \tag{2}$$

$$\mathbf{R}' = av([\mathbf{p_i} \times \mathbf{d_i}]) + [\mathbf{t}' \times av(\mu_i'\mathbf{d_i})] + av((\mathbf{R}' \bullet \mathbf{d_i})\mathbf{d_i}) \tag{3}$$

$$\mu_i' = -\mathbf{t}' \bullet (\mathbf{p_i} - [\mathbf{d_i} \times \mathbf{R}']) / (1 - (\mathbf{t}' \bullet \mathbf{d_i})^2)$$

In these equations k is a normalization factor that allows for $|\mathbf{t}'| = 1$, and $av(x) = 1/N \sum(x_i)$; i = 1,...,N. This set of coupled linear equations cannot be solved directly but only in an iterative procedure. As pointed out by Koenderink and van Doorn (1987) this offers a "best solution" in the sense that E is minimized. There is no way to approximate the measured flow vectors $\mathbf{p_i}$ better by other $\mathbf{p_i'}$ generated by egomotion in an environment of rigid and stationary objects whatever extraction algorithm for the egomotion components $\mathbf{R'}$, $\mathbf{t'}$ will be used. This is true for distributions of error vectors $\mathbf{p_i'}$-$\mathbf{p_i}$ which are independent of $\mathbf{d_i}$ and isotropic around $\mathbf{p_i}$. For non-isotropic distributions $\mathbf{p_i'}$-$\mathbf{p_i}$ have to be multiplied in E by proper weights.

In order to test the reliability of $\mathbf{R'}$, $\mathbf{t'}$ under various conditions we performed "numerical experiments" in the following way: We selected a combination of vectors $\mathbf{R}$ and $\mathbf{T}$, a visual field of the observer, and a certain structure for the environment, i.e. a set of parameters which describe a spherical or planar environment, the average angular density of markers, and the width of the Gaussian vector noise superimposed on the unperturbed image velocities. We call this set of parameters a visual configuration. For each visual configuration we selected 32 sets of randomly distributed markers. For each set of markers 32 sets of perturbed velocity fields were generated. Thus for each visual configuration 1024 simulations of a measured flow field were created. For each "measurement" we evaluated $\mathbf{R'}$ and $\mathbf{t'}$ and observed the size-scatter of $\mathbf{R'}$ and the angular scatter of $\mathbf{R'}$ and $\mathbf{t'}$. We selected $\mathbf{R}$ and $\mathbf{T}$ along one of the coordinate axes. Because equations (2), (3) are linear, results for any other orientation of $\mathbf{R}$ and $\mathbf{T}$ can be derived from this set of conditions.

3.2 "One shot" estimates and a matched filter for estimating egomotion parameters

In many visual configurations, iterative algorithms are time consuming. The short latencies of many animals' responses to pattern rotation and translation suggest that they are able to determine components of their own movement by way of a fast, almost instantaneous mechanism. Such an ability would also seem to be essential for visual control of robot navigation. To propose a suitable fast "matched" filter we reformulate equations (2) and (3):

$$t' = -v \left\{ I - av(\mu_i' \mathbf{d_i} \otimes \mu_i' \mathbf{d_i}) \right\}^{-1} (av(\mu_i' \mathbf{p_i}) + [\mathbf{R'} \times av(\mu_i' \mathbf{d_i})]) \qquad (2a)$$

$$R' = \left\{ I - av(\mathbf{d_i} \otimes \mathbf{d_i}) \right\}^{-1} (av([\mathbf{p_i} \times \mathbf{d_i}]) + [t' \times av(\mu_i' \mathbf{d_i})]) \qquad (3a)$$

(I is the unit matrix; $\otimes$ indicates the dyadic product; v is a normalization factor so that $|t'| = 1$).

The term $[\mathbf{R'} \times av(\mu_i' \mathbf{d_i})]$ in equation (2a) represents the "apparent translation" induced by rotation $\mathbf{R'}$, and the term $[t' \times av(\mu_i' \mathbf{d_i})]$ in equation (3a) represents the "apparent rotation" induced by translation t'. None of the terms depends on the actual image flow $\mathbf{p}_i$. Provided $\mathbf{D}_i$ are distributed sufficiently symmetrical the apparent terms become small. The matrices $\{...\}^{-1}$ depend only on the markers $\mathbf{d}_i$ and reflect their distribution in the solid angle of the visual field. With sufficiently many and symmetrically distributed markers, the equations can be further simplified, because the off-diagonal elements of the matrices practically disappear. Then equations (2a) and (3a) can be re-written:

$$\mathbf{t_0'} \sim \begin{array}{l} av(p_{x,i}/D_i)/av(\sin^2 \theta_{x,i}/D_i^2) \\[6pt] av(p_{y,i}/D_i)/av(\sin^2 \theta_{y,i}/D_i^2) \\[6pt] av(p_{z,i}/D_i)/av(\sin^2 \theta_{z,i}/D_i^2) \end{array} \qquad (2b)$$

$$\mathbf{R_0'} = \begin{array}{l} av(p_{y,i}d_{z,i} - p_{z,i}d_{y,i})/av(\sin^2\theta_{x,i}) \\[6pt] av(p_{z,i}d_{x,i} - p_{x,i}d_{z,i})/av(\sin^2\theta_{y,i}) \\[6pt] av(p_{x,i}d_{y,i} - p_{y,i}d_{x,i})/av(\sin^2\theta_{z,i}) \end{array} \qquad (3b)$$

where $\theta_{x,i}$, $\theta_{y,i}$, $\theta_{z,i}$ are the angles of $\mathbf{d}_i$ with the x-, y-, z-axis, respectively. The two equations (2b) and (3b) allow for an initial approximation of $\mathbf{t'}$ and $\mathbf{R'}$ in a single step calculation. In the iterative solution the resulting estimates are used as initial values for equations (2a) and (3a). To achieve a better estimate further iterations are performed that include the "apparent translation" and "apparent rotation" components. Iteration in general improves the estimates whenever the "apparent terms" deviate significantly from zero.

Equations (2b) and (3b) can be interpreted intuitively. Suppose, for instance, that we are interested in the rotational component $\mathbf{R_a'}$ about the axis $\mathbf{a}$. We then first construct a "template" field of unit vectors parallel to the flow field induced by a unit rotation about axis $\mathbf{a}$ at all markers $\mathbf{d}_i$:

$$\mathbf{U^R_{a,i}} = -[\mathbf{a} \times \mathbf{d}_i] / \sin \theta_i, \qquad (4)$$

where θ_i is the angle between $\mathbf{a}$ and $\mathbf{d}_i$. We project the actual flow $\mathbf{p}_i$ onto this vector field to evaluate the contribution of $\mathbf{p}_i$ to $\mathbf{R_a'}$:

$$m^R_i = \mathbf{p}_i \bullet \mathbf{u^R_{a,i}} = [\mathbf{p}_i \times \mathbf{d}_i] \bullet \mathbf{a} / \sin \theta_i \qquad (5)$$

In order to get the properly scaled contributions of all flow vectors $\mathbf{p}_i$ to $\mathbf{R_a'}$ we have to average $m^R_i / \sin \theta_i$. To arrive at equation (3b) and a quick, "best one shot", estimate of the rotation component about axis **a**, we have to take a weighted average of these contributions with the weighting factor

$$w^R_i = \sin^2\theta_i. \qquad (6)$$

Equation (2b) can be interpreted in an analogous way. When we are interested in the component $\mathbf{t_a}'$ of $\mathbf{t'}$ along the unit vector **a**, we regard the template of unit vectors parallel to the flow field induced at marker $\mathbf{d_i}$ by a unit translation **a**:

$$\mathbf{u}^T_{a,i} = - [[\mathbf{d_i} \times \mathbf{a}] \times \mathbf{d_i}] / \sin \theta_i . \qquad (7)$$

We project $\mathbf{p_i}$ onto $\mathbf{u}^T_{a,i}$:

$$m_i^T = \mathbf{p_i} \bullet \mathbf{u}^T_{a,i} = - \mathbf{p_i} \bullet \mathbf{a} / \sin \theta_i.$$

In order to get the proper contribution of all $\mathbf{p_i}$ to $\mathbf{t_a}'$ we have to average $D_i\, m_i^T / \sin \theta_i$. Comparison to equation (2b) tells us that we have to use a weighted average with the weighting factor

$$w^T_i = \sin^2 \theta_i / D_i^2 \qquad (8)$$

to find the "best one shot" estimate $\mathbf{t'}$.

3.3 What influences the accuracy of estimating egomotion parameters from optic flow?

In the following we demonstrate how various factors affect the accuracy of egomotion parameter estimates derived from the procedures described in the previous two sections.

3.3.1 The number of fiducial points

The reliability of $\mathbf{R'}$ and $\mathbf{t'}$ depends on the number of flow measurements, as demonstrated by the example of our simulation results in figure 1. For the visual configuration (see inset) in this case a rotation $\mathbf{R} = (0, 0, 1)$ around the vertical axis (yaw rotation) was combined with a forward translation $\mathbf{T} = (1, 0, 0)$, and the visual field consisted of two 160° wide cones that were centred symmetrically along the transverse axis. The markers were equally distributed in all directions, while their density was varied in this numerical experiment between 20 and 1000 points / 4π. Distances D_i were normally distributed around 1 with $\sigma = 0.1$ to generate a "spherical" environment. The perturbation vectors added to the original image flow to simulate errors of measurement were equally distributed in all directions. Their size was normally distributed with $\sigma = 0.1 |p_i|$. Figure 1a shows the mean and standard deviation of the angular deviation from the veridical rotation vector for R'_X ($\square$), R'_Y (O), t'_Y ($\triangle$), t'_Z (∇) as function of point density. Figure 1b illustrates the size scatter of $\mathbf{R'}$, given as fraction of the true rotation magnitude $|\mathbf{R}|$. It is apparent from these plots that the estimates of direction and size of egomotion from the fully iterative procedure are very reliable, both in terms of biases and random fluctuations that are due to noisy signals. Only when the density of fiducial points is reduced to less than $50/4\pi$, a large scatter and a 5% underestimation of rotation magnitude indicate a notable impairment of the performance (for a more extensive documentation, see Dahmen et al. 1997). For all further simulations we chose a marker density of $250/4\pi$ which corresponds to an average angle of about 10° between markers.

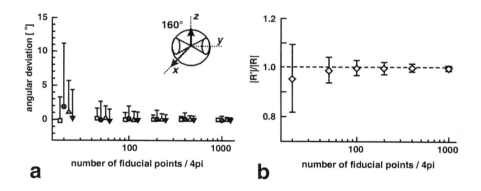

a   number of fiducial points / 4pi

b   number of fiducial points / 4pi

Fig. 1 Mean and standard deviation of the angular scatter of $\mathbf{R'}$ and $\mathbf{t'}$ (**a**) and of $|\mathbf{R'}|$ (**b**) versus the density of fiducial points. The visual configuration is indicated in the inset: rotation (black arrowhead) $\mathbf{R} = (0,0,1)$, translation (white arrowhead) $\mathbf{T} = (1,0,0)$, the visual field are two opposite cones, 160° wide, oriented along the +y and −y axis, distances D_i of fiducial points are normally distributed around 1 with range $\sigma = 0.1$ ("spherical" environment). The perturbation δp_i of the flow p_i was equally distributed in all directions and $|\delta p_i|$ was normally distributed with σ $(|\delta p_i|) = 0.1 |p_i|$. The angular error of $\mathbf{R'}$ and $\mathbf{t'}$ is indicated (in °) by the mean and standard deviation of the distribution of the two components of $\mathbf{R'}$ and $\mathbf{t'}$ orthogonal to $\mathbf{R}$ and $\mathbf{T}$: $\square = R'_x$, O $= R'_y$; $\triangle = t'_y$, $\nabla = t'_z$

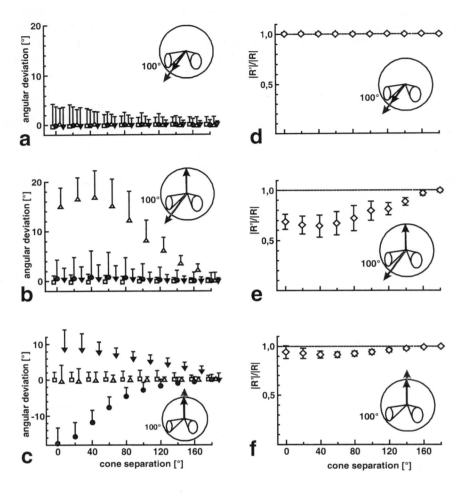

In **Fig. 2** the visual field were two 100° wide cones of variable angular separation. The cone axes pointed symmetrical to the left and right of the x-axis in the x-y plane. Results for three combinations of **R** and **T**, indicated in the insets, are presented. In the first column the panels show the angular distribution of **R'** and **t'** in an analogous way as in figure 1. In **a**: $\square$ = R'$_y$, $\bigcirc$ = R'$_z$; $\triangle$ = t'$_y$, ∇ = t'$_z$; in **b**: $\square$ = R'$_x$, $\bigcirc$ = R'$_y$, $\triangle$ = t'$_y$, ∇ = t'$_z$; in **c**: $\square$ = R'$_x$, $\bigcirc$ = R'$_y$, $\triangle$ = t'$_x$, ∇ = t'$_y$. In the second column mean and standard deviation of |**R'**| versus the angular separation of the two cone axes are reproduced. The density of fiducial points per solid angle was constant on the average (250/4π) in each cone, so that in the area of overlap of the two cones the density of markers was twice that of non-overlapping areas.

3.3.2 The angular separation of the cones in a two cone visual field

A number of recent studies of optomotor reflexes in insects, crabs, and birds show, that responses to pattern rotation are enhanced when contrasts are visible in opposite directions in the visual field (e.g. Nalbach 1990; Frost 1993; Kern et al. 1993; Blanke and Varjú 1995). To test whether there is a systematic reason for combining flow information from specific parts of the visual field, we restricted fiducial points to two cone-shaped segments of the visual field and varied their angular separation. The axes of the cones were located in the x-y plane, symmetrical to the x-axis. We asked how the accuracy of $\mathbf{R'}$ and $\mathbf{t'}$ depends on the angular separation Φ between the two visual cones for different types of egomotion. In figure 2 the results for specific combinations of $\mathbf{R}$ and $\mathbf{T}$ as indicated in the insets are presented for a cone width of 100°. The density of points in each cone was kept constant, so that in overlapping areas of the two cones the density of fiducial points was twice as high as in non-overlapping areas. In this way we excluded a possible influence of a variable number of points.

In the visual configurations of figure 2 we find in some cases a strong influence of the angular separation Φ on the accuracy of $\mathbf{R'}$ and $\mathbf{t'}$. For instance, small angular separations Φ lead to large systematic errors of t'_x and $|\mathbf{R'}|$ (Fig. 2b,e) and of R'_y, t'_y and $|\mathbf{R'}|$ (Fig. 2c,f). These errors practically disappear when the visual fields are oriented in opposite directions ($\Phi = 180°$). As might be expected, the accuracy also improves for enlarged visual fields (data not shown). In the special visual configuration of figure 2a,d (roll around the axis of translation) systematic errors are small. It can be concluded that in general it pays to analyse flow in opposite viewing directions to monitor egomotion. Since the estimation algorithm does not contain any specific interaction of information from opposite visual fields, this advantage is simply a consequence of geometry. The reason for this is that $av(\mu_i\mathbf{d_i})$ decreases when the fiducial points are located in opposite regions of the visual field, reducing the "apparent terms" in equations (2a) and (3a). In addition, the non-diagonal elements of the matrices in these equations are small under these conditions. For the special visual configuration of figure 2a,d the "apparent terms" are small for all Φ because $\mathbf{R}$, $\mathbf{T}$, and $av(\mu_i\mathbf{d_i})$ are nearly parallel independently of Φ. Hence the systematic angular error turns out to be negligible for the whole range of Φ.

In the following simulations we use two visual cones pointing either into the same direction along the x-axis simulating a visual field restricted to frontal regions, as in the human visual system, or in opposite directions along the y-axis as for instance in the laterally directed eyes of insects.

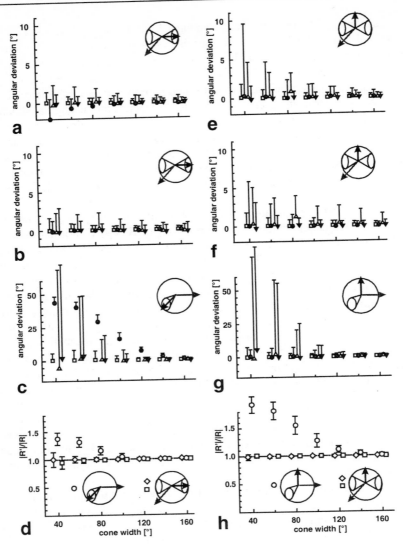

Fig. 3 Results for typical combinations of **R** and T in a "spherical" environment, illustrated in the insets, are presented as function of the width of the cones. In the first row angular errors for fully iterated estimates in opposite cone configurations are shown (conventions of symbols are analogous to Fig. 2). The second row reproduces angular errors of fast estimates (equations (2a),(3a) with **R'** and **t'** on the right side replaced by **R$_0$'** and **t$_0$'** (equations (3b),(2b) resp.) in the same configurations as row 1. The third row presents angular errors of fully iterated estimates in corresponding "frontal vision" configurations (2 parallel cones). The fourth row shows errors of size estimates |**R'**| in the same configurations as the previous three columns (opposite cones: ◇ = full iteration, □ = fast iteration; ○ = parallel cones). Opposite cones were oriented along the +y and -y axis, the two coaxial cones along the x axis (see insets). Note the different scales of angular deviation in the third row.

3.3.3 The size of visual cones and the relative orientation of R and T for iterated and one-shot estimates

We analysed how different orientations of **R** and **T** influence the errors of **R'** and **t'** by pointing both vectors either parallel or perpendicular to the axes of the cones. We also varied the width of the cones, keeping the density of markers constant. Because equations (2) and (3) are linear in **R** and **T**, estimates for all other orientations of **R** and **T** are linear combinations of the results for these orthogonal components.

We begin by considering two typical situations, namely a pitch (left column in figure 3; combinations of **R** and **T** are indicated in the insets) and a yaw rotation (right column in Fig. 3) during forward translation. The first row in figure 3 shows angular errors of fully iterated estimates, the second row those of one-shot estimates, the third row errors of fully iterated estimates with a double cone along the x-axis, and the fourth row the errors of size estimates |**R'**| for all visual the configurations considered here. For other combinations of **R** and **T** , in particular when they are oriented parallel to the cone axis, the estimates are generally in very good agreement to the veridical values, irrespective of the cone width. Fast estimates are the result of equations (2a) and (3a) with **R'** and **t'** on the right hand side replaced by the R_0' and t_0' of equations (3b) and (2b). It is quite obvious from the comparison of the first two rows and the corresponding results in the fourth row ($\diamond$ and $\square$) that for a spherical environment fast estimates are nearly as good as fully iterated ones. The disadvantage of a frontal vision system as far as egomotion estimates are concerned is obvious from the third row of this figure (note the different y-scales in Figs. 3c, g) and corresponding results in the fourth row ($\bigcirc$). In this visual configuration only large cone angles lead to reliable estimates.

Misjudgments are again due to the asymmetric distribution of the d_i in situations in which the angular separation of the two cones is small, inducing a large "apparent translation" component by rotation in equation (2a) and a large "apparent rotation" induced by translation in equation (3a). The one exception, for which no impairment of performance is expected from the theory, and actually not found in simulations, is the case where **R**, **T** and the cones are collinear. In this visual configuration $av(\mu_i d_i)$ is nearly parallel to **R** and **T**, and the "apparent" terms are very small.

3.3.4 The structure of the environment: moving relative to planes with or without "clouds"

Many animals and machines move on or parallel to the ground plane and the question arises what consequences environmental topography has for estimating egomotion parameters from optic flow. We thus analysed the principal limits of egomotion parameter extraction in a situation in which the fiducial points are located on a plane. To explore the role of depth cues, we simulated two parallel transparent planes at various distances from each other, since depth cues in the

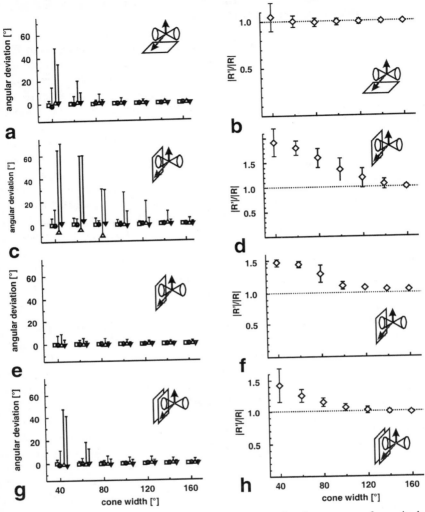

Fig. 4 Simulation results (angular deviation of **R'** and **t'** in left column, errors of magnitude of **R'** in right column) as function of cone width, for a typical combination of **R** and **T** and several relative orientations of a single plane environment. The distance of the planes from the origin is one unit; the distance of the second plane in g and h is two units. In **a, b** a horizontal plane is depicted; in **c, d** a vertical plane parallel to x-z; in **e, f** the same vertical plane as c and d but with 'clouds', and in **g, h** two vertical planes are depicted. The visual configuration is indicated in the insets. Symbol conventions are analogous to those in figures 1-3.

same viewing direction, if properly exploited, may allow the extraction of the heading direction even in the presence of rotation (cf. Longuet-Higgins and Prazdny 1980). Fiducial points were generated by piercing the plane(s) along straight lines through randomly distributed markers within the visual field. About

50% of the lines did not intersect the plane(s). In these cases we either assumed that there is visible contrast at infinity ($\mu_i = 0$, we call these distant contours "clouds") or that fiducial points are lacking in these directions (no "clouds"). In the presence of "clouds" or two transparent planes the number of fiducial points is, on the average, twice that of the single-plane environment. The effect of adding "clouds" in the hemisphere opposite to the plane is to add flow induced by rotation only, and to double the solid angle where flow is present. Adding a second transparent plane adds flow induced by translation only, but does not change the solid angle of visible flow compared to the single-plane environment.

Figure 4 shows the results we obtained for a yaw turn during forward translation and a number of environmental configurations. Figures 4a-d depict the situation for a "single plane without clouds", as seen in the ventral or lateral visual field, respectively, figures 4e and f for a "single plane with clouds", and figures 4g and h for "two planes". The distance of the first plane to the vantage point was one unit, that of the second one two units. For other distances of the second plane results were similar to those presented here and for other visual configurations they are generally better, mostly leading to a veridical representation of the egomotion parameters.

3.3.5 General rules for extracting egomotion parameters from optic flow

We have attempted to characterize the principal limits of extracting egomotion parameters from optic flow under the restriction of an isotropic error distribution. We considered the number of necessary flow measurements, the size of the visual field, the optimal directions in which flow is measured, and the distribution of contours in the environment. Our main results can be summarized as follows:

1. The precision with which the direction of **R'** or **t'** can be determined may be asymmetric: mean and variance of the two components orthogonal to **R** and **T** may be quite different (see for instance Figs. 2b,c; 3c; 4c).
2. With cone widths smaller than $100°$ egomotion estimates are unreliable especially in cases when a planar environment is seen through one cone only (compare Figs. 4a,b and 4c,d).
3. If both, **R** and **T**, are orthogonal to the cone axes egomotion estimates are impaired. Estimates are exceptionally good if both, **R** and **T**, are parallel to the cone axes (data not shown).
4. Adding "clouds" improves estimates of rotation more than adding depth cues. When the plane is seen through one cone, adding a set of distant points reduces the angular scatter of **R'**. Adding depth cues by a second depth plane is less effective (compare Figs. 4e and 4g).
5. One shot estimates are practically as good as fully iterated ones as long as distances do not play an important role (compare Figs. 3a,e and 3b,f). But in an environment, where distances are distributed anisotropically (plane environment) the fully iterated solutions, which include estimates μ_i', are (in some

visual configurations more than in others) superior to one shot estimates (not demonstrated here).

Our general conclusions are, firstly, that wide visual fields are of utmost importance for extracting reliable egomotion parameters, and secondly, that the "apparent" terms in equations (2a), (3a) should be small, i.e. flow should be measured in opposite directions of view.

4. Fly tangential neurones and matched filters for optic flow fields

4.1 Are tangential neurones "one shot" estimators for egomotion?

In the search of neural filters underlying the processing of optic flow, one of the most intriguing pieces of evidence comes from electrophysiological analyses of large integrating neurones in the fly visual system. As mentioned in the introduction, the local receptive field properties of some tangential neurones have been determined experimentally (Krapp and Hengstenberg 1997). When the resulting local sensitivities and preferred directions are plotted into a map of one visual hemisphere, they show a marked resemblance to the structure of optic flow fields generated by a unique egomotion component (Krapp and Hengstenberg 1996; Krapp et al. 1998). As we have seen in Section 3.2, this is exactly what we would expect in a fast filter for "one shot" estimates of egomotion: If we rewrite equations (2b) and (3b) using the equations (4)-(8), the response of the matched filter for the egomotion axis $\mathbf{a}$ is given by

$$t_{a,0}' = N \, av(\mu_i' \sin \theta_i \, \mathbf{p}_i \bullet \mathbf{u}^T_{a,i}) \tag{9}$$

$$R_{a,0}' = N \, av(\sin \theta_i \, \mathbf{p}_i \bullet \mathbf{u}^R_{a,i}) \tag{10}$$

with some suitable normalization factor N. The unit vectors $\mathbf{u}^T_{a,i}$ and $\mathbf{u}^R_{a,i}$ are parallel to the flow field generated by translation in direction $\mathbf{a}$ or rotation about $\mathbf{a}$ (cf. equations (1), (4) and (5)), just as one observes in the local preferred motion directions of the investigated tangential neurones. This suggests that equations (9) and (10) could provide a model for the egomotion preference of these neurones. However, the model predicts a local motion sensitivity following a $\sin \theta_i$ dependence which was not found in the measurements (Franz and Krapp, in press). Instead, the investigated neurones show a pronounced dorsoventral anisotropy in their receptive fields. For instance, the rotatory VS neurones are more sensitive to motion above than below the horizon. In contrast, the so-called Hx neurone is more sensitive below the horizon (cf. Figs. 5a,c,e). Thus, the model needs to be extended if we want to explain these anisotropic sensitivity distributions.

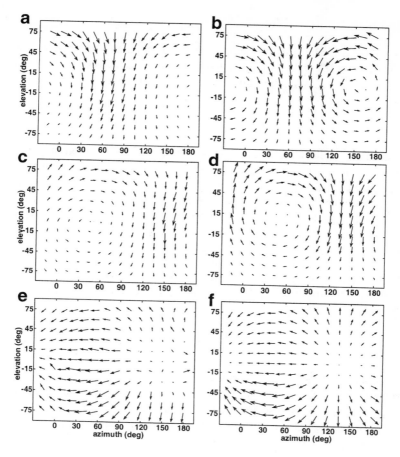

Fig. 5 Receptive fields and matched filters. **a** Averaged receptive field of VS4 neurones (N =5). The orientation of each arrow represents the local preferred direction this location and its length indicates the motion sensitivity normalized to the maximum response. The right visual hemisphere and the first meridian of the left hemisphere is mapped in this example. Positions are defined by azimuth and elevation. Azimuth and elevation of 0° correspond to the position frontal and in the equatorial plane of the visual field. The azimuth of 90° denotes the lateral, and the azimuth of 180° the caudal position in the visual field. The filter axis of the VS4 is aligned horizontally, corresponding roughly to the body axis of the fly. **c** The averaged VS10 receptive field (N = 5) shows a similar structure but its filter axis is shifted towards the frontolateral visual field (azimuth approx. 45° and elevation approx. 0°). **e** The receptive field of the Hx neurones is reminiscent to an optic flow field induced by translation along a horizontal axis pointing at an azimuth of about 135°. The sensitivities in the translatory receptive field are higher in the ventral visual field whereas the VS neurones are more sensitive to motion in the dorsal visual field. **b, d** and **f** show the corresponding matched filter structures as derived by combining the respective direction templates with local weights calculated from equation (13) for a specified world model and a particular distribution of translation directions (modified from Franz and Krapp 2000).

4.2 Optimized matched filters

The generalized model uses the basic structure of the matched filter as described above (equations (9) and (10)), i.e., a template field of unit vectors $\mathbf{u}_{a,i}$ parallel to the local velocity vectors induced by a particular egomotion:

$$e = N \, av(w_i f(\mathbf{p}_i \bullet \mathbf{u}_{a,i})). \tag{11}$$

As before, the detector signals are weighted by the local motion sensitivities w_i and summed up to give the filter output e. In addition, the model includes the velocity characteristic f of the fly motion detector which resembles an inverted U-shape. A model as expressed by equations (9) and (10) uses a linear response, i.e. $f(x) = x$. In the fly, this approximation is valid only for small image velocities (ca. 0 to 10°/s). At higher velocities, the detector response saturates in an extended, flat maximum ranging from ca. 20 to 200°/s before it decreases at higher speeds (Borst and Egelhaaf 1993). The behaviour near the maximal response can be modelled by dividing the current detector input by its absolute value. When the absolute value falls below a threshold P, we set the detector output to 0. Using $f(x) = x/|x|$ for $x > P$, equation (11) becomes

$$e = N \, av_{p,u > P} (w_i (\mathbf{p}_i \bullet \mathbf{u}_{a,i}) / | \mathbf{p}_i \bullet \mathbf{u}_{a,i} |). \tag{12}$$

In contrast to the model described in Section 4.1, a matched filter using such a mechanism cannot encode the magnitude of an egomotion parameter since it uses only the sign of the flow projection as input. The filter output rather indicates the presence and sign of the apparent egomotion component along its axis.

To obtain the local sensitivity distribution of the matched filter, we now adopt a different point of view: Instead of using a "one shot" version of an egomotion algorithm as in Section 3.2, we directly derive the w_i from an optimality criterion, namely by minimizing the variance of the filter output caused by noise and distance variability. The varying distance distribution in different scenes leads to variations of the translatory flow, even when the egomotion parameters remain exactly the same (cf. equation (1)). As a consequence, the filter output will be different for the same egomotion in different scenes. The variance in the filter output can be minimized by choosing appropriate w_i that assign less weight to detector signals with high noise content and distance variability. Such an optimized matched filter is able to maintain its output as consistent as possible between different scenes.

It can be shown that the optimal sensitivity distribution for equation (12), minimizing the output variance, is given by (Franz and Krapp 2000):

$$w_i = E(\mathbf{p}_i \bullet \mathbf{u}_{a,i})^2 / (\Delta t_i^2 + \Delta n_i^2) \tag{13}$$

where E denotes the expectation over all scenes, Δt_i^2 the local variance of the translatory flow caused by the distance variability and Δn_i^2 the noise variance of the motion detector signal. Δt_i^2 is especially high in viewing directions with small absolute distance, high distance variability, and a large component of $\mathbf{T}$ along $\mathbf{u}_{a,i}$ (cf. equation (1)).

4.3 Modelling anisotropies in distances and flight directions

A plausible start to explain the observed anisotropic sensitivity distributions is that they reflect the distribution of distances in the fly's environment. Distances to the ground usually tend to be smaller than to objects in the upper regions of the visual field. The variance of the translatory flow Δt_i^2 in equation (13) is therefore larger below the horizon. Flow above the horizon is thus more reliable for the determination of $\mathbf{R}'$ and, therefore, should be given a larger weight. Vice versa, flow below the horizon should be given a larger weight in estimating $\mathbf{t}'$ since the translatory flow projection is much larger in this part of the visual field (cf. Nalbach and Nalbach 1987). Since the statistics of distances and translations for an animal like the fly are not known, we have to make some reasonable assumptions about their distribution. As a crude approximation, we assume a simplified "world model" in which the mean distances below the horizon are smaller than those above the horizon. The distance scatter around the mean is assumed to be isotropic (see Fig. 6a). In addition, we assume that the fly is heading preferably forward. Thus, translations encountered during flight are modelled in a broad and unimodal distribution with a peak in the forward direction (see Fig. 6b).

Based on the assumptions about distances and translations, optimal matched filters can be computed according to equation (13). The resulting weight sets depend only on a few parameters. This allows us to apply a fitting procedure in which these parameters are varied until the best fit to the measured motion sensitivities of the tangential neurones is reached. The goodness-of-fit is measured by evaluating their χ^2-value. Weight sets with a χ^2-value corresponding to $p < 0.05$ are rejected.

We first tested weight sets for the linear range of motion detector response such as the ones described in equations (9) and (10), using our world model and the chosen translation distribution. All of the resulting weight distributions could be rejected with high probability. Provided that our assumptions capture essential elements of the fly's environment, this means that the tangential neurones under consideration are not optimized for direct "one shot" estimation of the current rotation or translation. The optimal weight sets from equation (13), however, produced a significant fit for some tangential neurones. For instance, there is a close correspondence between the measured response field of VS4 (Fig. 5a) and Hx (Fig. 5e) and the structure of an optimal matched filter for sensing a particular rotation (Fig. 5b) or translation (Fig. 5f), respectively. This suggests that these neurones rather act as detectors for the presence and sign of the rotatory flow around a particular set of axes instead of directly encoding egomotion.

Both matched filter models failed to explain the receptive field properties of a number of other VS neurones. The receptive fields of these neurones are very close to the model (13) in some regions, but in other regions where the model predicts a high motion sensitivity they respond only weakly to motion (e.g., VS10 in figure 5c and d which is not sensitive to motion in the frontal part of the visual field). This indicates that further, possibly anatomical or developmental constraints are at work in their design.

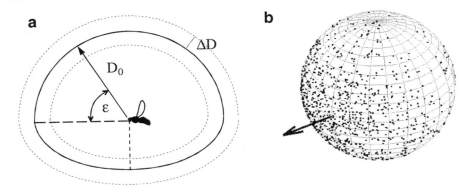

Fig. 6 a Simplified world model of the fly's environment. The distances are assumed to scatter around the mean distance D_O with a constant standard deviation ΔD. Above the horizon (elevation $\varepsilon > 0$) a constant average distance D_O is assumed. The environment is assumed to be flat below the horizon. **b** Distribution of translation directions encountered by the fly as assumed to generate filter sensitivity profiles. One thousand different directions were calculated using an unimodal two-dimensional von Mises distribution. (modified from Franz and Krapp 2000).

5. Discussion

We reviewed theoretical and neurophysiological studies aimed at understanding how egomotion parameters can be estimated in biological and artificial systems. In the wake of these considerations it is tempting to speculate about the optimal design of systems that solve a variety of tasks related to visual guidance.

5.1 Sampling the visual field for robust egomotion estimates

We note that the direction of **R** and **T** can be determined from optic flow in a rigid world to a precision of about 3°, provided that the flow can be measured with a relative error of 10% at 50 markers which are distributed over a whole spherical visual field. Under these conditions, the magnitude of **R** can be extracted to within

5 % (see Fig. 1). A system dealing with 50 markers, which are scattered across the whole sphere, will operate on signals at an average angular separation of about 22°. This can be compared to 1200 ommatidia in *Drosophila* or 6000 in *Calliphora* flies, or maximum spatial resolution of approximately 5° or 1°, respectively (Land 1997). Because this is thought to roughly correspond to the maximum number of local motion measurements available to the flies' visual system, there seems to be a considerable amount of redundancy in the sensory system reflecting the sparse distribution of contrasts in natural environments. For the extraction of egomotion parameters it therefore is not necessary to perform many (as compared to the potentially available number) and very precise local measurements. It is essential, however, to measure image motion over a large solid angle and to apply the "appropriate" integration. If these conditions are met, the estimates are robust against relatively large flow errors even when only a few image velocity measurements are available. This robustness is particularly relevant in environments where local contrast is not distributed homogeneously throughout the entire visual field.

Our investigation of an optical system equipped with two visual cones of variable angular separation and width was stimulated by results in visually induced behaviour in insects which showed that insects distinguish between rotational and translatory flow (Junger and Dahmen 1991). It was shown that the gain of compensatory head and body movements increases with the angular separation of two stripes rotating around an animal (Frost 1993; Kern et al. 1993; Blanke and Varjú 1995). Our theoretical analysis now shows that the errors in estimating egomotion parameters are minimized by pointing the receptive field axes into opposite directions (see Fig. 3a). As a simple consequence of geometry, the amount of error reduction depends on the visual configuration: for small visual fields it is often remarkable, for wide cones it is less pronounced (compare Figs. 3a and e with Figs. 3c and g). In contrast to humans, arthropods with their compound eyes benefit in this respect from their extended visual fields, an advantage which is increasingly being recognized in robotic applications (Nelson and Aloimonos 1988; Chahl and Srinivasan 1997; Nagle et al. 1997; Franz et al. 1998). In case animals have to operate on a restricted field of view, they can use knowledge about their typical translatory locomotion by orienting their cones forwards and backwards. This is somewhat counter-intuitive, because, particularly for small visual fields, the largest flow vectors are not visible in this case. However, the "apparent" terms in equations (2a), (3a) are reduced under such conditions, which allows for a more reliable estimation of translatory egomotion components. Vice versa, visual configurations with R and T being oriented orthogonal to the visual cone axes are most unfavourable, because the apparent terms tend to be large under such conditions.

5.2 Simplified estimation procedures and specific environments

Estimates **R'** and **t'** are good in a cluttered world where nearby contours can be seen in all directions, a situation, which we attempted to simulate with our "spherical" environment. "One shot" estimates **R'** and **t'** through equations (2b), (3b) are nearly as good as fully iterated ones for a spherical environment (compare Figs. 3a and e with Figs. 3b and f). In an environment in which the contrasts are confined to a plane, however, **R'** and **t'** estimates may be unreliable particularly for a "one-shot" algorithm. One fundamental limitation of the "one-shot" estimate appears to be the knowledge about the distribution of distances. The relative nearness μ_i and the weighting matrices $\{I\text{-av} (\mathbf{d_i} \otimes \mathbf{d_i})\}^{-1}$ and $\{I\text{-av} (\mu_i\mathbf{d_i} \otimes \mu_i\mathbf{d_i})\}^{-1}$ reflect the distribution of markers and distances of fiducial points. Knowledge about distance distributions can be incorporated by customizing these matrices to a given environment and to the optical system. Keeping this possibility in mind, we now discuss an example of biological implementation of egomotion estimation in the visual system of the fly.

5.3 Are tangential neurones "matched filters" for egomotion estimation?

Identifying the limits of accuracy for egomotion estimation allows us to judge how effective a biological or technical system may perform on this task. In the case of fly tangential neurones, some of these neurones fulfil the optimality criterion defined in Section 4.2. We found the best quantitative fit between the experimental data presented here and the model predictions under two major assumptions: (i) The receptive field organization of these cells is adapted to the distribution of distances and translation directions encountered by the fly. Thus the neurones reflect aspects of the animals' visual environment and their functional context. (ii) The elementary motion detectors feeding into these neurones do not appear to operate in the linear range, but in the plateau-like velocity range.

The observation that tangential neurones are best suited to indicate the presence and sign of a specific egomotion component rather than its magnitude is a consequence of the velocity characteristic of elementary motion detectors (EMDs) in our model. This restriction, however, may have little consequences for control performance as long as these neurones operate as part of a closed feedback loop with zero set-point. Further constraints may be imposed by the fact that EMDs do not compute the velocity of retinal image shifts (Reichardt 1987). Instead, their output signals are influenced by pattern properties like spatial frequency content or contrast (Egelhaaf and Borst 1993).

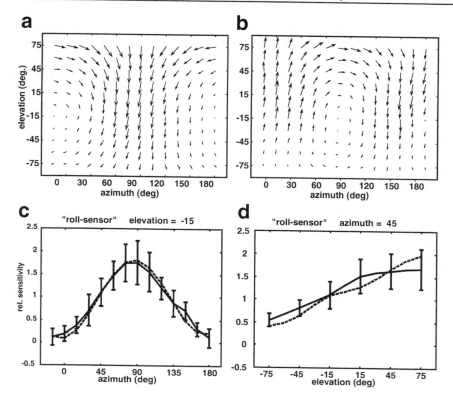

Fig. 7 Combined response fields to construct matched filters for self-motion estimation which cover one complete visual hemisphere. **a** Averaged response field of the neurones VS4 - VS7 (five neurones of each type). This combined filter has a preferred axis of rotation corresponding to the body axis of the fly. Thus it can be expected to responds best to roll-rotations of the animal ("roll-sensor"). **b** The response field shown was generated from the vector differences between the averaged response field of the neurones VS8-VS10 and the response fields of VS1 - VS3 (five neurones of each type). Such filter is designed to sense pitch-rotations around the transverse body axis ("pitch-sensor"). Note the relatively weak sensitivity to motion in the ventral visual field. **c, d** Comparison between the measured motion sensitivities (solid lines) of the roll-sensor and the theoretical weights w^R_i (dotted lines) derived from the model described in Section 4.2. Error bars indicate the SD of the measured sensitivities, the weights obtained from the matched filter model do not differ significantly from the experimental data (X^2-fitting procedure). The sensitivities and weights are plotted as a function of azimuth at an elevation of -15° and as a function of elevation at an azimuth of 45°.

Although VS neurones are quite obviously adapted to sense rotations, they are not insensitive to translatory flow components. Translatory components induce apparent rotations that are corrupting the rotation estimate (cf. equation 3a). The apparent rotation term can only be expected to vanish if the viewing directions are arranged in opposite directions so that the components induced by translation cancel each other out, and if the system operated in a world where contrast is homogeneously distributed. VS neurones with visual fields restricted to one hemisphere usually do not meet this criterion, and signals from appropriate pairs of VS neurones looking at either hemisphere would need to be combined at later processing stages. An example of another approach to generate an optimal fit between the predicted weight sets and the neurones' sensitivity profiles is shown in figure 7. Here, the response fields of several cells looking at one and the same hemisphere, are combined in such a way to completely cover one half of the visual field. The performance could then be further improved by binocular interactions which are known to exist for other subclasses of tangential neurones (Hausen 1984).

It is also tempting to speculate about the question why particular egomotion vectors seem not to be represented explicitly at the level of the visual system. The reason for this may be to maintain a higher flexibility with respect to the sensory-motor transformation (Oyster et al. 1972; Götz and Wandel 1984). The coordinates of the motor system and the sensory coordinates are not necessarily arranged in the same frame of reference. Thus, for efficiently driving the muscles involved in the optomotor and/or gaze stabilizing system, a specific combination of tangential neurones may be selected from the whole ensemble and converge at an appropriate integration stage.

5.4 Limitations of the present approach and outlook on future work

Whereas the simulation approach that was put forward in the present chapter provides an illuminating first approximation about the quality and performance limits of mechanisms to extract the parameters of egomotion, there are fundamental limitations when comparisons are to be made with actual implementations in biological systems. To start with, responses to combined stimuli in opposite parts of the visual field are by no means the linear superposition of the responses to the individual components (Frost 1993; Kern et al. 1993; Blanke and Varjú 1995), thus violating the assumption of linear summation of local response components. Therefore in attempting to understand exactly the neuronal mechanisms of egomotion extraction, it will be crucial to study quantitatively the behaviour and physiological foundations of flow field processing in more detail. The matched filters described in Section 3.2 and 4.2 include stages which compute the linear sum over all local estimates, but the response of the tangential neurones saturate with increasing pattern size at a level that depends on velocity (Borst et al. 1995). It is unclear as yet, what consequences this property has for the task of extracting egomotion information from the current optic flow under real time conditions (cf. Egelhaaf and Warzecha, this volume). To answer this important question the tangential neurones need to be studied in experiments involving realistic wide-field optic flow stimuli.

Acknowledgments

We would like to thank J. Zanker and J. Zeil for critically reading and discussing the manuscript and for language corrections. Helpful comments on the manuscript of an anonymous referee are appreciated. Many thanks to Karin Bierig for preparing some of the figures. The work was supported by the DFG (SFB 307) and grants of the Max-Planck-Gesellschaft.

References

Blanke H, Varjú D (1995) Visual determination of self motion components: Regionalization of the optomotor response in the backswimmer *Notonecta*. In: Elsner N, Menzel R (eds) Nervous systems and behaviour. Proc 23rd Göttingen Neurobiol Conf. Thieme, Stuttgart, p 265

Borst A, Egelhaaf M, Haag J (1995) Mechanisms of dendritic integration underlying gain control in fly motion-sensitive interneurons. J Comput Neurosci 2: 5-18

Buchner E (1976) Elementary movement detectors in an insect visual system. Biol Cybern 24: 85-101

Chahl JS, Srinivasan MV (1997) Reflective surfaces for panoramic imaging. Appl Optics 36: 8275-8285

Dahmen H (1991) Eye specialization in waterstriders: an adaptation to life in a flat world. J Comp Physiol A 169: 623-632

Dahmen H, Wüst RW, Zeil J (1997) Extracting egomotion parameters from optic flow: principal limits for animals and machines. In: Srinivansan MV, Venkatesh S (eds) From living eyes to seeing machines. Oxford University Press, Oxford, New York, pp 174-198

Egelhaaf M, Borst A (1993) Movement detection in arthropods. In: Miles FA, Wallman J (eds) Visual motion and its role in the stabilization of gaze. Elsevier, Amsterdam, London, pp 53-77

Franz MO, Krapp HG (2000) Wide-field, motion-sensitive neurons and matched filters for optic flow fields. Biol Cybern: in press

Franz MO, Schölkopf B, Mallot HA, Bülthoff HH (1998) Where did I take that snapshot? Scene-based homing by image matching. Biol Cybern 79: 191-202

Frost B (1993) Subcortical analysis of visual motion: Relative motion, figure-ground discrimination and self induced optic flow. In: Miles FA, Wallman J (eds) Visual motion and its role in the stabilization of gaze. Elsevier, Amsterdam, London, pp 159-175

Gibson JJ (1950) The Perception of the Visual World. Houghton Mifflin, Boston.

Götz KG, Hengstenberg B, Biesinger R (1979) Optomotor control of wing beat and body posture in *Drosophila*. Biol Cybern 35: 101-112

Götz KG, Wandel U (1984) Optomotor control of the force of flight in *Drosophila* and *Musca*. Biol Cybern 51: 135-139

Hausen K (1981) Monocular and binocular computation of motion in the lobula plate of the fly. Verh Dtsch Zool Ges 1981: 49-70

Hausen K (1982a) Motion sensitive interneurons in the optomotor system of the fly. I. The horizontal cells: Structure and signals. Biol Cybern 45: 143-156

Hausen K (1982b) Motion sensitive interneurons in the optomotor system of the fly. II. The horizontal cells: Receptive field organization and response characteristics. Biol Cybern 46: 67-79

Hausen K (1984) The lobula complex of the fly: structure, function and significance in visual behaviour. In Ali MA(ed) Photoreception and vision in invertebrates. Plenum, New York, London, pp 523-559

Hausen K (1993) The decoding of retinal image flow in insects. In: Miles FA, Wallman J (eds) Visual motion and its role in the stabilization of gaze. Elsevier, Amsterdam, London, pp 203-235

Hausen K, Egelhaaf M (1989) Neural mechanisms of visual course control in insects. In: Stavenga DG, Hardie RC (eds) Facets of Vision. Springer Verlag, Berlin, Heidelberg, pp 391-424

Heeger DJ, Jepson AD (1992) Subspace methods for recovering rigid motion I: Algorithim and implementaion. Int J Comp Vis 7: 95-117

Hengstenberg R (1981) Rotatory visual responses of vertical cells in the lobula plate of *Calliphora*. Verh Dtsch Zool Ges 1981: 180

Hengstenberg R (1982) Common visual response properties of giant vertical cells in the lobula plate of the blowfly *Calliphora*. J Comp Physiol A 149: 179-193

Hengstenberg R, Hausen K, Hengstenberg B (1982) The number and structure of giant vertical cells (VS) in the lobula plate of the blowfly *Calliphora erythrocephala*. J Comp Physiol A 149: 163-177

Junger W, Dahmen HJ (1991) Response to self-motion in waterstriders: visual discrimination between rotation and translation. J Comp Physiol A 169: 641-646

Kern R, Nalbach HO, Varjú D (1993). Interaction of local movement detectors enhance the detection of rotation. Optokinetic experiments with the rock crab *Pachygrapsus marmoratus*. Visual Neurosci 10: 643-52

Koenderink JJ (1986) Optic flow. Vision Res 26: 161-190

Koenderink JJ, van Doorn AJ (1987) Facts on optic flow. Biol Cybern 56: 247-54

Krapp HG, Hengstenberg R (1996) Estimation of self-motion by optic flow processing in single visual interneurons. Nature 384: 463-466.

Krapp HG, Hengstenberg, R (1997) A fast stimulus procedure for determining local receptive field properties of motion-sensitive visual interneurons. Vision Res 37: 225-234

Krapp HG, Hengstenberg B, Hengstenberg R (1998) Dendritic structure and receptive-field organization of optic flow processing interneurons in the fly. J Neurophysiol 79: 1902-1917

Land MF (1997) Visual acuity in insects. Ann Rev Entomol. 42: 147-177

Lappe M (1999) Neuronal processing of optic flow. Int Rev Neurobiol 44. Academic Press, San Diego

Lappe M, Bremmer F, van den Berg AV (1999) Perception of self-motion from optic flow. Trends Cog Sci 3: 329-336

Longuet-Higgins HC, Prazdny K (1980) The interpretation of a moving retinal image. Proc Roy Soc Lond B 208: 385-97

Miles FA, Wallman J (1993) Visual motion and its role in the stabilization of gaze. Elsevier, Amsterdam, London, New York, Tokyo

Nagle MG, Srinivasan MV, Wilson DL (1997) Image interpolation technique for measurement of egomotion in 6 degrees of freedom. J Opt Soc Am A 14: 3233-3241

Nalbach H-O (1990) Multisensory control of eye stalk orientation in decapod crustaceans. An ecological approach. J Crust Biol 10: 382-399

Nalbach H-O, Zeil J, Forzin L (1989) Multisensory control of eye-stalk orientation in space: Crabs from different habitats rely on different senses. J Comp Physiol A 165: 643-649

Nelson RC, Aloimonos J (1988) Finding motion parameters from spherical motion fields (or the advantage of having eyes in the back of your head). Biol Cybern 58: 261-218

Oyster CW, Takahashi ES, Collewijn H. (1972) Directional-selective retinal ganglion cells and control of optokinetic nystagmus in the rabbit. Vision Res 12: 183-193

Reichardt W (1987) Evaluation of optical motion information by movement detectors. J Comp Physiol A 161: 533-547

A Closer Look at the Visual Input to Self-Motion Estimation

John A. Perrone

Psychology Department, The University of Waikato, Hamilton, New Zealand

1. Introduction

In their Chapter, Dahmen et al. use an interesting approach to the self-motion (egomotion) estimation problem that combines traditional vector flow field decomposition schemes with a matched filter (template) scheme. In the field of visual self-motion perception these two approaches represent two different schools of thought about how biological systems solve the self-motion problem. The boundaries between these classes of models occasionally become blurred but there is one feature that distinguishes these two approaches and which can be used to assess their pros and cons: namely the type of input they are designed to process. In this section I examine the different input requirements of the two model types and identify some of the challenges that are faced by modellers in this area.

It has long been suggested that visual navigation through the environment relies on our ability to detect and respond to specific patterns of retinal velocity or optic flow (Gibson 1950). There is ample empirical evidence that the self-motion problem can be solved using just visual motion information (e.g., Rieger and Toet 1985; Cutting 1986; Warren and Hannon 1990; Stone and Perrone 1997). It is also well established that – *in theory* – the instantaneous retinal optic flow contains sufficient information for the recovery of information about our instantaneous retino-centric heading (3D translation direction), eye-body rotation, and the relative depth of points in the world (Koenderink and van Doorn 1975; Longuet-Higgins and Prazdny 1980; Zacharias et al. 1985; Heeger and Jepson 1992). These analyses demonstrated that the equations governing the projection of an environmental point (x, y, z) onto an image plane during movement of an observer, could be solved for the observer motion parameters, given a small number of image velocity vectors. Some of these mathematical considerations motivated models of human visual self-motion estimation.

2. Models of human visual self-motion estimation

2.1 Vector-based decomposition models

Most current models of self-motion estimation are inherently vector based (e.g., Koenderink and van Doorn 1975; Rieger and Lawton 1985; Heeger and Jepson 1992; Hildreth 1992; Lappe and Rauschecker 1993; Dyre and Andersen 1994; Royden 1997; Dahmen et al., this volume). Their algorithms assume that local velocity vectors have been derived from the retinal input imagery. At some stage, the input is assumed to be a distribution of velocity vectors ($\mathbf{x}$, $\mathbf{y}$) and the algorithms perform some calculation on these vectors (e.g., vector subtraction). The impetus for using the vectors is that the theoretical analyses of the flow fields have turned up some clever schemes that enable the local flow vectors arising from combined translation and rotation of the observer to be "decomposed" into a translation component and a rotation component. For example, by subtracting one vector from another in the same small local region, only the translation component remains because the rotation component is common to both vectors (Longuet-Higgins and Prazdny 1980; Rieger and Lawton 1985). This is very appealing because the observer heading direction can then – in theory – be derived from the vector flow field translation components even though the observer was rotating during the translation.

2.2 Template models

For primates at least, the current evidence casts some doubts on the suggestion that the clever but often complex algorithms underlying the decomposition schemes are used to estimate self-motion. Neurones in the Medial Superior Temporal area (MST) that have been identified as having a role in self-motion estimation (e.g., Saito et al. 1986; Tanaka et al. 1986; Duffy and Wurtz 1991a, b) do not display the critical properties one would expect from a system based on decomposition of the flow field (Perrone and Stone 1998). Their receptive field properties are more consistent with schemes that simply register the global patterns of image motion directly (e.g., Perrone 1992; Perrone and Stone 1994) without resorting to vector subtraction or other forms of vector algebra. Such a "template" based approach to the self-motion estimation problem was proposed as an alternative to decomposition models by a number of people. In the insect vision area, a number of large field motion sensitive neurones had been postulated as being involved in visual course stabilization and gaze control (Hengstenberg 1982; Hausen 1993). In primates, a group of Japanese researchers (Saito et al. 1986; Tanaka et al. 1986; Tanaka and Saito 1989) suggested that MT (Middle Temporal) neurones could be assembled into networks that responded best to the radial expansion patterns that occurred during forward translation. They suggested that MST neurones had the appropriate structure to fulfil this role. These specialized detector networks could be thought of as templates designed to match a particular pattern of image motion. The Saito and Tanaka groups did not provide the specific details of the template construction but other heading models that followed used a similar design and provided more specific connectivity rules (Perrone 1987, 1990; Glünder 1990; Hatsopoulos and Warren 1991).

These original template models of human self-motion estimation could not process the inputs from combined translation and rotation self-motion. They were basically just expansion detectors (cf. Regan and Beverley 1978) and produce incorrect heading estimates when confronted with combined translation-rotation scenarios. Although they were biologically inspired and consistent with the known physiology, they could not compete with vector-based decomposition schemes that could solve the general self-motion problem. This deficit was overcome eventually when a model that could process translation and rotation was proposed (Perrone 1992). This model uses rotation detector networks to first detect the rotation visually. The output from these detectors is used to modify the heading templates to compensate for the rotation. The translation and rotation components of the flow field are never "separated out" locally as is the case in decomposition models. Since then, more elaborate template models which also incorporate eye-movement signals have been proposed (Beintema and van den Berg 1998).

A common general complaint against template models is that they require large numbers of detectors to handle the varieties of self-motion that can occur. We have, however, demonstrated that template models do not necessarily require large numbers of templates to operate successfully (Perrone and Stone 1994). In addition we have shown that the templates in our model share many of the properties of neurones in MST (Perrone and Stone 1998) and that heading can be directly encoded by individual MST neurones acting as templates. There is no need to postulate complex decomposition algorithms based on vector inputs.

In their analysis of visual interneurone properties in the fly (Krapp and Hengstenberg 1996; Dahmen et al., this volume) discovered a structure that is consistent with the basic template approach but they refer to the special detectors they found as "matched filters". They also propose that the function of the filters is to implement a form of vector decomposition. In this sense, their model does not fit cleanly into the category of template models discussed above. One way to evaluate models such as the Dahmen, Franz and Krapp model is to examine the types of inputs they require.

3. Vector flow fields: The bane of self-motion estimation theorists

The majority of current models of retino-centric instantaneous self-motion estimation assume that a velocity vector flow field is available to the navigating organism. While the notion of isolated points and velocity vectors is a useful theoretical construct for demonstrating that visual self-motion estimation is possible from retinal optic flow, it is far removed from the realities of everyday human visual navigation derived from retinal image motion. The two dimensional image motion generated from self-motion rarely arises from the movement of isolated points – instead surfaces and edges make up the major part of our visual environments and edges exist at a variety of orientations. A vector flow field consists of magnitude and direction measures at a number of locations in the visual field. I will examine both of these measures in detail in an attempt to convince the reader that vector flow fields are not realistic inputs to self-motion estimation in biological systems.

3.1 Speed (vector magnitude)

Vector-based algorithms

If an edge moves over a particular retinal location at 2°/s, for instance, then vector-based algorithms require the value "2" to be passed onto the next stage of processing. If the image speed increases to 8°/s then some coded value that is 4 times greater than the first speed must be sent to the self-motion processing unit. It requires a neurone that can provide a direct (e.g., rate coded) signal proportional to the speed. Evidence for such metrical coding of image speed has yet to be found in biological visual systems. For example, the most popular candidate for motion detection in insects, the Reichardt detector, produces an output that is influenced by the contrast and spatial structure of the input and which is not monotonically related to the image speed (Egelhaaf and Borst 1993). What we do find, however, is ample evidence for neurones in primate visual systems that are *tuned* to particular speeds.

Figure 1 shows re-plotted average speed tuning data from 109 MT cells collected by Maunsell and van Essen (1983). Similar data can be seen in a study by Lagae et al. (1993, their Fig. 9A). It is clear that at the level of MT there is no simple one-to-one relationship between the neurone's output and the speed of the edge moving across it. Some neurones do show low-pass or high-pass tuning in respect to image speed (see Lagae et al., their Fig. 9B) but these are in the minority and can hardly be described as exhibiting linear output properties. This means that in order to obtain the vector magnitude required for the processing of self-motion, the decomposition models must assume the existence of a processing level *above* MT that uses some form of population code to derive the image speed. (e.g., Lappe et al. 1996).

The decomposition models cannot assume that neurones in cortical area MT of the primate brain, are producing the speed part of the velocity vectors they require in their computations. The MT neurones feed into area MST which has neurones with properties suitable for self-motion estimation (Perrone and Stone 1998). But the MST neurones appear to simply integrate the signals being generated by the MT neurones (Tanaka and Saito 1989; Perrone and Stone 1998), without any intervening speed coding or "vector processing". Given that a signal proportional to the edge speed has yet to be found at any stage of the motion pathway, the decomposition models are faced with the problem of not having an obvious physiological correlate to the speed estimation part of their algorithms.

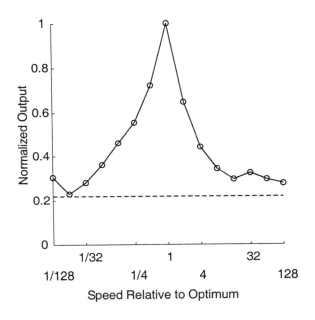

Fig. 1 Replotted mean speed tuning data from Maunsell and Van Essen (1983, their Fig. 6b). Horizontal dashed line corresponds to the average normalized background rate of firing of their sample.

Template models. Speed tuning versus speed estimation

How do template models avoid the "vector magnitude" problem outlined above? The simple answer is that these models make use of speed tuning, not speed estimation, and are built directly around the properties of MT neurones. Therefore the physiological counterpart to their input stage is clearly specified. The templates are based on the idea that if there is a good match between the velocity preference of the MT units making up the template and the image flow that is occurring on the retina, then a large signal will be generated in that template. If the match is not so good, the response will be less. The basic mechanism relies on speed and direction tuning as found in MT neurones (Maunsell and van Essen 1983; Albright

1984). These neurones produce a large output when a feature moves across their receptive field at the correct speed and direction and less of an output when the velocity does not match, i.e., they have exactly the properties required by the template models (which is not surprising since the models evolved from them). For vector-based schemes the non-linear relationship between speed and neurone output is a problem (see above). Rather than being an integral part of their mechanism – as is the case for template models – the speed tuning is an obstacle that must be overcome by the vector-based models.

It should be pointed out, that template models of the type proposed by Perrone (1992) and Perrone and Stone (1994) cannot function properly with a single MT motion sensor at each location. Because it is not possible to anticipate the exact structure of the 3D environment, and hence the speed of the retinal image that will occur at a particular location, a number of different speed tunings must be included at each location. The range of speeds can be minimized by assuming particular environment layouts are more common than others (Perrone and Stone 1998; Dahmen et al., this volume) but usually more than one is required. A particular retinal image speed will activate one of the MT units at that location more than the others. Using a winner-takes-all scheme, the *output* of this MT neurone is passed onto the next stage. This is where the template models differ from the vector-based models. The speed preference (e.g., 2°/s) of the winning MT neurone is not needed for later computations. It only becomes relevant when the structure of the environment needs to be determined (see Perrone and Stone 1994) and even then it is just the relative responses that are used, not the actual speeds.

The input to our template model is the neural activity from direction and speed tuned motion sensors (e.g., MT neurones). Examination of the figures in the early papers discussing the template model (Perrone 1992; Perrone and Stone 1994, 1998) may lead to some confusion in this regard. The inputs to the model are described and depicted as velocity vector fields. Because we did not have a model of MT neurones available that can be applied to image sequences, these template models have been implemented using velocity vectors as the initial input. The vectors are first passed through a stage involving idealized MT speed tuning curves and direction tuning curves to determine the simulated neurone output. We are currently attempting to bypass this unnecessary step by working with image sequences and models of MT neurones rather than idealized vector flow fields (Perrone 1997).

3.2 Direction

Obtaining the direction of the vector would seem straightforward for the visual system since there is ample evidence of neurones tuned for particular directions of motion (e.g., Middle Temporal neurones; Albright 1984). Nevertheless the problem is far from trivial.

As was the case for speed estimation discussed above, there still needs to be a stage which converts the direction tuning into a measure of actual directions and the locus of this operation needs to be specified. In addition to this coding problem, there is an even bigger obstacle facing theorists who rely on knowledge of the velocity vector direction. For many self-motion scenarios, the image motion varies considerably from one retinal location to the next. This is because most scenes contain a range of objects at different distances from the moving observer and important information concerning the layout of the scene often occurs at nearby adjacent regions of the visual field. In order to register these small spatial variations, the analysis of the retinal image motion needs to be reasonably localized and invariably some of the 2D motion sensors performing the analysis will have receptive fields that contain a single moving edge. Once this occurs, the visual system is faced with the aperture problem (Wallach 1935) – motion in the direction parallel to the edge is invisible and only the component normal to the edge can be detected by the motion sensors. Figure 2 illustrates the problem and shows two flow fields, one where the aperture problem is assumed to not exist (a) and the other where it is present (b). The structure of the flow field is changed radically by the presence of the aperture problem.

Vector-based approaches and the aperture problem

The aperture problem has been largely ignored in the field of self-motion perception (see however, Perrone 1990, 1992; Heeger and Jepson; 1990). Many of the vector-based techniques rely on the detection of small vector differences that would be heavily masked by the large perturbations in speed and direction that arise from the aperture problem. Some self-motion estimation models claim resistance to the aperture problem because they incorporate spatial integration over large areas of the visual field (e.g., Perrone 1992; Heeger and Jepson 1992). This resistance however can only work up to a point and usually relies on certain assumptions about the distribution of edges in the scene. The analysis carried out by Dahmen et al is typical in that they simulate "noise" such as the aperture problem by perturbing the flow vector directions and speeds using a normal noise distribution that shifts the direction equally about the correct one. The noise is assumed to be statistically independent at neighbouring locations. Figure 2 demonstrates a weakness with this assumption because the direction perturbations can often be asymmetrically distributed and the image speeds are systematically distorted in certain areas of the visual field, not just increased or decreased randomly across the field. Humans routinely navigate safely in environments similar to that depicted in figure 2b, but most vector-based models of egomotion would have serious problems when faced with the input shown in this figure.

The aperture problem has mainly been addressed in the context of rigid two-dimensional motion in the image plane. Techniques for overcoming the aperture problem under uniform 2D motion conditions (e.g., Adelson and Movshon 1982;

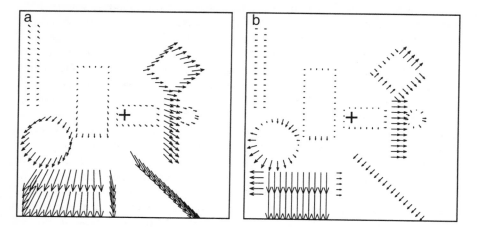

Fig. 2 Demonstration of the aperture problem. **a** This is the theoretical image motion that would be generated for an observer moving in a direction up and to the left of the viewing direction (marked with a cross). **b** Image motion for the same heading direction of the observer but it shows the result if only the motion normal to the edges in the scene could be detected.

Hildreth 1983; Yuille and Grzywacz 1988; Sereno 1993; Simoncelli and Heeger 1998) cannot be applied directly to the 3D self-motion situation. These techniques assume rigid 2D object motion and were not designed to accommodate the ambiguous speeds and directions endemic to 3D flow fields. You cannot "smooth" or average the vectors in figure 2b without introducing large errors because in many cases adjacent vectors come from different objects separated in depth. More elaborate solutions to the aperture problem have been proposed which can deal with multiple objects, transparency and partial occlusion (e.g., Nowlan and Sejnowski 1995) although it is not clear how well such models can deal with the multi-speed, multi-direction motion inputs that result from observer motion. This model also requires a stage above MT in order to obtain the final image velocity from the activity of the "velocity units" computing the motion. It therefore suffers from the same problems outlined under speed tuning in Section 3.1.

Template models and the aperture problem

If one assumes that MT neurones are able to solve the aperture problem, then in theory the aperture problem should not cause template models any problems because MT neurones form the main input to these models. However, because the exact mechanisms for how MT neurones actually overcome the aperture problem are currently unknown, the template models cannot include this stage as part of their processing chain. We have begun exploring different possible methods for solving the aperture problem in the context of self-motion estimation and the template approach. Each heading template in our model of self-motion estimation is made up of a network that connects together particular sets of MT-like 2D motion sensors. For a particular preferred heading direction (α, β) coinciding with retinal image location (x, y), the direction preferences of the motion sensors are radially aligned around (x, y). If (α, β) coincides with the actual heading direction, then the true image motion of an isolated point would be along one of these radial directions (ϕ). Assume that the speed of the point is V (°/s) and, for simplicity, that the template 2D motion sensor speed tuning matches this speed. If the point is replaced with an edge, then depending on its orientation, the motion of the edge normal could lie anywhere in a 180° semi-circle centred around ϕ. The speed of the edge will be reduced (relative to that of the point), by the cosine of the angle between the edge normal direction and ϕ. This 180° of possible directions could be sampled with a modest number of additional 2D motion sensors given the relatively broad directional tuning of MT units (Albright 1984). At each image location, the motion sensor tuned to [V, ϕ] could be augmented by sensors tuned to [0, ±90°], [Vcos(60), ±60°] and [Vcos(30), ±30°].

One solution for template models is therefore to include an additional set of MT-like sensors into their detector networks which are tuned to a subset of the extra possible image motions that can arise as a result of the aperture problem. Initial tests of this concept using "artificial" inputs such as those in figure 2b have been successful, but a thorough testing of the aperture problem solution cannot be undertaken until an image-based version of the model is implemented. Natural images contain features at a variety of spatial scales, contrasts and orientations and the motion "information" is much more complex than that depicted in figure 2b.

4. Conclusion

The vector flow field has become an integral part of many treatments of the visual self-motion estimation problem. Over the years researchers have either assumed its existence or sought after it as though it was some sort of Holy Grail – although I think a better analogy is that of a Siren luring sailors onto the rocks. The current physiological data better supports the view that primate visual systems do not use the vector flow field but instead have developed a solution around speed- and direction-tuned motion sensors. Template models have followed this lead and thus avoid many of the problems associated with vector flow field inputs. Even so, they still have a long way to go before they can provide an adequate description of biological self-motion estimation. The starting point for self-motion perception is the changing patterns of light falling onto the retinae of the eyes. Therefore the true input for self-motion estimation is a two-dimensional image sequence. Template models currently assume that MT neurone activity has been derived from the image motion. The challenge is to remove this assumption by simulating the MT stage as well. The next generation of self-motion models must be designed to work with natural image sequences and the vector flow field can be laid to rest.

Acknowledgements

Supported by NASA-Ames Grant NAG 2-1168

References

Adelson EH, Movshon JA (1982) Phenomenal coherence of moving visual patterns. Nature 300: 523-525

Albright TD (1984) Direction and orientation selectivity of neurons in visual area MT of the Macaque. J Neurophysiol 52: 1106-1130

Beintema JA, van den Berg AV (1998) Heading detection using motion templates and eye velocity gain fields. Vision Res 38: 2155-2179

Cutting JE (1986) Perception with an eye for motion. Bradford, Cambridge

Duffy CJ, Wurtz RH (1991a) Sensitivity of MST neurons to optic flow stimuli. I. A continuum of response selectivity to large-field stimuli. J Neurophysiol 65: 1329-1345

Duffy CJ, Wurtz RH (1991b) Sensitivity of MST neurons to optic flow stimuli. II. Mechanisms of response selectivity revealed by small-field stimuli. J Neurophysiol 65: 1346-1359

Dyre BP, Andersen GJ (1994) Statistical moments of retinal flow may be used to determine heading. Invest Ophthalmol Vis Sci 35: S 1269

Egelhaaf M, Borst A (1993) Movement detection in arthropods. In: Miles FA, Wallman J (eds) Visual motion and its role in the stabilization of gaze. Elsevier, Amsterdam, pp 53-77

Glünder H (1990) Correlative velocity estimation: Visual motion analysis, independent of object form, in arrays of velocity-tuned bilocal detectors. J Opt Soc Am A 7: 255-263

Gibson JJ (1950) The perception of the visual world. Houghton Mifflin, Boston

Hatsopoulos N, Warren WH (1991) Visual navigation with a neural network. Neural Networks 4: 303-317

Hausen K (1993) Decoding of retinal image flow in insects. In: Miles FA, Wallman J (eds) Visual motion and its role in the stabilization of gaze. Elsevier, Amsterdam, pp 203-235

Heeger DJ, Jepson AD (1992) Subspace methods for recovering rigid motion I: Algorithm and implementation. Int J Comp Vision 7: 95-177

Hengstenberg R (1982) Common visual response properties of giant vertical cells in the lobula plate of the blowfly *Calliphora*. J Comp Physiol A 149: 179-193

Hildreth EC (1983) The computation of the velocity field. Proc Roy Soc Lond B 221: 189-220

Hildreth EC (1992) Recovering heading for visually-guided navigation. Vision Res 32: 1177-1192

Koenderink JJ, van Doorn AJ (1975) Invariant properties of the motion parallax field due to the movement of rigid bodies relative to an observer. Opt Acta 22: 773-791

Krapp HG, Hengstenberg R (1996) Estimation of self-motion by optic flow processing in single visual interneurons. Nature 384: 463-466

Lagae S, Raiguel S, Orban GA (1993) Speed and direction selectivity of macaque middle temporal neurons. J Neurophysiol 69: 19-39

Lappe M, Rauschecker JP (1993) A neural network for the processing of optic flow from ego-motion in man and higher mammals. Neural Comput 5: 374-391

Lappe M, Bremmer F, Pekel M, Thiele A, Hoffmann KP (1996) Optic flow processing in monkey STS: A theoretical and experimental approach. J Neurosci 16: 6265-6285

Longuet-Higgins HC, Prazdny K (1980) The interpretation of moving retinal images. Proc Roy Soc Lond B 208: 385-387

Maunsell JHR, van Essen DC (1983) Functional properties of neurons in the middle temporal visual area of the Macaque monkey. I. Selectivity for stimulus direction, speed, orientation. J Neurophysiol 49: 1127-1147

Nowlan SJ, Sejnowski TJ (1995) A selection model for motion processing in area MT of primates. J Neurosci 15: 1195-1214

Perrone JA (1987) Extracting 3-D egomotion information from a 2-D flow field: A biological solution? Opt Soc Am Tech Digest Series 22: 47

Perrone JA (1990) Simple technique for optical flow estimation. J Opt Soc Am A 7: 264-278

Perrone JA (1992) Model for the computation of self-motion in biological systems. J Opt Soc Am A 9: 177-194

Perrone JA (1997) Extracting observer heading and scene layout from image sequences. Invest Ophthalmol Vis Sci 38: S 481

Perrone JA, Stone LS (1994) A model of self-motion estimation within primate extrastriate visual cortex. Vision Res 34: 2917-2938

Perrone JA, Stone LS (1998) Emulating the visual receptive field properties of MST neurons with a template model of heading estimation. J Neurosci 18: 5958-5975

Regan D, Beverley KI (1978) Looming detectors in the human visual pathway. Vision Res 18: 415-421

Rieger JH, Lawton DT (1985) Processing differential image motion. J Opt Soc Am A 2: 354-360

Rieger JH, Toet L (1985) Human visual navigation in the presence of 3D rotations. Biol Cybern 52: 377-381

Royden CS (1997) Mathematical analysis of motion-opponent mechanisms used in the determination of heading and depth. J Opt Soc Am A 14: 2128-2143

Saito H, Yukie M, Tanaka K, Hikosaka K, Fukada Y, Iwai E (1986) Integration of direction signals of image motion in the superior temporal sulcus of the Macaque monkey. J Neurosci 6: 145-157

Sereno ME (1993) Neural computation of pattern motion: modeling stages of motion analysis in the primate visual cortex. MIT Press, Cambridge, Mass

Simoncelli EP, Heeger DJ (1998) A model of neuronal responses in visual area MT. Vision Res 38: 743-761

Stone LS, Perrone JA (1997) Human heading estimation during visually simulated curvilinear motion. Vision Res 37: 573-590

Tanaka K, Hikosaka K, Saito H, Yukie, M, Fukada Y, Iwai E (1986) Analysis of local and wide-field movements in the superior temporal visual areas of the macaque monkey. J Neurosci 6: 134-144

Tanaka K, Saito H (1989) Analysis of the motion of the visual field by direction, expansion/contraction, and rotation cells clustered in the dorsal part of the medial superior temporal area of the macaque monkey. J Neurophysiol 62: 626-641

Wallach H (1935) Über visuell wahrgenommene Bewegungsrichtung. Psychol Forsch 20: 325-380

Warren WH, Hannon DJ (1990) Eye movements and optical flow. J Opt Soc Am A 7: 160-168

Yuille AL, Gryzywacz NM (1988) A computational theory for the perception of coherent visual motion. Nature 333: 71-74

Zacharias GL, Caglayan AK, Sinacori JB (1985) A visual cueing model for terrain-following applications. J Guidance 8: 201-207

Visual Navigation: The Eyes Know Where Their Owner is Going

Mandyam V. Srinivasan

Visual Sciences Group, Research School of Biological Sciences, Australian National University, Canberra, Australia

Vision serves us in many important ways, one of which is to convey to us a precise sense of our own motion in the world. In a racing car video game, for example, the view of the moving landscape through the virtual windscreen provides a compelling and accurate impression – or, rather, illusion – of the car's motion, even though our vestibular senses are not stimulated by the visual display.

Humans, and a wide variety of animals use visually-derived information on egomotion for several purposes.

First, hovering creatures stabilize their position in space by using egomotion signals to sense unwanted displacements. This is probably how hoverflies, for example, manage to stay virtually locked in mid-air (Collett 1980), how hawkmoths (Pfaff and Varju 1991) and probably hummingbirds maintain a stable hovering position relative to the flower from which they are feeding, and how guard bees of the tropical genus *Tetragonisca* maintain vigilant stations in front of their hive entrances (Kelber and Zeil 1997).

Second, information on egomotion can be used to execute motion in a straight line, or along any desired trajectory. The strategy would be to apply the motor control that is needed to produce the pattern of image motion (optic flow) that corresponds to the desired trajectory. Deviations from the desired trajectory could be detected in terms of deviations in the pattern of optic flow, which can be used to generate corrective motor commands. Thus, in flying insects, the well-studied "optomotor response" is believed to stabilize yaw disturbances by detecting the apparent motions of the environment that are induced by such yaws and generating appropriate counter-turns to promote flight in a straight line (reviews

Reichardt 1969; Egelhaaf et al. 1988). Similar mechanisms are probably involved in the stabilization of roll (Srinivasan 1977), as well as pitch.

Third, information on translatory egomotion can be used to regulate the speed of locomotion. Flies (David 1982) and bees (Srinivasan et al. 1996) control the speed of flight by holding the velocity of self-induced image motion approximately constant. This visuomotor strategy cannot be used to achieve a desired absolute speed of flight, because the speed of image motion that is experienced will depend not only on flight speed but also upon the distance and bearing of the objects in the environment. However, keeping the overall image velocity constant at some prespecified value automatically ensures that flight speed is adjusted to the height above the ground (Srinivasan et al. in press) and that narrow passages are negotiated at slower, safer speeds. This provides a simple means of controlling flight speed without having to explicitly measure the distances to objects in the environment by using, say, complex stereo mechanisms, which most insects do not possess (Srinivasan 1993).

Fourth, visually derived egomotion signals can be used to estimate how far one has travelled. It has recently been shown that honeybees estimate the distance they have flown by integrating, over time, the self-induced optic flow that they experience en route to the destination (Esch and Burns 1996; Srinivasan et al. 1996, 1997, 2000).

Finally, humans and many animals are capable of detecting other moving objects in the environment, even whilst they are themselves in motion. The ability of predatory animals to detect, chase and capture their prey is a clear testimony to this capacity. Clearly, humans are able to detect moving objects effortlessly, regardless of whether they themselves are walking, running or driving a car. In the insect world, one striking example of moving object detection is provided by the male housefly, which detects females during its patrolling manoeuvres and chases them in a rapid and impressive display of aerobatics (Land and Collett 1974).

Computationally, the task is not as trivial as it might seem at first glance, because when an animal is in motion, the image of the environment (i.e. the background) is constantly in motion on the retina, as is the image of the moving object. In principle, one way in which a moving animal could detect other moving objects would be, firstly, to estimate its egomotion from the global pattern of image motion that is induced on the retinae. This computation would be based on the assumption that the bulk of the image motion that is experienced by the eyes is due to the animals' own motion in the (largely) stationary environment. The next step would be to compute (i.e. predict) the pattern of optic flow that is expected to be experienced by the eye on the basis of the inferred egomotion, assuming that the environment is entirely stationary. Finally, objects that are moving in the environment would be detected by sensing regions of the image in which the optic flow differs from that expected on the assumption of a stationary environment. At least, this is the standard, computer-vision approach to the problem. It is interesting in this context that flies sometimes "shadow" other flies by moving in such a way as to appear as though they are part of the stationary environment (Srinivasan and Davey 1995).

As we have seen above, computation of egomotion is likely to be an essential first step in a number of important navigational tasks. Indeed, neurophysiological experiments in a number of animal species, such as monkeys (review: Wurtz 1998), pigeons (review: Frost and Sun 1997) and flies (review: Hausen 1993; Krapp and Hengstenberg 1996) have revealed the existence of motion-sensitive neurones that appear to play an important role in the analysis of egomotion. These neurones typically possess large visual fields and are sensitive to the patterns of optic flow generated by particular types of egomotion, such as rotation and translation about specific axes.

The Chapter by Dahmen et al. describes how egomotion can be computed visually by using a set of motion-sensitive neurones that are each sensitive to a rotatory or a translatory motion along a specific axis. One of the striking findings of the study is that the canonical patterns of flow-field sensitivities that are predicted by their model are very similar to the patterns of sensitivity displayed by real neurones in the lobula plate of the fly's brain. Thus, it appears that the visual system of the fly may indeed carry neural circuitry for determining egomotion in three dimensions, involving six degrees of freedom: three in rotation, and three in translation. This solution for computing egomotion appears to be a rather general one in that it does not require severely restrictive assumptions about the nature of the environment in which the animal moves. The "world model" of Dahmen et al. only assumes that (a) objects or surfaces below the horizon are nearer than those above the horizon and (b) on average, the fly is most likely to move in the forward direction.

While this is obviously a very satisfying finding, it is worth noting that many navigational tasks may not need a complete solution to the egomotion problem as described by Dahmen et al.. This is because the structure of the natural environment sometimes offers simpler solutions. Furthermore, in some instances, the problem can be greatly simplified if the perceptual problem is not addressed in isolation, but in conjunction with the behavioural role that it is meant to fulfil. Some possibilities are sketched below.

An insect flying outdoors under a clear sky can, at least in principle, use the reference direction provided by the sun not only to maintain a straight course, but also to stabilize yaw, pitch and roll. Even when the sun is hidden by a cloud, the pattern of polarized light produced by the sun in the sky, and the spatial variations in the spectrum of the light from different parts of the sky can be used instead of the sun for the same purpose (Rossell and Wehner 1986). In bees, ants and crickets, there is clear evidence, behaviourally, anatomically and physiologically, for the existence of a visual subsystem that analyses the orientation of the pattern of polarized light in the sky. Behavioural experiments on ants and bees show that this system is crucial for navigation (Wehner 1997). There is evidence that bees and ants even allow for the movement of the sun in the sky through the course of the day by combining sun-compass information with information from their circadian clock (review Wehner 1992). The advantage of using cues from the sky for navigation is that, since they are derived from sources (or objects) that are effectively infinitely far away, they provide information on rotational movements (yaw, pitch and roll) that is uncontaminated by the animal's translatory movements (see, for example, Nalbach and Nalbach 1987). An additional advantage of using such information from the sky is that there is no need to compute optic flow in order to maintain a straight course or to stabilize yaw, pitch and roll. It is only necessary to ensure that the spatial pattern of intensity, colour or polarized light from the sky is held stable on the eye during flight. Optic flow information provided by the ground and surrounding vegetation can then be used secondarily for regulating flight speed, inferring distance travelled, and for other manoeuvres such as landing or avoiding obstacles. Even when the sky is not visible – as when flying in dense vegetation – a straight course can be maintained, and yaw and pitch stabilized, by fixating two features in the environment, one in the direction of the intended course and another in the opposite direction (see, for example, Warren 1998). If the fixated features are extended, rather than point-like, roll can be stabilized as well. Here again, there would be no need to measure optic flow everywhere in the visual field: it is only necessary to fixate the features in the two directions. Of course, the fixated features would have to be modified or updated as the flight progresses. There is evidence that hoverflies may indeed use such a strategy for moving in a straight line (Collett 1980). Clearly, this would be a very simple navigational strategy for an animal with compound eyes that provide nearly panoramic vision.

It is possible that natural visual systems even solve the difficult problem of detecting a moving object whilst in motion, without computing egomotion as a first step. A recent psychophysical study has examined the ability of human observers to detect a moving object in the presence of optic flow patterns that would have been generated by their own motion in the environment (Hoffman and Zanker 1997). The results show that the primary cue for detecting the moving object under such conditions is the extent to which the motion of the image of the object differs from that of the image of its immediate surround on the retina. In other words, if the object appears to move in the same direction as the background, it is taken to be stationary, if it moves in a different direction, it is assumed to be moving. If this is indeed the basis on which moving animals detect moving objects, then the underlying computation is a relatively simple, local one that does not require determination of egomotion as a first step. Indeed, there are neurones in the visual systems of a number of animal species that do not react to homogeneous motion within their receptive field, but do respond when motion in one part of the receptive field is different from that in another (flies: review Egelhaaf et al. 1988; pigeons: review Frost and Sun 1997; monkeys: Eifuku and Wurtz 1998). Such neurones could potentially form part of a circuit that enables their moving owner to detecting moving objects in the environment, without having to first compute egomotion.

In conclusion, it is reassuring that, at least in some instances, natural visual systems seem to have converged on the kinds of solutions that might have been arrived at by tackling the problem from a "first principles" standpoint, and treating it as a purely sensory one. This seems to be the case with egomotion computation by the lobula plate of the fly, as shown elegantly in the chapter by Dahmen et al.. However, we see from the above discussion that, at least in some situations, surprisingly simple, "alternative" solutions can be postulated. To arrive at these solutions, one has to (a) take into account the environmental conditions under which animals operate and (b) consider perception and action as part of a single closed loop, rather than as separate entities. Further research should uncover whether, and to what extent animals take advantage of these potential "short cuts".

References

Collett TS (1980) Some operating rules for the optomotor system of a hoverfly during voluntary flight. J Comp Physiol A 138: 271-282

David CT (1982) Compensation for height in the control of groundspeed by *Drosophila* in a new Barber's Pole wind tunnel. J Comp Physiol 147: 495-493

Egelhaaf M, Hausen K, Reichardt W, Wehrhahn C (1988) Visual course control in flies relies on neuronl computation of object and background motion. Trends Neurosci 11: 351-358

Eifuku S, Wurtz RH (1998) Response to motion in extrastriate area MST1 – center-surround interactions. J Neurophysiol 80: 282-296

Esch H, Burns J (1996) Distance estimation by foraging honeybees. J Exp Biol 199: 155-162

Frost BJ, Sun J (1997) Visual motion processing for figure/ground segregation, collision avoidance, and optic flow analysis in the pigeon. In: Srinivasan MV, Venkatesh S (eds) From living eyes to seeing machines. Oxford University Press, Oxford, pp 80-103

Hausen K (1993) Decoding of retinal image flow in insects. In: Miles FA, Wallman J (eds) visual motion and its role in the stabilization of gaze. Elsevier, Amsterdam, pp 203-235

Hoffmann MJ, Zanker JM (1997) Detection of moving objects in optic flow fields. In: Elsner N, Wässle H (eds) Göttingen Neurobiology Report 1997. Thieme, Suttgart, p 1002

Kelber A, Zeil J (1997) *Tetragonisca* guard bees interpret expanding and contracting patterns as unintended displacements in space. J Comp Physiol A 181: 257-265

Krapp H, Hengstenberg R (1996) Estimation of self-motion by optic flow processing in single visual interneurons. Nature 384: 463-466

Land MF, Collett TS (1974) Chasing behaviour of houseflies (*Fannia canicularis*). J Comp Physiol 89: 331-357

Nalbach H-O, Nalbach G (1987) Distribution of optokinetic sensitivity over the eye of crabs: its relation to habitat and possible role in flow-field analysis. J Comp Physiol A 160: 127-135

Pfaff M, Varju D (1991) Mechanisms of visual distance perception in the hawk moth *Macroglossum stellatarum*. Zool Jb Physiol 95: 315-321

Reichardt W (1969) Movement perception in insects. In: Reichardt W (ed) Processing of optical data by organisms and insects. Academic Press, New York, pp 465-493

Rossel S, Wehner R (1986) Polarization vision in bees. Nature 323: 128-131

Srinivasan MV (1977) A visually-evoked roll response in the housefly: open-loop and closed-loop studies. J Comp Physiol 119: 1-14

Srinivasan MV (1993) How insects infer range from visual motion. In: Miles FA, Wallman J (eds) Visual motion and its role in the stabilization of gaze. Elsevier, Amsterdam, pp 139-156

Srinivasan MV, Davey M (1995) Strategies for active camouflage of motion. Proc Roy Soc Lond B 259: 19-25

Srinivasan MV, Zhang SW, Chahl JS, Barth E, Venkatesh S (2000): How honeybees make grazing landings on flat surfaces. Biol Cybernetics (in press)

Srinivasan MV, Zhang SW, Lehrer M, Collett TS (1996) Honeybee navigation en route to the goal: Visual flight control and odometry. J Exp Biol 199: 237-244

Srinivasan MV, Zhang SW, Bidwell N (1997) Visually mediated odometry in honeybees. J Exp Biol 200: 2513-2522

Srinivasan MV, Zhang SW, Altwein M, Tautz J (2000) Honeybee navigation: Nature and calibration of the "odometer". Science 287: 851-853

Warren WH, Jr. (1998) Visually controlled locomotion: 40 years later. Ecol Psychol 10: 177-219

Wehner (1992) Arthropods. In: Papi F (ed) Animal homing. Chapman and Hall, London, pp 45-144

Wehner R (1997) Insect navigation: Low-level solutions to high-level tasks. In: Srinivasan MV, Venkatesh S (eds) From living eyes to seeing machines. Oxford University Press, Oxford, pp 158-173

Wurtz RH (1998) Optic flow - a brain region devoted to optic flow analysis. Curr Biol 8:554-556

Part IV

Motion Vision in Action

The Role of Inertial and Visual Mechanisms in the Stabilization of Gaze in
Natural and Artificial Systems
Giulio Sandini, Francesco Panerai and Frederick A. Miles

Gaze Control: A Developmental Perspective
Janette Atkinson and Oliver Braddick

Does Steering a Car Involve Perception of the Velocity Flow Field?
Michael F. Land

The Role of Inertial and Visual Mechanisms in the Stabilization of Gaze in Natural and Artificial Systems

Giulio Sandini[1], Francesco Panerai[2] and Frederick A. Miles[3]

[1]Laboratory for Integrated Advanced Robotics, Department of Communication, Computers and Systems Science, University of Genova, Genova, Italy; [2]Laboratoire de Physiologie de la Perception et de l'Action, Collège de France, Paris, France; [3]Laboratory of Sensorimotor Research, The National Eye Institute, Bethesda, USA

Contents

1. Abstract

Vision is arguably our premier navigational aid, allowing us to map out and actively explore our surroundings. However, we view the world from a constantly shifting platform and some visual mechanisms function optimally only if the images on the retina are reasonably steady. As we go about our everyday activities, visual and vestibular mechanisms help to stabilize our gaze on particular objects of interest by generating eye movements to offset our head movements. The general picture that has emerged of gaze stabilization in primates during motion is of two vestibulo-ocular reflexes, the RVOR and TVOR, that compensate selectively for rotational and translational disturbances of the head, respectively, each with its own independent visual backup mechanisms. A major objective of this chapter is to review recent work on low-level, pre-attentive mechanisms that operate with ultra-short latencies and are largely independent of conscious perception. Recent advances in the field of robotics and, particularly, in the domain of active vision, provide a complementary view of the uses and associated problems of visuo-inertial integration for the stabilization of gaze. Much like biological systems, robots have to comply with physical constraints imposed by the environment and/or by the need to coordinate their sensori-motor components in an efficient way. In contrast with biological systems, however, the experimental variation of implementation parameters and control strategies allows, among other things, a comparison of the different hypotheses and implementations. The goal of this chapter is to draw parallels between the results from biology and from a robot which uses inertial and visual information to stabilize its cameras/eyes. The Chapter is organized as follows. Section 2 describes the peculiarities of the patterns of retinal motion (optic flow) experienced by an observer moving through the environment. The appropriate compensatory eye movements required to stabilize gaze are described in Section 3, introducing the distinction between vergence and version eye movements. Section 4 describes the main characteristics of the vestibular system from a biological and artificial perspective and the distinction between "rotational" and "translational" components of the vestibulo-ocular reflex (VOR). The magnitude of the eye movements required for complete compensation depends on various kinematic parameters such as the position of the eyes in the head, the inter-ocular distance, as well as the distance to the fixation point. Section 5 deals with the integration of visual and inertial information for gaze stabilization. This Section builds upon the concept of "translational" and "rotational" components of the VOR and highlights the differences between "version" and "vergence" control of compensatory eye movements. In particular the role of the radial component of optical flow in the feed-forward control of eye movements is compared with the feed-back loop mediated by binocular disparity. In Section 6, the contribution of inertial and visual information in gaze stabilization is discussed with reference to the different latencies and processing power required by the two modalities. The advantage of the integration of visual and inertial data is discussed from a biological and robotics perspective in the concluding Section.

2. Optic flow: the visual consequences of moving through the environment

The pattern of retinal image motion, or optic flow, resulting from motion of the observer can be highly complex and biologists tend for convenience to treat rotational and translational disturbances independently. The latter is in part a historical accident but evidence is beginning to accumulate which suggests that the brain also tends to parse optic flow into these two components. Of course, the patterns of optic flow associated with rotations and translations are distinctly different. A passive observer who undergoes pure rotation experiences *en masse* motion of her entire visual world, the direction and the speed of the optic flow at all points being dictated solely by the observer's rotation. The overall pattern of optic flow resembles the lines of latitude on a globe (see Fig. 1a) but, of course, the observer's restricted field of view means that only a portion will be visible at any given time (e.g., Fig. 1b).

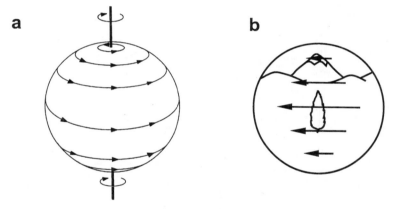

Fig. 1 Patterns of optic flow experienced by a (passive) rotating observer. **a** The retinal optic flow can be considered to be distributed over the surface of a sphere and created by projection through a vantage point at the centre. Here, the observer rotates about this vantage point and the pattern of flow resembles the lines of latitude on a globe. In reality things are never as simple as this, voluntary head turns occurring about an axis some distance behind the eyes so that the latter always undergo some slight translation. Such second-order effects are ignored here (but see Miles et al. 1991). **b** A cartoon showing the observer's limited field of view and the kind of motion experienced during rotation about a vertical axis as the observer looks straight out to the side. The speed of optic flow is greatest at the centre ("equator") and decrements as the cosine of the angle of latitude. However, both the pattern and the speed of the optic flow at all points are determined entirely by the observer's motion - the 3D structure of the scene is irrelevant (Miles 1997).

In principle, appropriate compensatory eye movements could completely offset the visual effects due to rotational disturbances so that the entire scene would be stabilized on the retina. This ignores the second-order translational effects due to the eccentricity of the eyes with respect to the usual axis of head rotation which, as we shall see later, are of consequence only for close viewing. If compensation is less than adequate, which is often the case, the *speed* of flow is reduced and the overall *pattern* of flow is largely preserved, provided the compensatory eye movements are in the correct direction. When the passive observer undergoes pure translation, the optic flow consists of streams of image flow vectors emerging from a focus of expansion straight ahead and disappearing into a focus of contraction behind, the overall pattern resembling the lines of longitude

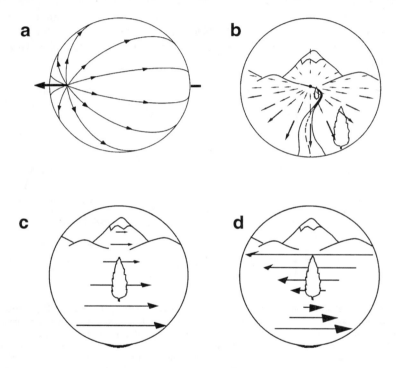

Fig. 2 Patterns of optic flow experienced by a translating observer. **a** The pattern of optic flow resembles the lines of longitude on a globe (Miles et al. 1991). **b** A cartoon showing the centrifugal pattern of optic flow experienced by the observer who looks in the direction of heading - the black dot at the foot of the mountain (Busettini et al. 1997). **c** The optic flow experienced by the moving observer who looks off to the right but makes no compensatory eye movements so that the visual scene appears to pivot about the distant mountains (effective infinity). The speed of image motion is inversely proportional to the viewing distance. **d** Again, the observer looks off to one side but here attempts to stabilise the retinal image of a particular object in the middle ground (tree), necessitating that she track to compensate for her own motion, thereby reversing the apparent motion of the more distant objects and creating a swirling pattern of optic flow. The scene now appears to pivot about the tree (c, d after Miles et al. 1992).

on a globe (see Fig. 2a). As with rotational disturbances, the *direction* of flow at any given point depends solely on the motion of the observer but, in contrast, the *speed* of the flow at any given point depends also on the *viewing distance* at that location: nearby objects move across the field of view much more rapidly than more distant ones, a phenomenon that is called *motion parallax* (Gibson 1950; Gibson 1966). Again, in the observer's restricted field of view the pattern of motion actually experienced depends very much on where the observer chooses to look. If the observer looks straight ahead, as when driving a car, for example, she sees an expanding world (see Fig. 2b) whereas, off to one side, as when looking out from a moving train onto a landscape, the sensation is of the visual world pivoting around the far distance (Fig. 2c).

3. Compensatory eye movements

Appropriate compensatory eye movements can almost eliminate the visual conse-quences of head *rotations*, but this is not the case with *translations* if the scene has 3D structure because of the dependence on viewing distance. During translation, eye movements can stabilize only the images in one particular depth plane and we shall see that the problem confronting the system here is how to make that "plane of stabilization" coincide with the plane of fixation. In the case of the observer looking out from the train and making no attempt to compensate for the motion (Fig. 2c), only the images of the most distant mountains are stable. If the observer transfers gaze to the tree in the middleground, then it is reasonable to assume that priority should now go to stabilizing the image of the tree, which requires that the observer now compensates for the motion of the train. If the observer succeeds in this then her visual world will now pivot about the tree (Fig. 2d). The optic flow here is a combination of translational flow due to the motion of the train in our example, and rotational flow, due to the subject's compensatory eye movements. Of course, many other combinations of translational and rotational flow are possi-ble in everyday situations. From the implementation perspective, a robot in similar situations has to solve exactly the same problems and in particular has to select a "pivot" – the tree in the previous example – to maintain a stable view of that part of the visual field. A possible solution, as we shall see, is based on the decom-position of optical flow information into behaviourally significant components, sometimes utilizing binocular information to achieve successful parsing, together with inertial information that can reduce the response time.

3.1 Version and vergence

The ocular compensations during rotations, as well as during translation when the subject looks off to one side, are conjugate, i.e. in the same direction for both eyes. In contrast, if the translating observer looks in the direction of heading, her two eyes must move in opposite directions (that is, towards the nose) if she is to keep them both aligned on the object of interest as it gets nearer. The required compensatory eye movements are thus disconjugate. There is considerable evidence that the brain controls conjugate and disconjugate movements of the eyes largely independently and in many laboratory experiments it is now usual to compute the conjugate components – so-called *version* – by averaging the movements of the two eyes, and to compute the disconjugate components – so-called *vergence* – by subtracting the orientation of one from the other. In fact, version and vergence provide a complete (binocular) representation of eye movements from which the movements of each of the two eyes can be reconstructed, and we think it perhaps more indicative of the way that eye movements are encoded in many brain areas. It is worth noting, however, that ocular compensation for motion of the observer in most everyday situations requires a combination of version and vergence. Vergence movements are required whenever the distance between the fixation point and the observer changes. Only in the unlikely circumstance that viewing distance changes along the cyclopean line of sight, version movements are *not* required, but oscillatory motion of the head generated by locomotion makes this virtually impossible. On the other hand, depending upon the distance of the fixation point and/or the motion of the observer one of the two components may become, in practical terms, predominant. For instance, vergence movements become negligible, when the observer is looking sideways at a far point. It is worth stressing that the control of vergence and version for the purpose of gaze stablization must utilise different components of image velocity and possibly different combinations of visual and inertial information. This lends support to the idea that vergence and version movements, although concurrently controlled, are indeed processed in different brain areas.

4. Inertial stabilization mechanisms

4.1 Vestibulo-ocular reflexes

The primate vestibular system senses motions of the head through two kinds of end organ that are embedded in the base of the skull, the semicircular canals and the otoliths, which are selectively sensitive to angular and linear accelerations, respectively (Goldberg and Fernandez 1975). These two kinds of sensors support two vestibulo-ocular reflexes: the canals provide the information to compensate for rotations (RVOR) and the otoliths provide the information to compensate for translations (TVOR). The corresponding design of an inertial stabilization mechanism for a robot requires firstly the development of an appropriate sensory system, and secondly the synthesis of effective eye control strategies. A prototype "vestibular system" for a robot, designed at the Laboratory of Integrated Advanced Robotics (LIRA) at Genova, is able to measure two angular velocities and two linear accelerations. The device is composed of independent sensing elements, positioned as indicated in figure 3.

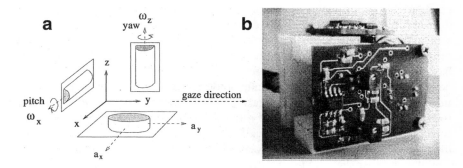

Fig. 3 The artificial vestibular system of the LIRA robot head. **a** Schematic of the spatial arrangement of two rotational and two translational sensing elements. Such configuration enables, in principle, to measure and therefore compensate disturbances along the pitch and yaw axes and fronto-lateral accelerations. **b** A side view of the corresponding prototype assembly combining two artificial otolith modules and two canal modules. Overall dimensions: 5x4x4 cm.

The electronics were customized to characterize precisely the dynamic response of the sensing elements. When integrated into the structure of the robot head, the device provides inertial signals in response to pitch and yaw rotations, and linear frontal and lateral accelerations in the horizontal plane. The device thus operates like an artificial "vestibular system" that senses rotational as well as translational robot movements to generate appropriate compensatory eye reflexes. In the case of the RVOR, ignoring for the moment the eccentricity of the eyes with respect to the axis of head rotation, perfect compensation would require simply that the output (eye rotation) match the input (head rotation), in which case the gain would be unity. However, for the TVOR to be optimally effective, its gain should depend on the proximity of the object of interest, nearby objects necessitating much greater compensatory eye movements than distant ones in order for their retinal images to be stabilized during translation. In fact, to stabilize an image off to one side (as in Figs. 2c and d), the gain of the TVOR should be inversely proportional to the viewing distance, and this has been shown to be the case for primates (Paige 1989; Schwarz et al. 1989; Paige and Tomko 1991b; Schwarz and Miles 1991; Busettini et al. 1994b; Bush and Miles 1996; Gianna et al. 1997; Telford et al. 1997). In addition, the compensatory eye movements generated by the TVOR depend on the direction of gaze with respect to the direction of heading, consistent with the idea that the system attempts to stabilize the central image regions in the plane of fixation (Paige and Tomko 1991a, b). Accordingly, when gaze is in the direction of heading, so that the object of interest is directly ahead and getting closer, the TVOR converges the two eyes to keep both foveas aligned on the object (pure vergence). If gaze is eccentric with respect to the direction of heading during the forward motion then the responses include conjugate (version) components to increase the eccentricity of gaze exactly in accordance with the local pattern of optic flow. For example, if the observer's gaze is directed downwards during the forward motion then his/her compensatory eye movements have a downward component, while compensatory eye movements have a rightward component if gaze is directed to the right of the direction of heading, and so forth. Thus, the oculomotor consequences of vestibular stimulation are here contingent upon the gaze position. Another complication is that the rotational axis of the eye and head often do not coincide; therefore even a pure rotational movement of the head usually causes both a rotation and a translation of the eyes, so that compensation for rotation of the head must also include a component that is dependent on viewing distance (Viirre et al. 1986). Of course, an artificial robot is subject to these same physical challenges and so requires information about viewing distance through a range-finding mechanism, as well as "gaze" eccentricity with respect to the direction of heading, together with implied knowledge of head kinematics and optical geometry.

4.2 The kinematics of inertial stabilization

A formal description of the eye-head kinematic parameters is given in the following, including their dependence on fixation distance, head rotation and translation. This formalism should not be taken to suggest that biological systems explicitly compute direct or inverse kinematics. In fact, we think that this is not the case. However the formalism makes explicit the relative roles of inertial and visual information in different situations and helps in highlighting the advantages of visuo-inertial integration and the asymmetries of eye-control commands.

4.2.1 Rotational movements

Figure 4 shows the schematic geometry of a binocular system for stabilization around the vertical axis and indicates the relevant parameters: the inter-ocular distance (or baseline) b, the distance a between the rotational axis of the head and the baseline, and the viewing distance d. The analytical relation among these parameters can be derived by considering the kinematics of this model, and imposing the constraint that the eye $\mathbf{E}$ maintains gaze at point $\mathbf{P}$ when the head rotates. Consider two vectors, $\vec{V}_g$ and $\vec{V}_b$, on the ZX plane which connect the eye position $\mathbf{E}$ respectively with gaze point $\mathbf{P}$ and mid-baseline point $\mathbf{B}$. Simple vectorial rules and differentiation with respect to time leads to the following expression of angular velocity $\dot{\omega}_e$ (Panerai and Sandini 1998):

$$\dot{\omega}_e = \left[\frac{d(d - Z_L)}{d^2 - 2dZ_L + \left(a^2 + b^2/4\right)} \right] \dot{\omega}_h \tag{1}$$

where the auxiliary expression $Z_L = (b/2 \sin \omega_h + a \cos \omega_h)$ represents the Z-coordinate of the left eye. Equation (1) determines for any given head velocity, $\dot{\omega}_h$, the relationship between eye velocity, $\dot{\omega}_e$, the geometrical parameters of the eye-head system b and a and the distance d of the fixation point **P**.

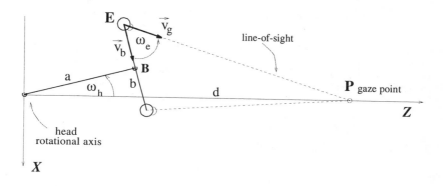

Fig. 4 Geometry of the eye-head system showing the parameters a, b, d relevant to inertial and visual measures.

In a robot vision system different choices of the a and b parameters (i.e. geometric configurations) determine different shapes of the eye-head velocity relationship. Equation (1) and figure 5 also show the inverse dependence upon distance, the considerable influence that distance has on the eye-to-head velocity ratio. The RVOR of primates does indeed exhibit such dependence on distance (Biguer and Prablanc 1981; Viirre et al. 1986; Hine and Thorn, 1987; Snyder and King 1992; Crane et al. 1997; Telford et al. 1998). This finding indicates that in order to synthesize an efficient ocular compensation in response to rotational movements, the distance parameter must play an important role in the close range domain. It is worth noting that the eye velocity $\dot{\omega}_e$ required to maintain fixation on near objects can be as much as twice the value of $\dot{\omega}_h$ and, for fixation distances in the range 25-200 cm, the optimally effective amount of ocular compensation needed to obtain gaze stabilization can change rapidly. One point worth stressing here is the fact that the range over which fixation distance has a strong effect on inertial stabilization, may not be very relevant for locomotion. On the other hand this range overlaps entirely with manipulation workspace and, in this respect, justifies appropriate control circuits.

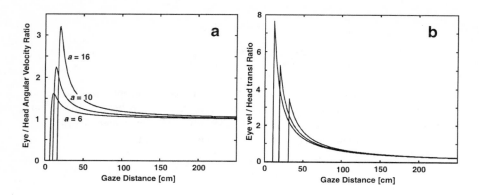

Fig. 5 Theoretical eye-head velocity ratios plotted for rotations and translations of the head. **a** Rotational movements of the head: note the influence of the geometrical parameter a and of distance d of the fixation point (see eye-head geometry in figure 4). **b** Translational movements of the head: again, the velocity ratio as a function of fixation distance d for different values of a. In both cases baseline b is fixed (6 cm) and eye-to-neck distance a increases (from 6 to 16 cm).

4.2.2 Translational movements

The same formalism is used to derive the analytical expression of the eye velocity, $\dot{\omega}_e$, required to maintain fixation on the object, **P**, positioned at distance, d, for translational movements or disturbances in the fronto-perpendicular direction of a robot head. In figure 6 we sketch the situation of a binocular system fixating an object at distance d translating with instantaneous velocity, T_x, along the x-axis. This gives:

$$\dot{\omega}_e = \left[\frac{(d-a)}{\left(\frac{b}{2}-x\right)^2+(d-a)^2} \right] T_x \qquad (2)$$

Figure 5 represents graphically the gain required for perfect compensation, $\dot{\omega}_e/T_x$, as a function of distance, d. It shows clearly that the ocular compensation required to fixate near objects (i.e., 20-150 cm) can be quite demanding, but decreases inversely with distance; therefore, an object at infinite distance does not require, in principle, any ocular compensation, irrespective of translation speed. Moreover, when the eye is fixating an object in the range 50-200 cm, the required gain (i.e. required eye velocity per unit linear translation) changes dramatically with distance: from a value of 0.5 at a distance of 150 cm, the gain raises to about 2.0 at 50 cm. In terms of ocular velocity, this means that changing fixation from a point at 150 cm distance to another at 50 cm, while translating at velocity T_x, requires a four-fold increase in the eye velocity if fixation is to be maintained. As indicated above, the gain of the primate TVOR varies inversely with viewing distance, although compensation is often less than complete. The modulation of the TVOR gain is subject to instantaneous changes in fixation distance and it can be modelled using a linear relationship with binocular vergence angle, though vergence is not the only parameter used for range-finding (Schwarz and Miles 1991; Shelhamer et al. 1995).

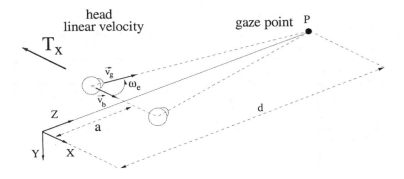

Fig. 6 Geometry of the binocular responses required to maintain the alignment of the two eyes during lateral translations of the head.

4.2.3 Eccentric gaze

When looking at an eccentric target in near space, the compensatory eye movements required to maintain binocular alignment during head rotations are different for the two eyes (Hine and Thorn 1987). From a kinematic point of view, the origin of this asymmetry is clear if one compares the analytical expressions for the angular velocities of the left and the right eye during rotation of the head while maintaining fixation. The two required eye angular velocities for the left and the right eye are:

$$\dot{\omega}_l = \left[\frac{d(d - Z_l)}{d^2 - 2dZ_l + \left(a^2 + \frac{b^2}{4}\right)} \right] \dot{\omega}_h \tag{3}$$

$$\dot{\omega}_r = \left[\frac{d(d - Z_r)}{d^2 - 2dZ_r + \left(a^2 + \frac{b^2}{4}\right)} \right] \dot{\omega}_h \tag{4}$$

where $Z_l = (b/2 \sin \omega_h + a \cos \omega_h)$ and $Z_r = (-b/2 \sin \omega_h + a \cos \omega_h)$ are, respectively, the Z-coordinates of the left and right eye. Note that the expressions are identical except for a couple of signs that reflect the difference in the eye positions with respect to the centre of the baseline. Figure 7 represents in polar coordinates the gain function, $\dot{\omega}_e / \dot{\omega}_h$, defined by equations (3) and (4). The gains are plotted with respect to head angular position, ω_h, and for a given distance of fixation, d.

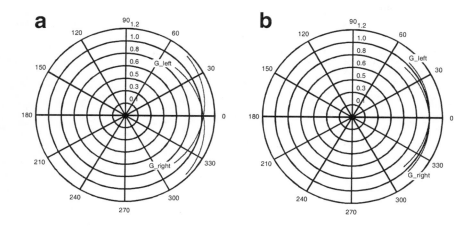

Fig. 7 Theoretical polar plots of the gains of the two eyes required to maintain binocular fixation of a target during head rotations: fixating a target at a distance of 30 cm (**a**), and at a distance of 70 cm (**b**). It is evident how the optimal gains vary with head position or equivalently with eccentricity of gaze.

In the case of an object at 30 cm distance (Fig. 7a), an angular deviation from the frontal direction of 30° introduces a relative gain difference between the two eyes of about 0.2; for example, with a head velocity of 200°/s, deviating 30° from the frontal direction gives a relative angular differential velocity of 40°/s. Thus, the angular velocities of the two eyes can be rather different in the near space. Although in humans there is clear evidence that during compensatory eye movements binocular alignment is not strictly maintained (Collewijn and Erkelens 1990), this constraint might be more important for a robot vision system, especially if the system uses binocularly-derived cues to control camera movements (Capurro et al. 1997).

4.2.4 Binocular gaze

The kinematics description given in the Sections 4.2.1 and 4.2.2 shows that a constant gain would be optimal only for constant fixation distances. With eyes fixed on objects at close range, rotational as well as translational head movements require continuous non-linear corrections of the amount of counter-rotational eye speed (see Fig. 5). The block diagram in figure 8 illustrates the basic idea of modulating the performance of the inertial ocular compensation, controlled by the block tagged "stabilization module", using distance and gaze direction information. This scheme represents one possible solution to account for the additional "contextual" information needed to perform visuo-inertial integration appropriately. It introduces concept of system state, which allows to adapt oculomotor control for gaze/camera stabilization in real-time to different contextual situations by using information available internally to the system (Panerai et al. 2000). Robust distance information is easy to compute in a binocular system provided the system is able to control vergence dynamically (Capurro et al. 1997). This type of control keeps both eyes/cameras pointing at the object of interest, and dynamically changes their orientation to maintain the object centred in the fovea, when moving in depth. The estimate of fixation distance can be updated continuously on the basis of instantaneous eye/camera orientations.

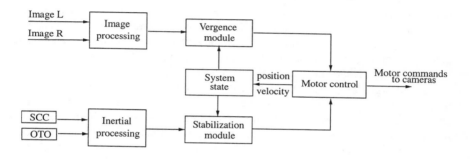

Fig. 8 Block diagram showing the integration of vergence and stabilization in the LIRA robot. By combining these two subsystems all the required signals (i.e. eye position, fixation distance) to adapt RVOR and TVOR appropriately are internally available. SCC: semicircular canals; OTO: otolith organs.

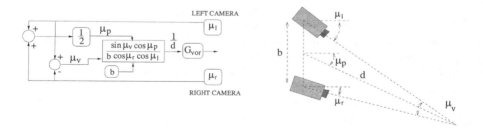

Fig. 9 Block diagram showing the modulation of gain in the LIRA robot. Note that angular eye position information appropriately combined leads to an estimate of distance of fixation point. The resulting distance-dependent G_{vor} gain controls the amount of counter rotational eye movement to optimally stabilize gaze.

In fact, with reference to figure 9, the distance d of the object being fixated can be derived from the vergence (μ_v) and version (μ_p) angles, using the following equation:

$$d = b\frac{\cos(\mu_r)\cos(\mu_l)}{\sin(\mu_v)\cos(\mu_p)} \tag{5}$$

where $\mu_p = 1/2\ (\mu_l + \mu_r)$ and $\mu_v = (\mu_l - \mu_r)$.

5. Visuo-inertial stabilization

None of the inertial stabilization mechanisms – biological or robotic – is perfect, hence bodily motions must often be associated with some residual image motion which brings visual stabilization mechanisms into operation. The primate vestibular system's decomposition of head movements into rotational and translational components results directly from the physical properties of the end organs in the labyrinth. However, there is no such decomposition of the optic flow by the visual end organ or camera: these see *all* visual disturbances and any decomposition must be achieved by image processing. Traditionally, visual stabilization in animals has been considered only in relation to the RVOR and compensation for rotational disturbances of the observer. Indeed, the usual visual stimulus of choice involves placing the subject inside a cylindrical enclosure that has patterned walls which can be rotated around the subject to simulate the visual consequences of shortcomings in the horizontal RVOR. The rotating cylinder elicits a pattern of tracking eye movements, termed *optokinetic nystagmus (OKN)*, which has two independent components distinguished by their dynamics: OKNe with brisk dynamics and OKNd with sluggish dynamics (Cohen et al. 1977). It has recently been suggested that these two components of the primate optokinetic response are in fact generated by mechanisms that evolved independently to deal with rotational (OKNd) and translational (OKNe) disturbances of the observer and provide the drives for visual backups to the RVOR and the TVOR, respectively (Schwarz et al. 1989; Busettini et al. 1991; Miles et al. 1991, 1992; Miles and Busettini 1992; Miles 1993, 1995, 1997, 1998). The block diagrams in figure 10 illustrate the two hypothetical visuo-vestibular mechanisms that deal independently with rotational and translational disturbances. Note that each visual mechanism shares a gain element with its vestibular counterpart. In the case of the rotational mechanism, the gain element mediates adaptive gain control in the RVOR, and in the case of the translational mechanism it mediates the TVOR's dependence on viewing distance.

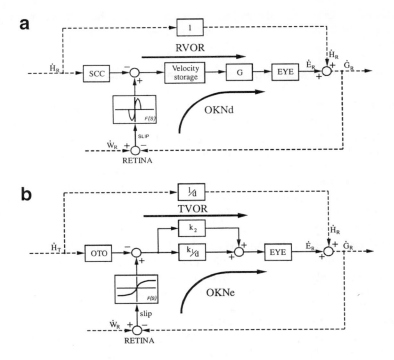

Fig. 10 Block diagrams showing the proposed linkages between the visual and vestibular reflexes stabilizing gaze. **a** The open-loop RVOR and the closed-loop OKNd generate eye movements, $\dot{E}_R$, that compensate for rotational disturbances of the head, $\dot{\omega}_e$. These reflexes share (a) a velocity storage element, which is responsible for the slow build-up in OKN and the gradual decay in RVOR with sustained rotational stimuli, and (b) a variable gain element, G, which mediates long-term regulation of RVOR gain. SCC: semicircular canals. The characteristics F(s) indicates that the visual input is sensitive to low slip speeds only (Miles et al. 1992). **b** The open-loop TVOR and the closed-loop OKNe generate eye movements that compensate for translational disturbances of the head, $\dot{H}_T$, which affect gaze in inverse proportion to the viewing distance, d. These reflexes share (a) a variable gain element, k_1/d, which gives them their dependence on proximity, and (b) a fixed gain element, k_2, which generates a small response irrespective of proximity. OTO: otolith organs (Schwarz et al. 1989). Dashed lines represent physical links: $\dot{H}_T$, head velocity in linear coordinates; $\dot{H}_R$, $\dot{E}_R$, $\dot{G}_R$, $\dot{W}_R$, velocity of head, eyes (in head), gaze and visual surroundings, respectively, in angular coordinates.

The proposed sharing of pathways and gain elements by the visual and vestibular mechanisms comes from the following observations: (1) changes in the gain of the RVOR resulting from exposure to telescopic spectacles are associated with parallel changes in the gain of OKNd but not in the gain of OKNe (Lisberger et al. 1981); (2) changes in the gain of the TVOR resulting from changes in the viewing distance are associated with parallel changes in the gain of OKNe, often also termed *ocular following* (Schwarz et al. 1989; Busettini et al. 1994b). The translational visual mechanisms represented by ocular following (OKNe) operate with machine-like consistency and have latencies of less than 60 ms in monkeys (Miles et al. 1986) and <85 ms in humans (Gellman et al. 1990).

5.1 Compensations calling for conjugate (version) responses

All animals with mobile eyes seem able to generate conjugate eye movements to help compensate for rotational disturbances of the head using canal-ocular and optokinetic reflex mechanisms, indicating that these systems emerged very early in evolution. Recent experiments indicate that primates also have evolved mechanisms for dealing with the visual problems posed by translation when the observer looks off to one side, as in figures 2c and d. The visual task confronting the visual stabilization mechanisms here is to single out the motion of particular elements in the scene – such as the mountain in figure 2c and the tree in figure 2d – and ignore all of the competing motion elsewhere. One way to achieve this would be to use attentional focussing mechanisms to spotlight the target of interest. Such mechanisms exist and are used by the so-called *pursuit system*. They have the limitation that they require high-level executive decisions to select the image to be tracked which requires time (Keller and Khan 1986; Kimmig et al. 1992). The ocular following system solves this problem more expeditiously using low-level stereo mechanisms that perform rapid parallel processing of binocular images, effectively sorting them on the basis of the depth plane that they occupy. This stereo algorithm utilizes the fact that we have two eyes with slightly differing viewpoints, as illustrated in figure 11, which is a "binocular" elaboration of the cartoons in figures 2c and d. The object on which the two eyes are focussed (the

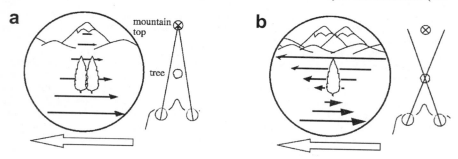

Fig. 11 The optic flow experienced by a translating observer with binocular view. **a** As in figure 2c, except that in the binocular view the mountain in the plane of fixation is seen as single and the nearer tree is seen double (disparate). A plan view of the observer and the two objects is shown to the right. **b** As in figure 2d, except that, with binocular viewing, the tree is placed in the plane of fixation and so is seen as single whereas the distant mountain is now seen as double (disparate). Again, the plan view is shown to the right. Note that the dimensions of the eyes and their separations have been exaggerated to illustrate the disparity more clearly. In fact, disparity is much more evident with near viewing, which is also associated with the most vigorous optic flow and requires the most vigorous tracking from the observer to compensate. All of the laboratory experiments used near viewing (Busettini et al. 1996a).

mountain in figure 11a or the tree in figure 11b) resides in the plane of fixation and is imaged at corresponding positions on the two retinae; the object is therefore perceived as a single, fused image.

In contrast, objects that are nearer or farther than the plane of fixation have images that occupy non-corresponding positions on the two retinae – referred to as "binocular disparity" – and are seen as double (the tree in figure 11a and the mountain in figure 11b). Clearly, a highly reliable algorithm for stabilizing gaze on objects of particular interest would track only those objects whose images occupy corresponding positions on the two retinae: objects in the plane of fixation. Recent experiments indicate that the fastest conjugate tracking responses show a preference for binocular images that lack disparity (Busettini et al. 1996b). Thus, the ocular following system generates conjugate eye movements that help to stabilize gaze on objects of interest not by selecting a particular one, but by stabilizing the image of any object that happens to lie close to the plane of fixation, an implicit assumption, therefore, being that this plane contains the objects likely to be of most interest. Note that the time-consuming process of selecting the object of interest therefore rests with the oculomotor subsystems that bring images into the plane of fixation – that is, the saccadic system working in concert with the vergence system. These latter systems redirect gaze to objects using higher-level criteria whereas ocular following relies on low-level rapid parallel filters. Thus, the general concept is of low-level reflex systems stabilizing whatever image components the high-level systems happen to bring into the plane of fixation. In robot vision, the idea of using disparity information to separate the fixated object from the background was first developed by Coombs and Brown (1993). In their implementation the computation of binocular disparity, followed by a zero-disparity filter was used to isolate points belonging to the *horopter*[1] in order to simplify a tracking procedure. The requirement for real-time performance constrained the authors to adopt a windowing strategy to reduce the implicit computational load of the image processing techniques adopted.

A different approach was followed by Capurro et al. (1997) who proposed the use of space variant sensing, as an alternative imaging geometry for robot vision systems. Interestingly enough, the choice of this geometry reduces the amount of visual information to be processed without constraining the visual field size nor the resolution, and allows for more simplified image processing techniques. Space-variant sensors intrinsically weight the central fovea much more than the peripheral areas. As a result the choice of the sub-image size for the computation of disparity becomes less critical to the extent that the overall image can be used, as will be shown later (figure 12 presents a direct comparison between different sampling schemes). In other words the space-variant approach implic-

---

[1]In the human system, any point in space lying on a circle which also contains the fixation point and the two retinae (*horopter*) will have a constant angular disparity (Griswold et al. 1992). Here we are referring to the horopter as the zero disparity locus.

itely enhances objects that happen to lie close to the fixation point and through this provides a pre-categorical, fast selection mechanism which requires no additional computation. Among space-variant sensing approaches the log-polar one is receiving increasing attention (Weiman and Juday 1990; Sandini et al. 1993; Tunley and Young 1994).

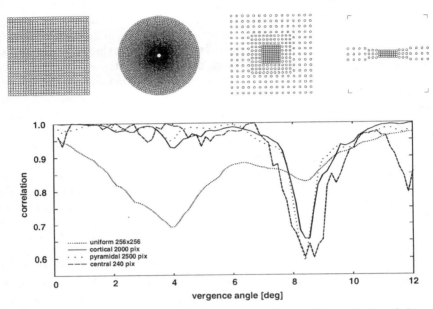

Fig. 12 Image tessellations and computational advantages. Top: spatial organization of elementary image elements (i.e. pixels) in four different tessellations, from left to right, uniform (standard), log-polar (cortical), log-Cartesian (pyramidal), horizontal log-Cartesian (central) produce more or less efficient numerical measures for the control of vergence. Bottom: space variant tessellations characterize uniquely correct vergence configurations. A cross-correlation technique (normalized correlation, NC) applied to pairs of stereo images of identical tessellation produces the curves shown in the plot (1-NC is actually plotted). On the x-axis the vergence angle [deg], on the y-axis the normalized cross-correlation index. For space variant layouts the correlation index produces robust minima when the robot correctly converges its cameras (see plot between 8° and 9°). On the other hand, in the uniform tessellation (standard) incorrect matches may occur in the periphery producing false minima as the one shown on the left side, at about 4°.

The initial analytical formulation of the log-polar transformation based on biological data is due mainly to Schwartz (1977). His model described the projection from the retina to the cortex via the lateral geniculate nucleus (Daniel and Whitteridge 1961; Cowey 1964; Allman and Kaas 1971; Hubel and Wiesel 1977) as a mapping between a polar plane (ρ, θ) (retinal plane) and a Cartesian plane (η, ξ) (log-polar or cortical plane). From the robotics and image processing perspective the peculiarities of log-polar mapping have been studied for many years

(Weiman and Chaikin 1979; Sandini and Tagliasco 1980). At present, the LIRA-Lab binocular vision system performs the log-polar transformation at frame rate using re-mapping software routines. Hardware re-mappers (Fisher and Juday 1988; Rojer and Schwartz 1990) and prototypes of space variant CCD's have recently been designed and a compact camera using C-MOS technology is already available (Ferrari et al. 1995; Sandini et al. 1998). Once the foveated images have been separated from the background, their relative motion with respect to the observer can be retrieved by analysing the dynamics of displacements in the fused optic array. Normal flow computation (Horn 1986) is sufficient to derive a linear *affine* model of the motion of the light patterns in the fused image sequence. The affine formulation (which is based on the assumption that the flow pattern is generated by a planar surface) is described on the basis of four quantities: image *translation*, image *rotation*, *divergence* and *shear* (Koenderink and van Doorn 1991). The components of the affine motion relevant for image stabilization are *translation* and *divergence*. Translation components, in particular, have been used to implement an "ocular following" servo that proved successful in tracking moving objects located in the plane of fixation (Capurro et al. 1996).

5.2 Compensations calling for disconjugate (vergence) responses

5.2.1 Feedforward mechanisms using radial optic flow

In the previous section we considered the visual challenge an observer is confronted with during rotation, or during translation while looking off to one side. The observer who undergoes translation but looks in the direction of heading experiences the radial pattern of optic flow featured in figures 2a and b and she must converge her eyes if an object of interest in the scene ahead is to stay imaged on both foveae. Of course, the amount of convergence required to maintain binocular alignment is inversely related to the viewing distance, hence the greatest challenge comes with viewing close objects. Centrifugal flow, which signals a forward approach and hence a decrease in the viewing distance, has recently been shown to elicit increased convergence, while centripetal flow, which signals the converse, elicits decreased convergence (Busettini et al. 1997). This and other experiments indicate that the human brain is able to extract the radial pattern of flow (cf. Part III of this volume) and to infer from this that there has been a change in viewing distance. However, such radial-flow-induced vergence is purely transient and operates solely as a velocity feedforward mechanism. In robot vision, image rate of expansion/contraction (also called divergence) has been proposed as a measure of depth change and is being used as a feedforward mechanism to drive the initial part of vergence movements (Capurro et al. 1997). Measuring image divergence seems most suited for guiding camera vergence, for the following reasons:

1. Divergent flow discriminates between fixated objects that are approaching (expansion) or moving away (contraction).
2. It also provides a measure of "how rapidly" depth is changing, a useful parameter for the vergence control loop that can improve the dynamics.
3. This information, at least in principle, is monocular and, as such, does not require a binocular matching procedure. Integrating signals from both eyes can be used to increase robustness.
4. Image divergence can be easily computed, particularly in log-polar space and is ideal for fast feedback-loop control.

On the other hand, a binocular visual measurement is necessary to distinguish correct vergence configurations from incorrect ones. To this end, a cross-correlation technique between pairs of stereo images can combined with the dynamic measurement described above. This correlation index, called binocular fusion, and the image divergence measurement are shown in figure 13 for an object passing through the plane of fixation.

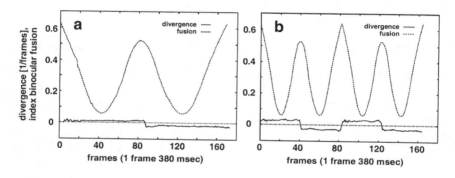

Fig. 13 Binocular fusion and image divergence measures computed by the robot for two target speeds (**a**) 1.6 cm/s (or 0.61 cm/frame) and (**b**) 3.2 cm/s (or 1.22 cm/frame) The two indices characterize uniquely the movement of an object passing through the robot's fixation plane. In this experiment, the robot maintains the vergence angle constant while fixating a point in three-dimensional space at 110 cm distance from the baseline. The object moves back and forth crossing the fixation plane. The target motion is symmetric. The x-axis indicates time in frames (frame period: 380 ms). The two w-shaped curves showing the fusion index reach minima when the object is at the correct fixation depth. The divergence index generates squared profiles that change sign when the object inverts the direction of movement: a positive value indicates the object is moving toward the robot and vice versa.

5.2.2 Feedback mechanisms using binocular disparity

When a moving observer looks in the direction of heading, radial optic flow is only one of several cues which indicate the direction and speed of heading. Another cue, which is very potent at generating vergence at ultra-short latencies is binocular disparity. If the observer were to move forward without converging her eyes adequately then the object of regard would be overtaken by the plane of fixation and would be imaged at non-corresponding positions on the two retinas with so-called "crossed disparity". Recent experiments have demonstrated that when random-dot patterns are viewed dichoptically and small binocular mis-alignments are suddenly imposed, corrective vergence eye movements are elicited at latencies of less than 60 ms in monkeys (Busettini et al. 1996a), and less than 85 ms in humans, (Busettini et al. 1994a), values closely comparable with those for ocular following and radial flow vergence. Crossed disparity steps elicited increased convergence and uncrossed steps decreased convergence, exactly as expected of a depth-tracking servo mechanism driven by disparity. However, the range of disparities over which the system behaves like a servo mechanism, that is, the disparity range over which increases in the disparity vergence error result in roughly linear increases in the vergence response, is less than $2°$. Thus, this vergence mechanism can correct only small misalignments of the two eyes, com-mensurate with a mechanism that performs only local stereo correlations and merely attempts to bring the nearest salient images into the plane of fixation. During forward locomotion this mechanism will help to prevent images from leaving the plane of fixation. Note that this disparity vergence mechanism is in a somewhat different category from ocular following and radial flow vergence insofar as its *primary* function is to eliminate small vergence errors. Evidence comes from the observation that it also operates in the vertical plane using vertical disparity, which is unrelated to depth and translation *per se* (Busettini, Masson and Miles, unpublished observations). While the specific involvement with vergence errors resulting from locomotion is clear, this is only a *secondary* function. Once more, we have a mechanism that functions as a low-level automatic servo and is not involved in high-level operations like the transfer of fixation to new images in new depth planes, which requires time-consuming target selections, and (often) the decoding of large disparity errors (above $10°$) that require solving the cor-respondence problem. In the same context, the idea of using space-variant image geometry in a binocular robot opened the view on the new concepts of a global *index of fusion* and a binocular *fusion filter* to segment targets from the back-ground. Figure-ground discrimination at this stage is in fact an extremely effective pre-processing step that enables further motion analysis to be performed more selectively and efficiently. To show the advantage of a simple stereo correlation technique in the space variant domain, the *fusion index* has been measured for different image tessellations during a smooth vergence movement of the cameras with a person standing in front of the vision system. In figure 12 we plot the fusion index as a function of the vergence angle for the four different sensor array geometries, demonstrating that space variant mappings are unaffected by false minima resulting from incorrect matching in the periphery of the visual field. In addition, as the cameras approach the correct vergence configuration the degree of correlation between the stereo pair becomes higher, as indicated by the lower index value in space variant mappings. With respect to the biological findings mentioned previously, the computation of the index of fusion effectively provides a measure of disparity by applying a local correlation filter to the entire image.

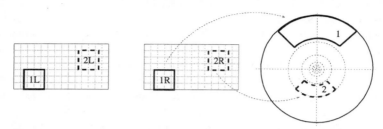

Fig. 14 A correlation index is computed between corresponding patches in the cortical plane (1L and 1R, 2L and 2R and so on, left panels). This corresponds to applying space variant masks in retinal coordinates (right panel).

Similarly, a "fusion filter" can be implemented by applying a local mask of constant size in the log-polar plane, (see Fig. 14) to isolate the fused portions of the stereo pair. This assumes that image regions with a high degree of correlation in their spatially-variant neighbourhood have very low disparity. Partial results of these processing steps and a comparison of the same algorithms with a Cartesian stereo pair are shown in figure 15. It is worth noting that in contrast to other methods proposed for figure-ground segmentation and tracking (Nordlund and Uhlin 1995), this approach is binocular. A fast, pre-attentive binocular mechanism based on disparity computation in the space variant domain was developed by Bernardino and Santos-Victor (1996). Following the idea of the global *fusion index* extracted from the whole images, they derived the concept of a global *disparity index*. They showed that disparity estimation can be considered in the log-polar domain as a minimum search over a set of different candidate disparities.

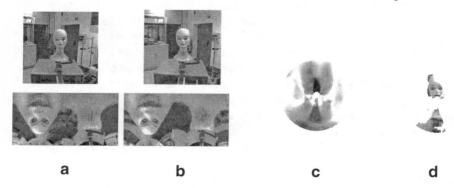

a   **b**   **c**   **d**

Fig. 15 Log-polar binocular fusion filtering. **a** and **b** are the original log-polar stereo pair (the face is located in the horopter); **c** is the result of the computation of the *fusion* map (fused areas are represented by dark values on a grey scale) and **d** the fused image. Images are all remapped for convenience in retinal coordinates.

6. Low-level, pre-attentive, parallel processing?

In terms of latencies, the primate vestibulo-ocular responses are ultra-rapid, the latency for the RVOR being less than 10 ms (Tabak et al. 1997), and the latency for the TVOR being less than 20 ms (Bush and Miles 1996). The shortest latency of purely visual responses are longer than this, less than 50 ms in monkeys (Miles et al. 1986) and 80 ms in humans (Gellman et al. 1990), but it seems likely that the eyes start moving well before the subject is even aware that there has been a disturbance. This is in accord with the general finding that vestibular reflexes have high-pass dynamics which are nicely complemented by visual reflexes with low-pass dynamics (Paige 1983; Telford et al. 1997). Moreover, there is accumulating evidence that the visual reflexes are mediated by the dorsal stream of cortex (Ungerleider and Mishkin 1982) where motion is processed. In regard to ocular following, chemical lesions in MST result in impairments of even the earliest responses (Kawano et al. 1997) and single unit recordings in this region indicate the presence of many directionally selective neurones that discharge in close relation to the large-field, high-speed motion stimuli that are optimal for eliciting ocular following (Kawano et al. 1994). Many motion-selective neurones in MT (Maunsell and van Essen 1983a), and MST (Roy and Wurtz 1990; Roy et al. 1992) also appear to be selective for binocular disparity (cf. Part II of this volume), but stimuli optimal for ocular following were not tried in these studies. There is extensive evidence that area MST contains neurones that are selectively sensitive to radial optic flow patterns such as those which evoke vergence eye movements at ultra-short latencies (Saito et al. 1986; Tanaka and Saito 1989; Tanaka et al. 1989; Duffy and Wurtz 1991a; Duffy and Wurtz 1991b; Lagae et al. 1994; Duffy and Wurtz 1995; Lappe et al. 1996; Pekel et al. 1996). In fact, MST is the *first* stage in this dorsal pathway at which *global* flow is encoded at the level of single cells: at earlier stages, such as MT, individual cells have much smaller receptive fields and encode only local motion (van Essen et al. 1981; Maunsell and van Essen 1983b; Albright and Desimone 1987; Komatsu and Wurtz 1988; Albright 1989; Lagae et al. 1994). In regard to the neural mediation of disparity vergence we are presumably concerned with disparity selective neurones that have no particular motion preference, except perhaps for motion in depth, and that discharge to *non-zero* disparities, thereby effectively encoding vergence error. Many such neurones have been described in visual cortex as early as V1 (see Poggio (1995) for recent review), and in the dorsal stream, including MT (Maunsell and van Essen 1983a) and MST (Takemura et al. 1997). Up to MT, these neurones encode a limited and sometimes very restricted range of depth. However, in MST some of the neurones that discharge in close relation to the large-field binocular stimuli used to elicit short-latency disparity vergence have much more global disparity tuning curves that exactly match the disparity tuning curves of the vergence motor responses. Once more, it seems that we have a short-latency oculomotor response that relies on signals that first occur in their entirety at the level of single cells in MST. The above discussion indicates that there are neurones or networks that act like templates to detect specific patterns of optic flow and disparity, rapidly generating appropriate oculomotor responses to serve the needs of visual stabilization. Latencies are so short that the system must depend on pre-attentive, parallel processing to get started. Recent experiments (Masson et al. 1997) have shown that the short-latency vergence responses can also be elicited by disparity stimuli applied to dense (50%) anti-correlated binocular patterns, in which each black dot in one eye is matched to a white dot in the other eye. The important point here is that such patterns do not give rise to percepts of depth: In two-alternative-forced-choice tests, subjects could readily discriminate between crossed and uncrossed disparities when applied to the usual correlated patterns but not when applied to these anti-correlated patterns (Masson et al. 1997). This is consistent with the idea that these short-latency vergence responses derive their visual input from an early stage of cortical processing prior to the level at which depth percepts are elaborated. We suspect that the same is true of all of these rapid visual reflexes.

Artificial systems using inertial and visual stabilization mechanisms exploit the same difference in the dynamics of the inertial and visual contributions to image stabilization. Consider the simple block diagram of a monocular system sketched in figure 16, which shows the basic structure of the visuo-inertial stabilization mechanism implemented in the LIRA robot's head.

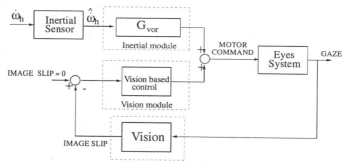

Fig. 16 Block diagram of the LIRA robot control system: the inertial sensory information is processed in open loop acting in a parallel, synergistic way with to the closed loop visual feedback subsystem.

Purely visual stabilization performance is restricted because computational resources in a robot system are limited and processing of visual information is very demanding, dealing with large amounts of two-dimensional data and computationally expensive algorithms. Moreover, all algorithms adopted for optical flow estimation (i.e. to estimate image slip) have a limited range of image velocities that can be reliably measured (i.e. saturation effect). Inertial data range in quantity some orders of magnitude below the visual data and, hence, can be processed much more expeditiously. For example, with 3 rotational and 3 linear accelerometers sampled at 1000 Hz, the amount of data that needs to be processed per second is almost negligible with respect to the amount of visual data that need to be processed for reliable optical flow estimation. The LIRA robot, for example, is able to maintain a stable "gaze" at 1 m distance in response to transient rotations of the "head" at 10°/s, necessitating that the eye/camera rotate at 12°/s for full compensation; the inertial mechanism ($G_{vor} = 1$) accounted for 10°/s of this compensation, leaving the visual mechanism to deal with a mere 2°/s. The use of inertial information enables more efficient use of the visual computational resources. Another direct advantage is that sensory integration also extends the dynamic range of motions or external disturbances the system can effectively deal with, and without increasing system complexity (cf. Hengstenberg 1993). As a direct consequence, gaze stabilization based on visuo-inertial sensory information, is more responsive: the bandwidth of sensory sources is increased, but to some extent, the computational requirements are optimized, and as a whole, reduced. Furthermore, the inertial data do not depend on visual processing (and vice versa) and their integration does indeed add a completely new and independent data source, which in turn increases system stability. Some of the examples presented in the previous sections are described in greater details in Capurro et al. (1997), Panerai and Sandini (1998) and Panerai et al. (2000). They show that the integration of low-level inertial and visual information allows the system to operate more efficiently, not only because of the short delays of the control loops but also because the kind of visual information that is strictly necessary and the most appropriate form of representation is made explicit in such an approach.

Closing remarks

Natural and artificial systems that share the same environment may adopt similar solutions to cope with similar problems. Neurobiologists are interested in finding *the* solutions adopted by natural systems and roboticists are interested in which of the technologically feasible solutions are optimal for implementing such systems in hardware. Occasionally, the biologist probing the natural system and the engineer seeking to build an artificial one that fulfills similar functions encounter similar problems and their thinking converges on similar solutions. This was the case with the idea that radial optic flow produces convergence eye movements. As it happens, this particular discovery was made independently by the biologists and engineers at about the same time, but there must be times when one discipline is going to have the edge on the other: at such times a dialogue between biologists and engineers would be especially fruitful. One example of this is the engineer's recent use of binocular disparity – a well-established concept in neurobiology – to allow a robot to segregate moving images from the textured background and thence to track them. Examples going the other way are legion as engineers have provided the major conceptual tools for characterizing biological control systems, so that biologists are no longer intimidated by terms like "velocity feedforward" and "high-pass dynamics". The challenge is to find a common ground on which reciprocal interactions are the immediate mutual advantage, and the stabilization of gaze seems to be a good example of a model system that rightfully engages the interest of biologists and engineers alike. The biological approaches to the problem of gaze stabilization have reached a stage of quantification that allows formal modelling with sufficient rigor to engage the interest of the engineer. An example of this is the recent discovery of gaze-stabilization mechanisms in primates that deal with the problems created by translational disturbances of the observer: no doubt robots too would benefit from inertial sensors that encode the linear as well as the angular accelerations of the head just as the human oculomotor system does. On the other hand, in order to construct robots that work in real time the engineer must tackle motor coordination problems that also challenge the human sensorimotor control systems and thus engineers might be able to offer insights into how the latter operate. An example of this is the use of the log-polar representation in the LIRA robot head, a computationally efficient way of encoding visual inputs with advantages for extracting correlations between binocular images without the need to derive disparity explicitly. The human visual system, which has a similar fovea-based anisotropy, might rely on such a representation, for example to perform rapid, pre-attentive extraction of vergence error. In the present chapter, the biological and the artificial aspects are not blended as well as we would have liked. This is partly due to some differences in the aspects investigated in natural systems and those synthesized in artificial systems, but the limits to the topical coherence at present is also due to communication problems: too often biologists and engineers become aware of their common interests only during chance encounters. The ideal solution, although currently not supported by many funding agencies, would be to establish and support multidisciplinary research teams where engineers could fully understand the difficulties of biological experiments and neuroscientists could appreciate the inspirational difference between theoretical models and real-time control of physical systems.

References

Albright TD (1989) Centrifugal directional bias in the middle temporal visual area (MT) of the macaque. J Vis Neurosci 2: 177-188

Albright TD, Desimone R (1987) Local precision of visuotopic organization in the middle temporal area (MT) of the macaque. Exp Brain Res 65: 582-592

Allman JM, Kaas JH (1971) Representation of the visual field in striate and adjoining cortex of the owl monkey (*Aotus trivirgatus*). Brain Res 35: 89-106

Bernardino A, Santos-Victor J (1996) Vergence control for robotic heads using log-polar images. Proc of IROS 96, Osaka, Japan, pp 1264-1271

Biguer B, Prablanc C (1981) Modulation of the vestibulo-ocular reflex in eye-head orientation as a function of target distance in man. In: Fuchs AF, Becker W (eds) Progress in oculomotor research, Elsevier, Amsterdam, pp 525-530

Busettini C, Masson GS, Miles FA (1996b) A role for stereoscopic depth cues in the rapid visual stabilization of the eyes. Nature 380: 342-345

Busettini C, Masson GS, Miles FA (1997) Radial optic flow induces vergence eye movements at ultra-short latencies. Nature 390: 512-515

Busettini C, Miles FA, Krauzlis RJ (1994a) Short-latency disparity vergence responses in humans. Soc Neurosci Abstr 20: 1403

Busettini C, Miles FA, Krauzlis RJ (1996a) Short-latency disparity vergence responses and their dependence on a prior saccadic eye movement. J Neurophysiol 75: 1392-1410

Busettini C, Miles FA, Schwarz U (1991) Ocular responses to translation and their dependence on viewing distance. II. Motion of the scene. J Neurophysiol 66: 865-878

Busettini C, Miles FA, Schwarz U, Carl JR (1994b) Human ocular responses to translation of the observer and of the scene: dependence on viewing distance. Exp Brain Res 100: 484-494

Bush GA, Miles FA (1996) Short-latency compensatory eye movements associated with a brief period of free fall. Exp Brain Res 108: 337-340

Capurro C, Panerai F, Sandini G (1996) Vergence and tracking fusing log-polar images. Proc Intern Conf Pattern Recog, Vienna, pp 740-744

Capurro C, Panerai F, Sandini G (1997) Dynamic Vergence using log-polar images. Int J Computer Vision 24: 79-94

Cohen B, Matsuo V, Raphan T (1977) Quantitative analysis of the velocity characteristics of optokinetic nystagmus and optokinetic after-nystagmus. J Physiol 270: 321-344

Collewijn H, Erkelens CJ (1990) Binocular eye movements and the perception of depth. In: Kowler E (ed) Eye movements and their role in visual and cognitive process: review of oculo-motor research. Elsevier, Amsterdam, pp 213-261

Coombs D, Brown C (1993) Real-time binocular smooth pursuit. Int J Computer Vis 11: 147-164

Cowey A (1964) Projection of the retina on to striate and prestriate cortex in the squirrel monkey (*Saimiri sciureus*). J Neurophysiol: 266-293

Crane BT, Viirre ES, Demer JL (1997) The human horizontal vestibulo-ocular reflex during combined linear and angular acceleration. Exp Brain Res: 304-320

Daniel M, Whitteridge D (1961) The representation of the visual field on the cerebral cortex in monkeys. J Physiol 159: 203-221

Duffy CJ, Wurtz RH (1991a) Sensitivity of MST neurons to optic flow stimuli. I. A continuum of response selectivity to large-field stimuli. J Neurophysiol 65: 1329-1345

Duffy CJ, Wurtz RH (1991b) Sensitivity of MST neurons to optic flow stimuli. II. Mechanisms of response selectivity revealed by small-field stimuli. J Neurophysiol 65: 1346-1359

Duffy CJ, Wurtz RH (1995) Response of monkey MST neurons to optic flow stimuli with shifted centers of motion. J Neurosci 15: 5192-5208

van Essen DC, Maunsell JHR, Bixby JL (1981) The middle temporal visual area in the macaque: myeloarchitecture, connections, functional properties and topographic organization. J Comp Neurol 199: 293-326

Ferrari F, Nielsen J, Questa P, Sandini G (1995) Space variant imaging. Sensor Rev 15: 17-20

Fisher TE, Juday RD (1988) A programmable video image remapper. Proc SPIE Conf Pattern Recog Signal Processing, Orlando, pp 122-128

Gellman RS, Carl JR, Miles FA (1990) Short latency ocular-following responses in man. Vis Neurosci 5: 107-122

Gianna CC, Gresty MA, Bronstein AM (1997) Eye movements induced by lateral acceleration steps. Effect of visual context and acceleration levels. Exp Brain Res 114: 124-129

Gibson JJ (1950) The perception of the visual world. Houghton Mifflin, Boston

Gibson JJ (1966) The senses considered as perceptual systems. Houghton Mifflin, Boston

Goldberg JM, Fernandez C (1975) Responses of peripheral vestibular neurons to angular and linear acceleration in the squirrel monkey. Acta Otolaryngol 80: 101-110

Griswold NC, Lee JS, Weiman CFR (1992) Binocular fusion revisited utilizing a log-polar tessellation. Comp Vis Image Proc 92: 421-457

Hengstenberg R (1993) Multisensory control in insect oculomotor systems. In: Miles FA, Wallman J (eds) Visual motion and its role in the stabilization of gaze. Elsevier, Amsterdam, pp 285-298

Hine T, Thorn F (1987) Compensatory eye movements during active head rotation for near targets: effects of imagination, rapid head oscillation and vergence. Vision Res: 1639-1657

Horn BKP (1986) Robot vision. MIT Press, Cambridge, USA

Hubel DH, Wiesel TN (1977) Functional architecture of macaque monkey cortex. Proc Roy Soc Lon: 1-59

Kawano K, Inoue Y, Takemura A, Kitama T, Miles FA (1997) A cortically mediated visual stabilization mechanism with ultra-short latency in primates. In: Thier P, Karnath H (eds) Parietal lobe contributions to orientation in 3D space. Springer Verlag, Heidelberg, pp 185-199

Kawano K, Shidara M, Watanabe Y, Yamane S (1994) Neural activity in cortical area MST of alert monkey during ocular following responses. J Neurophysiol 71: 2305-2324

Keller EL, Khan NS (1986) Smooth-pursuit initiation in the presence of a textured background in monkey. Vision Res 26: 943-955

Kimmig HG, Miles FA, Schwarz U (1992) Effects of stationary textured backgrounds on the initiation of pursuit eye movements in monkeys. J Neurophysiol 68: 2147-2164

Koenderink J, van Doorn J (1991) Affine structure from motion. J Opt Soc Am 8: 377-385

Komatsu H, Wurtz RH (1988) Relation of cortical areas MT and MST to pursuit eye movements. I. Localization and visual properties of neurons. J Neurophysiol 60: 580-603

Lagae L, Maes H, Raiguel S, Xiao DK, Orban GA (1994) Responses of macaque STS neurons to optic flow components: a comparison of areas MT and MST. J Neurophysiol 71: 1597-1626

Lappe M, Bremmer F, Pekel M, Thiele A, Hoffmann KP (1996) Optic flow processing in monkey STS: a theoretical and experimental approach. J Neurosci 16: 6265-6285

Lisberger SG, Miles FA, Optican LM, Eighmy BB (1981) Optokinetic response in monkey: underlying mechanisms and their sensitivity to long-term adaptive changes in vestibuloocular reflex. J Neurophysiol 45: 869-890

Masson GS, Busettini C, Miles FA (1997) Vergence eye movements in response to binocular disparity without the perception of depth. Nature 389: 283--286

Maunsell JHR, van Essen DC (1983a) Functional properties of neurons in middle temporal visual area of the macaque monkey. I. Selectivity for stimulus direction, speed, and orientation. J Neurophysiol 49: 1127-1147

Maunsell JHR, van Essen DC (1983b) Functional properties of neurons in middle temporal visual area of the macaque monkey. II. Binocular interactions and sensitivity to binocular disparity. J Neurophysiol 49: 1148-1167

Miles FA (1993) The sensing of rotational and translational optic flow by the primate optokinetic system. In: Miles FA, Wallman J (eds) Visual motion and its role in the stabilization of gaze. Elsevier, Amsterdam, pp 393-403

Miles FA (1995) The sensing of optic flow by the primate optokinetic system. In: Findlay JM, Kentridge RW, Walker R (eds) Eye movement research: mechanism, processes and applications. Elsevier, Amsterdam, pp 47-62

Miles FA (1997) Visual stabilization of the eyes in primates. Curr Opinion Neurobiol 7: 867-871

Miles FA (1998) The neural processing of 3-D visual information: evidence from eye movements. Eur J Neurosci 10: 811-822

Miles FA, Busettini C (1992) Ocular compensation for self motion: visual mechanisms. In: Cohen B, Tomko DL, Guedry FE (eds) Sensing and controlling motion: vestibular and sensorimotor function. Ann NY Acad Sci 656, pp 220-232

Miles FA., Busettini C, Schwarz U (1992) Ocular responses to linear motion. In: Shimazu H, Shinoda Y (eds) Vestibular and brain stem control of eye, head and body movements. Japan Scientific Societies Press, Tokyo, pp 379-395

Miles FA, Kawano K, Optican LM (1986) Short-latency ocular following responses of monkey. I. Dependence on temporospatial properties of the visual input. J Neurophysiol: 1321-1354

Miles FA, Schwarz U, Busettini C (1991) The parsing of optic flow by the primate oculomotor system. In: Gorea A (ed) Representations of vision: trends and tacit assumptions in vision research. Cambridge University Press, Cambridge, pp 185-199

Miles FA, Schwarz U, Busettini C (1992) The decoding of optic flow by the primate optokinetic system. In: Berthoz A, Graf W, Vidal PP (eds) The head-neck sensory-motor system. Oxford Univerity Press, New York, pp 471-478

Nordlund P, Uhlin T (1995) Closing the loop: pursuing a moving object by a moving observer. Proc 6th Int Conf on Computer Analysis of Images and Patterns, pp 400-407

Paige GD (1983) Vestibuloocular reflex and its interactions with visual following mechanisms in the squirrel monkey. I. Response characteristics in normal animals. J Neurophysiol 49: 134-168

Paige GD (1989) The influence of target distance on eye movement responses during vertical linear motion. Exp Brain Res 77: 585-593

Paige GD, Tomko DL (1991a) Eye movement responses to linear head motion in the squirrel monkey. I. Basic characteristics. J Neurophysiol 65: 1170-1182

Paige GD, Tomko DL (1991b) Eye movement responses to linear head motion in the squirrel monkey. II. Visual-vestibular interactions and kinematic considerations. J Neurophysiol 65: 1183-1196

Panerai F, Metta G, Sandini G (2000) Visuo-inertial stabilization in space-variant binocular systems. Robotics and Autonomous Systems 30: 195-214

Panerai F, Sandini G (1998) Oculo-motor stabilization reflexes: integration of inertial and visual information. Neural Networks 11: 1191-1204

Pekel M, Lappe M, Bremmer F, Thiele A, Hoffmann KP (1996) Neuronal responses in the motion pathway of the macaque monkey to natural optic flow stimuli. NeuroReport 7: 884-888

Poggio GF (1995) Mechanisms of stereopsis in monkey visual cortex. Cerebral Cortex, 5: 193-204

Rojer AS, Schwartz EL (1990) Design considerations for a space-variant visual sensor with complex-logarithmic geometry. Proc 10th ICPR IEEE Comp Soc, Atlantic City, 2: 278-285

Roy JP, Wurtz RH (1990) The role of disparity-sensitive cortical neurons in signalling the direction of self-motion. Nature 348: 160-162

Roy JP, Komatsu H, Wurtz RH (1992) Disparity sensitivity of neurons in monkey extrastriate area MST. J Neurosci 12: 2478-2492

Saito H, Yukie M, Tanaka K, Hikosaka K, Fukada Y, Iwai E (1986) Integration of direction signals of image motion in the superior temporal sulcus of the macaque monkey. J Neurosci 6: 145-157

Sandini G, Alaerts A, Dierickx B, Ferrari F, Hermans L, Mannucci A, Parmentier B, Questa P, Meynants G, Sheffer D (1998) The project SVAVISCA: a space-variant color CMOS sensor. Bernard, Thierry M (eds) Advanced Focal Plane Arrays and Electronic Cameras II, Zürich, SPIE 3410: 34-45

Sandini G, Gandolfo F, Grosso E, Tistarelli M (1993) Vision during action. In: Aloimonos Y (ed) Active perception. Lawrence Erlbaum Associates, London, pp 151-190

Sandini G, Tagliasco V (1980) An anthropomorphic retina-like structure for scene analysis. Comp Vis Graphics Image Proc 14: 365-372

Schwartz EL (1977) Spatial mapping in the primate sensory projection: Analytic structure and relevance to perception. Biol Cybern 25: 181-194

Schwarz U, Busettini C, Miles FA (1989) Ocular responses to linear motion are inversely proportional to viewing distance. Science 245: 1394-1396

Schwarz U, Miles FA (1991) Ocular responses to translation and their dependence on viewing distance. I. Motion of the observer. J Neurophysiol 66: 851-864

Shelhamer M, Merfeld DM, Mendoza JC (1995) Effect of vergence on the gain of the linear vestibulo-ocular reflex. Acta Otolaryngol Suppl 520: 72-76

Snyder LH, King WM (1992) Effect of viewing distance and location of the axis of head rotation on the monkey's vestibuloocular reflex. I. Eye movement responses. J Neurophysiol 67: 861-874

Tabak S, Collewijn H, Boumans LJ, van der Steen J (1997) Gain and delay of human vestibulo-ocular reflexes to oscillation and steps of the head by a reactive torque helmet. I. Normal subjects. Acta Otolaryngol 117: 785-795

Takemura A, Inoue Y, Kawano K, Miles FA (1997) Short-latency discharges in medial superior temporal area of alert monkeys to sudden changes in the horizontal disparity. Soc Neurosci Abstr 23: 1557

Tanaka K, Saito H (1989) Analysis of motion of the visual field by direction, expansion/contraction, and rotation cells clustered in the dorsal part of the medial superior temporal area of the macaque monkey. J Neurophysiol 62: 626-641

Tanaka K, Fukada Y, Saito H (1989) Underlying mechanisms of the response specificity of expansion/contraction and rotation cells in the dorsal part of the medial superior temporal area of the macaque monkey. J Neurophysiol 62: 642-656

Telford L, Seidman SH, Paige GD (1997) Dynamics of squirrel monkey linear vestibuloocular reflex and interactions with fixation distance. J Neurophysiol 78: 1775-1790

Telford L, Seidman SH, Paige GD (1998) Canal-otolith interactions in the squirrel monkey vestibulo-ocular reflex and the influence of fixation distance. Exp Brain Res: 115-125

Tunley H, Young D (1994) First order optical flow from log-polar sampled images. Proc 3rd European Conf Comp Vision (ECCV), Stockholm, pp 132-137

Ungerleider LG, Mishkin M (1982) Two cortical visual systems. In: Ingle DJ, Goodale MA, Mansfield RJW (eds) Analysis of visual behavior. MIT Press, Cambridge, pp 549-586

Viirre E, Tweed D, Milner K, Vilis T (1986) Re-examination of the gain of the vestibulo-ocular reflex. J Neurophysiol 56: 439-450

Weiman CFR, Chaikin G (1979) Logarithmic spiral grids for image processing and display. Comp Graphic and Image Process 11: 197-226

Weiman CFR, Juday RD (1990) Tracking algorithms using log-polar mapped image coordinates. Int Conf Intelligent Robots Computer Vision VIII: Algorithms and techniques, Philadelphia, SPIE 1192: 843-853

Gaze Control: A Developmental Perspective

Janette Atkinson and Oliver Braddick

Visual Development Unit, University College London, London, UK

1. Introduction

The chapter by Sandini et al. provides good examples of how the working of bio-logical visual systems can help us to understand the design problems of an engi-neered system, and vice versa. However, it is important to bear in mind that the organization of a biological system has arisen through processes that are rather different from the design and manufacture of an engineered system. The biological system has to be understood as the end point of an evolutionary and a develop-mental process.

2. Modularity and the evolution of the visual system

At the present state of engineering, it does not appear a sensible objective to design a general purpose robot. Thus artificial vision systems usually have rather specific and limited goals. Furthermore, a modular approach makes their design a more tractable problem, so if there are different subgoals, they are treated fairly independently in the design process. Thus the present chapter treats gaze stabili-zation as a self-contained problem.

The modular approach to neural systems has also made them more tractable to understand, and the long evolutionary history of gaze stabilization encourages us to believe that there are special purpose mechanisms, which may have been conserved in evolution, to perform this task. However, these mechanisms have co-evolved with other aspects of a system that achieves many diverse visual goals, such as guiding orienting actions and recognizing objects. The constraints and tradeoffs that might apply to the gaze stabilization problem in isolation do not necessarily apply when it is considered as part of this wider context.

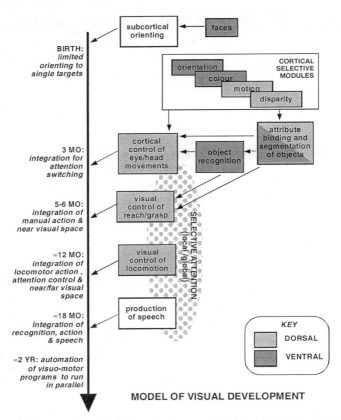

MODEL OF VISUAL DEVELOPMENT

Fig. 1 Model of the developing visual system in human infants. The different integration processes, involving functioning of different action systems, are shown on the left hand side of the vertical line. Several different action systems become functional at different ages of development in the first two years of life. In temporal order of development these are (a) the newborn subcortically controlled orienting system of the head and eyes, involving reflexive eye movement systems; (b) cortical action systems controlling both saccadic and smooth pursuit eye movements, involving integration of subcortical and cortical (parietal, frontal) circuits. These circuits are also dependent on information from cortically selective modules for different visual attributes of objects, i.e. shape, colour, size, and those responsible for attribute binding and segmentation (parietal, temporal) which develop in the first months of life. (c) action modules for reaching and grasping in nearby space (parietal-frontal, temporal); (d) action modules for controlling locomotion, accompanied by mechanisms for attentional shifting between different scales of representation of space at different distances (parietal, frontal); (e) action modules for production of speech and integration of information from modules for object recognition, motor actions and speech; (f) action systems for automating visuo-motor programs and providing parallel processing across modules.

This problem is illustrated by the Sandini et al.'s discussion of "pre-attentive" processing. At several points they refer to target selection and attentional processes as being a "waste of resources" in gaze stabilization. The relative cost of different resources in computer engineering is not necessarily the same as that in the richly parallel, noisy environment of the brain. But in any case, if other functions of the visual system (e.g. the initiation of saccades) depend strongly on selective processes, then the resources are being used anyway. Thus selective processes might play a role in human gaze stabilization even if they did not make "economic" sense for achieving the goal of stabilization on its own. Work on single units in primates (e.g. Desimone and Duncan 1995; Desimone 1998; Seidemann and Newsome 1999), and human brain imaging studies (e.g. Tootell et al. 1998; Gandhi et al. 1999; Martinez et al. 1999) have shown that effects of selective attention are remarkably pervasive in visual areas of the cortex. It is likely that this information feeds into gaze stabilization mechanisms, although it may be possible to dissociate it from more basic computations underlying stability by exploiting the differential latency of different effects, as Sandini et al. suggest. This is an empirical question, which cannot be prejudged in terms of the notional resource demands of neural processing.

3. The developmental context

A second important aspect of biological vision is that its capabilities are achieved through a developmental process. At intermediate stages of development, different aspects of the system (e.g. acuity, binocularity, motion integration) are nearer or further from their mature state. The developing system therefore has to function with a different relation between its components from that we see in the mature system.

4. Development of "space variant geometry"

One example of this is the role of what the engineers call "space variant sensing" – in biological terms, the inhomogeneity of the retina. The mature human visual system contains much denser retinal sampling at the fovea than in surrounding regions, like Sandini et al.'s "log-polar" tessellation. However, in infancy the human fovea is poorly differentiated in terms of receptor density, and photoreceptors migrate considerably across the visual field in the course of development to achieve the final foveal packing (Youdelis and Hendrickson 1986). Thus, the kind of mechanism proposed by Sandini et al., which weights central vision because of its higher sampling density, would not achieve the right result in a human infant. We do not know whether this difference has any effects on the operation of gaze stabilization mechanisms in infancy. We do know, however, that even newborn infants' fixation behaviour operates to crudely bring significant targets (to which we assume the infant is attending) to the fovea, even though that fovea does not yet have the anatomical specialization that gives it higher resolution than surrounding areas. This illustrates that the space-variant organization of human vision does not operate solely through differences in pixel density. Presumably the fovea-centred system is already embodied in cortical topography, and this is linked to oculomotor control at a very early stage of development.

The immaturity of the receptor distribution, coupled with oculomotor systems that are adapted to the not-yet-developed specialization of the fovea, would be expected to lead to a different functional balance from the mature state. This implies that the role of space-variant organization, suggested by Sandini et al., might lead to some interesting developmental predictions.

The orienting mechanism, bringing an object of interest onto the fovea for stable viewing by means of head and eye movements, which has been referred to above, is the first system in a series of visual action mechanisms developing in human infancy. A schematic diagram of this developmental sequence of action systems, including stabilizing systems, and its timescale is shown in figure 1 (after Atkinson, 2000).

Each of these action streams are likely to involve the integration of many cortical and subcortical areas. A scheme of the major action systems, which we have proposed, based on studies of non-human primates and neurological studies of adult patients (e.g. Jeannerod 1988; Milner and Goodale 1995; Rizzolatti et al. 1997), is shown in figure 2 (after Atkinson, 2000).

Here we see both separate and overlapping subcortical-cortical circuitry for saccadic eye movements and smooth pursuit movements, but there is also overlap in the adult systems with action modules for reaching and grasping.

5. Developmental action systems

The newborn orienting system and its relation to the model of Sandini et al. has already been briefly discussed. In the newborn infant this system functions sub-optimally initially, with hypotonic saccades being made to fixate an initially peripheral target of interest (Aslin and Salapatek 1975). However, there is some debate as to just how hypotonic these saccades are when the infant is free to move both the head and eyes together to orient, rather than when the head is held in a fixed position, as in many studies.

Differences have also been found between one month and three months olds in the extent and accuracy of smooth pursuit, which implies at least in human development that there are different stabilizing mechanisms for providing information to enable targets to be smoothly tracked to those for saccadic movements in orienting. Many of the most detailed developmental studies of these stabilizing mechanisms in infants have been made by Claus von Hofsten and his colleagues (von Hofsten and Rosander 1996, 1997). They recorded the presence of smooth pursuit eye movements even in newborns, if targets of sufficient size and contrast were used and their velocity was kept relatively slow. They have also looked at the development of the initial coupling of eye and head movements, as the infants develop these stabilizing mechanisms. Fairly accurate coupling and a mature vestibular ocular response is achieved in the first few months of life, although many of the tracking eye movements observed in everyday situations in this period are saccadic, rather than continuous smooth pursuit. An understanding of the limitations and parameters in the developmental context should make it possible to extend and test the models of Sandini et al. for gaze stabilization in the human infant.

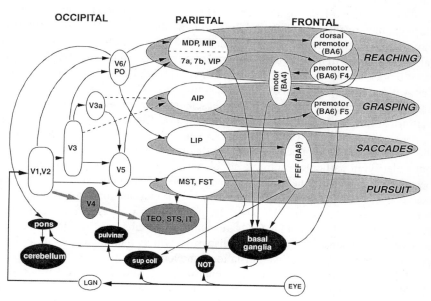

Fig.2 A schematic account of the areas contributing to different visuo-motor action streams in the primate brain (based on neurophysiological and neuropsychological findings). Four action streams are outlined; these are for reaching, grasping, saccadic eye movements and pursuit eye movements. BA: Brodmann's area; V1, V2, etc.: visual area 1, visual area 2, etc.; F4, F5: frontal areas; LIP: lateral intraparietal; V5 = MT: middle temporal, V6 = PO: parietal occipital; AIP: anterior intraparietal; MIP: medial intraparietal; MDP: mediodorsal parietal; VIP: ventral intraparietal; MST: medial superior temporal; FEF: frontal eye fields; TEO: posterior inferior temporal; IT: inferior temporal; STS: superior temporal sulcus; NOT: nucleus of the optic tract; sup coll: superior colliculus. (Reference source mainly: Rizzolatti et al. 1988; Milner and Goodale 1995; Jeannerod 1997).

6. Binocularity and stabilization

Sandini et al. give a very important role to binocular interactions, especially in maintaining stabilization against translational movements. As we can see from the schematic developmental model of figure 1, in human development binocular interaction is not one of the earliest features of cortical organization; systems sensitive to binocular disparity start to be come functional around three to four months of age (Birch 1993; Braddick 1996). Optokinetic and vestibulo-ocular reflexes for gaze stability certainly operate well before this age, as we have discussed above. However, their organization is different from the mature system. Up to three months, monocular OKN can be driven only by the temporal-to-nasal direction of movement for each eye (Atkinson and Braddick 1981). This has been attributed to the operation of a direct pathway from the retina to the contralateral midbrain nucleus of the optic tract (NOT) since the left NOT responds only to stimulus movement in the leftward direction and the right NOT only to rightward movement (Hoffmann and Schoppmann 1975; Hoffmann 1981). Nasal-to-temporal OKN appears to depend on development of a descending pathway to the NOT from binocular cells in cortex, including area MT/MST (Ilg and Hoffman 1993). Gross disruption or absence of the cortical input to NOT, e.g. in infants who have had early hemispherectomy to alleviate seizures due to developmental cortical malformations, leads to a loss of even the temporal-to-nasal response (Braddick et al. 1992; Morrone et al. 1999). However, even if the cortex on both sides is intact, the development of symmetrical OKN is disrupted by developmental problems of binocularity, in particular esotropic strabismus. It is not yet known whether this is because only binocular neurones can carry information to the NOT from the ipsilateral eye, or whether binocularity as such is important for normal optokinetic function. It is possible that the binocular pathway, as well as providing the basis for a symmetrical monocular OKN response, also subserves the modulation of optokinetic responses by binocular disparity, which Sandini et al. describe as providing stabilization against translational movements.

The approach through computational design issues, taken by Sandini et al. can therefore give insight into states of the developing system as well as the mature system. However, the developmental perspective also provides important tests of its relevance to the human brain.

References

Aslin RN, Salapatek P (1975) Saccadic localization of targets by the very young human infant. Percept Psychophys 17: 293-302

Atkinson J, Braddick OJ (1981) Development of optokinetic nystagmus in infants: an indicator of cortical binocularity? In: Fisher DF, Monty RA, Senders JW (eds) Eye movements: cognition and visual perception. Lawrence Erlbaum Associates, Hillsdale, N J, pp53-66

Atkinson J (2000) The developing visual brain. Oxford University Press, Oxford

Birch E (1993) Stereopsis in infants and its developmental relation to visual acuity. In: Simons K (ed) Early visual development: normal and abnormal. Oxford University Press, New York, pp 224-236

Braddick O, Atkinson J, Hood B, Harkness W, Jackson G, Vargha-Khadem F (1992) Possible blindsight in babies lacking one cerebral hemisphere. Nature 360: 461-463

Braddick O (1996) Binocularity in infancy. Eye 10: 182-188

Desimone R (1998) Visual attention mediated by biased competition in extrastriate visual cortex. Phil Trans Roy Soc B 353: 1245-1255

Desimone R, Duncan J (1995) Neural mechanisms of selective visual attention. Ann Rev Neurosci 18: 193-222

Gandhi SP, Heeger DJ, Boynton GM (1999) Spatial attention affects brain activity in human primary visual cortex. Proc Natl Acad Sci USA 96: 3314-3319

Hoffmann K-P (1981) Neuronal responses related to optokinetic nystagmus in the cat's nucleus of the optic tract. In: Fuchs A, Becker W (eds) Progress in oculomotor research. Elsevier, New York, pp 443-454

Hoffmann K-P, Schoppmann A (1975) Retinal input to the direction sensitive cells of the nucleus tractus opticus of the cat. Brain Res 99: 359-366

von Hofsten C, Rosander K (1996) The development of gaze control and predictive tracking in young infants. Vision Res 36: 81-96

von Hofsten C, Rosander K (1997) Development of smooth pursuit tracking in young infants. Vision Res 37: 1799-1810

Ilg UJ, Hoffmann K-P (1993) Functional grouping of the cortico-pretectal projection. J Neurophysiol 70: 867-869

Jeannerod M (1988) The neural and behavioural organization of goal directed movements. Oxford University Press, Oxford

Jeannerod M (1997) The cognitive neuroscience of action. Blackwell, Oxford

Martinez A, Anllo-Vento L, Sereno MI, Frank LR, Buxton RB, Dubowitz DJ, Wong EC, Hinrichs H, Heinze HJ, Hillyard SA (1999) Involvement of striate and extrastriate visual cortical areas in spatial attention. Nature Neurosci 2: 364-369

Milner AD, Goodale MA (1995) The visual brain in action. Oxford University Press, Oxford

Morrone MC, Atkinson J, Cioni G, Braddick OJ, Fiorentini A (1999) Developmental changes in optokinetic mechanisms in the absence of unilateral cortical control. NeuroReport 10: 2723-2729

Rizzolatti G, Fogassi L, Gallese V (1997) Parietal cortex: from sight to action. Current Opinion Neurobiol 7: 562-567

Rizzolatti G, Camarda R, Fogassi L, Gentilucci M, Luppino G, Matelli M (1988) Functional organization of area 6 in the macaque monkey II. Area F5 and the control of distal movements. Exp Brain Res 71: 491-507

Seidemann E, Newsome WT (1999) Effect of spatial attention on the responses of area MT neurons. J Neurophysiol 81: 1783-1794

Tootell RB, Hadjikhani N, Hall EK, Marrett S, Vanduffel W, Vaughan JT, Dale AM (1998) The retinotopy of visual spatial attention. Neuron 21: 1409-1422

Youdelis C, Hendrickson A (1986) A qualitative and quantitative analysis of the human fovea during development. Vision Res 26: 847-855

Does Steering a Car Involve Perception of the Velocity Flow Field?

Michael F. Land

Sussex Centre for Neuroscience, School of Biological Sciences, University of Sussex, Brighton, UK

1. Introduction

In 1950 Gibson introduced the idea of a flow-field – the pattern of velocity vectors in the field of view that results, in particular, from an organism's own motion. Since then there has been a continuing debate, evident from the chapters in this book, concerning the use of different cues in the guidance of locomotion. In crude terms these are viewed as being of two kinds. There are those based on the changing positions of identifiable features: thus "keep to the right of the row of trees and keep the church tower straight ahead". And there are those based on the pattern of velocity vectors in the flow-field. For example, the current direction of the traveller's heading – important for guiding locomotion – is represented by the location of the pole, or "focus of expansion" of the flow-field pattern, and the distances of objects in the surroundings can be recovered from their velocities across the retina. In the task I consider here, steering, the flow-field view of how we aim the vehicle was stated very clearly by Gibson (1950): "The behavior involved in steering an automobile, for instance, has usually been misunderstood. It is less a matter of *aligning the car with the road* than it is a matter of *keeping the focus of expansion in the direction one must go.*"

This view was explored further by Lee and Lishman (1977), who showed that the correspondence between the edges of the road and the direction of motion on the retina (the locomotor flow lines) indicate whether or not the vehicle is on course. On the other hand, Land and Lee (1994) found that on winding roads drivers look at rather specific regions of the road, notably the "tangent point" on the inside of each bend (this is the point where the driver's line of sight is tangential to the road edge or centre line, and it moves around the bend with the driver. Its important attribute is that - like the focus of expansion on a straight road – it does not move laterally in the visual field provided the road curvature remains constant). This suggests that drivers may be using road *features* to steer by, rather than patterns of optic *flow*. In what follows I review what is known about the information drivers need and use when steering, with a view to resolving the vexed question of whether it is feature displacement or flow that is used in loco-motor guidance (rev. Vishton and Cutting 1995).

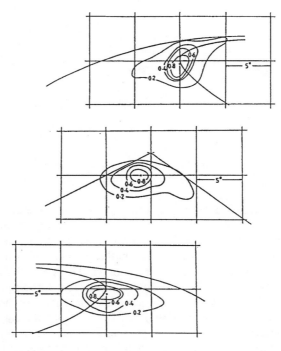

Fig. 1 Contour plots showing the location of fixations made on right and left-hand bends, and on straight road sections where no tangent point is visible. The contours give the density of fixations relative to the maximum (approximately 0.12 fixations / deg^2s^1). Measurements on bends were made from the tangent point and on the straight road from the vanishing point. The 0.2 contour includes about 65% of all fixations. Data from three 1km drives by different drivers. Note that on bends the highest fixation densities are within 1° of the tangent points. (Land and Lee 1994)

2. Where do drivers look?

Land and Lee (1994) used a head-mounted eye movement camera to determine the direction of drivers' gaze when driving on a winding road (Queen's Drive round Arthurs Seat in Edinburgh). The outcome was very clear; drivers looked at the region around the tangent point on the inside of each bend (Fig. 1) much more than anywhere else, and at the beginning of each bend they looked at it for about

80% of the time (Land and Lee 1994). There are a number of reasons why the tangent point might be particularly valuable to drivers, but perhaps the most important is that its direction, relative to the driver's current heading, gives a particularly simple measure of the curvature of the bend (Fig. 2). Since required steering-wheel angle is determined by bend curvature, this measurement provides an ideal visual input for any steering control system. The accuracy of this method depends to some extent on the driver keeping a constant, known, distance from the road edge, because this is a term in the curvature equation (Fig. 2), and so the tangent point direction needs to be supplemented by position-in-lane information, if it is to result in accurate steering. The other qualification is that the tangent point is some distance ahead, and provides information about the curvature between the driver and the tangent. For example, on a bend of 100 m radius with a driver 2 m from the lane edge, the tangent point is 20.1 m ahead of the driver, and so a delay is required. Land and Lee (1994) found that during the Edinburgh drives the maximum correlation between visual direction and steering-wheel angle occurred after a delay of 0.75 s, which is approximately the delay needed at moderate speeds.

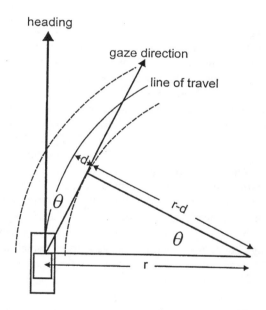

Fig. 2 Geometry of tangent-point steering. The curvature of the bend ($1/r$) can be obtained from the gaze angle θ, using the geometry of the right angled triangle. Here, $\cos\theta = (r-d)/r$, where d is the distance of the driver from the lane-edge. However, the expansion of the cosine gives $\cos\theta \approx 1 - \theta^2/2$. Substitution for $\cos\theta$ then gives: $1/r \approx \theta^2/2d$

As we will see later, it is not necessary for a driver to look at the tangent point, nor even to have the tangent point visible in the field of view. However, some part of the more distant region of the road does need to be visible, and provided that its distance is known (there are several cues that can be used for this) its direction relative to the driver's heading can be used in much the same way as the tangent point direction (Land 1998). This suggests that perhaps the significance of the tangent point is only partly to obtain curvature information, and that it may have another role. One thing that comes to mind is that in the region around the tangent point there no lateral motion of the flow-field, and in fact rather little vertical motion, as most of the structures in the region – lines and kerbs – are also oriented vertically in the flow field at this point. The tangent point is thus a place that the eyes can rest, without being "dragged around" by the optokinetic consequences of moving but irrelevant regions of the flow-field.

3. Simulator studies: which parts of the road does a driver need?

Although we have shown where drivers direct their gaze on bends, it is almost impossible to say where they are actually attending. We can, however, ask a related question: where should drivers look in order to get the best information to steer by? If this coincides with where they actually look, then this gives us grounds for thinking that they are attending there as well.

The method involved a simple simulator (Land and Horwood 1995), in which subjects drove round a skeletal version of the Queen's Drive road. Only the road edges were present, plus a horizon and a sketch of the car bonnet, but no other scenery. The drive could be run at a variety of constant speeds, from 12.5 to 19.7 m/s (28 to 44mph). Drivers found this similar to night driving, and had no difficulty negotiating the bends of the simulated road. The main measure of performance was the standard deviation of the subject's position in lane, taken over the whole 1km drive; under ideal conditions this was between 0.1 and 0.2m. After a few trials with the whole road outline visible, the view was restricted to either one or two 1° high segments of the road edge (Fig. 3) which could be located at varying positions between 1 and 10° below the horizon, corresponding to distances between 63 m and 6 m from the vehicle. These segments behaved exactly as they would had the whole road been present.

In the first set of experiments only one segment was visible The principal result was that at each speed there was an optimum (vertical) position of the segment that gave the best steering performance. It was nearer to the vehicle at slow speeds and further at high speeds, but in terms of "time ahead" it was close to 0.7 s at all speeds. Except for the slowest speed, however, the performance was not as good as when the whole road edge was present, and at faster speeds the difference was big enough to make the vehicle stray from its lane. This implies that any one segment of the road is not capable of providing the whole of the required control signal. Another disconcerting feature of this study was that the part of the road that provided the best performance was somewhat closer to the vehicle than the region containing the tangent point, and closer than the region where drivers usually looked when the whole simulated road was present.

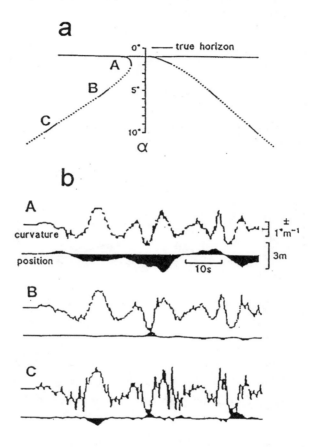

Fig. 3 Driving on a simulator with only parts of the road edge visible. **a** Appearance of simulated road, showing the angular scale used in Fig. 4 and the three road segments used in b, each subtending 1° vertically. **b** Differences in driver behaviour depending on the position of the visible segment. The upper trace in each pair gives the curvature of the road and the vehicle's track, with differences between them appearing as a solid black region. The lower trace shows position relative the road midline, with solid black indicating the extent of error. With only a distant segment visible (A) curvature matching is smooth and reasonably accurate, but position-in-lane accuracy is very poor. With only a near segment (C) curvature matching becomes jerky and unstable, although lane position is more accurate than in A. An intermediate distance (B) gives the best result. (From Land and Horwood 1995)

We also found that drivers behaved quite differently to near and far regions of the road. When only the far part of the simulated road was visible, drivers matched curvature well, but their lane keeping performance was poor; and when only the near part was visible lane keeping was better, but steering was unstable and jerky (Fig. 3b). The drivers' control system had changed from smooth to "bang-bang" (Land and Horwood 1995). This suggested that far and near regions contribute to the overall control system in different ways.

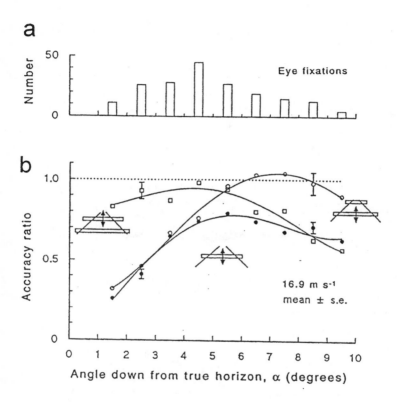

Fig. 4 Driver performance with two segments of road visible. **a** Distribution of vertical gaze direction when viewing the whole road. Abscissa as in b. The peak is 4-5° down from the horizon. (3 drivers) **b** With a single visible road segment (filled circles) the apparent optimum is 5.5° down from the horizon, but it is not as accurate as when the whole road is present (dotted line). Adding a second segment to the near part of the road (squares) greatly improves performance, but only when the first segment is in the far part of the road; when a second segment is added to the far part of the road (open circles) performance is enhanced for segment positions in the near part of the road. With two appropriately spaced segments, performance is as good as with the whole road edge visible. 5 drives each by 3 drivers. Abscissa is the angle down from horizon (α, Fig. 3a); Ordinate is the ratio of accuracy (1/s.d.) of the vehicle's position-in-lane under each condition, compared with the accuracy when the whole road edge is visible. (Land and Horwood 1995)

In a second series of simulations we added a second segment of road, and this improved performance provided that the two segments were well separated (Fig. 4). Taking out the "middle distance" region of the road, but leaving the more distant and nearer parts not only improved performance, but it made it indistinguishable from having the whole road present. In these simulations one visible segment was either at the most distant region (1-2° below the horizon) or the nearest (9-10° down), and the position of the other segment was varied. Broadly, the result shows that adding a second segment dramatically improves performance if and only if it is added to the opposite end of the road: the far segment if the near one is present, and vice versa. With the near-road segment present the optimum location of the far-road segment was about 4° down from the horizon (at 16.9 m/s), and with the far-road segment, the best location of the near-road segment was about 7° down.

The explanation of this is that the near and far regions of the road supply different and complementary information. The distant region (including tangent points where present) supplies feed-forward information about the future curvature of the road, and the near region supplies feedback information about position in lane. (A two-component model of this kind was originally proposed by Donges in 1978). Both are necessary. If only the far region is visible the driver may steer a course with a good approximation to the right road curvatures, but with no guarantee that he actually stays on the road. With only the near-road visible he can keep between the lane-edges easily at low speeds, but as speed increases and the lead time becomes short compared with his reaction time, the feedback becomes unstable. It is like driving too fast in fog. However, if the far-road feed-forward mechanism has already done most of the work, the near-road mechanism can work at low gain, and it is stable again.

4. Conclusions: features and flow

The locations of the edges of the road in the field of view appear to provide the principal visual cues for steering. They are necessary and sufficient, and the rest of the flow-field does not seem to be involved in any very direct or essential way. On poorly marked roads the differences in texture and motion at the road edge may substitute for discrete lines, but most drivers would agree that white or yellow lines provide a much better cue, and that when road markings are absent it is much harder to steer. However, "cross country" driving, which is normally undertaken at low speed, must involve other types of cue.

Are the edges of the road properly regarded as features, or components of the flow-field? They seem to have attributes of both. On a straight road, or on a road of constant curvature, their appearance and location in the visual field are constant, which makes them features. The direction of motion of their texture coincides with their orientation, so the "flow" appearance is weak, and is in any case irrelevant to the task of steering. However, when road curvature changes, or the vehicle strays from the lane centre, the lateral movements of the road edges in the field of view are of crucial importance. At that point one could regard them either as moving features, or as lateral components of the flow field. Perhaps the question should be: is it their instantaneous position or their velocity that matters? Again the answer is likely to be both. In this paper and its predecessors I have concentrated on the relation of the position of the road edge as an input to the human control system that turns the steering wheel, but in any practical control system the addition of a velocity input invariably improves performance, and that is undoubtedly also the case here. In a way the feature/flow argument is a sterile one. Features have both positions and velocities, and both are important control variables.

Another argument that is still active concerns the role of the focus of expansion of the flow field in providing moving observers with the direction of their heading (e.g. Gibson 1950; Cutting 1986, chap. 10). Drivers do need to know their heading, so that they can determine the angle between that heading and, say, the tangent point, in order to obtain a signal to steer by. A problem with this is that if drivers are actually looking at the tangent point (Fig. 1), their eyes will be rotating with the curve, and this means that there will be no focus of expansion, since this requires the eye to be in linear translational motion. On a curved trajectory the locations of the stationary points in the flow-field vary with distance, generating a curved line across the ground plane, not a single focus of expansion (see Raviv and Herman 1993). Thus to detect the direction of instantaneous heading in general requires the decomposition of the flow-field back into translational and rotational components. It seems this can be done in laboratory conditions, but with some difficulty. Warren et al. (1991) showed that under appropriate circumstances subjects can extract *circular* heading (i.e. their future curved path) from the sort of combined rotational and translational flow field that would result from driving on a curving road. However it isn't clear that this is actually relevant to the driving task, and the assumption that heading (linear or circular) needs to be obtained by visual means is very questionable. For a driver belted to the seat, the orientation of the trunk axis *is* the direction of the vehicle's instantaneous heading, and the measurement required for steering is simply the angle between the trunk and the tangent point, if that is the feature being used. This angle is the sum of two physiologically available measurements, the head/trunk angle and the eye/head angle, if the driver is actually looking at the tangent point (Fig. 1). Thus it is hard to find any information required for the task of steering that *must* come from the velocity flow-field.

Acknowledgment

I am grateful to the BBSRC (UK) and to the Gatsby Foundation (UK) for funding the research that led to this review.

References

Cutting JE (1986) Perception with an eye for motion. MIT Press, Cambridge, Mass
Donges E (1978) A two-level model of driver steering behavior. Human Factors 20: 691-707
Gibson JJ (1950) The perception of the visual world. Houghton Mifflin, Boston
Land MF (1998) The visual control of steering. In: Harris LR, Jenkin H (eds) Vision and action. Cambridge University Press, Cambridge UK, pp 172-190
Land MF, Lee DN (1994) Where we look when we steer. Nature 369: 742-744
Land MF, Horwood J (1995) Which parts of the road guide steering. Nature 377: 339-340
Lee DN, Lishman JR (1977) Visual control of locomotion. Scand J Psychol 18: 224-230
Raviv D, Herman M (1993) Visual servoing from 2-D image cues. In: Aloimonos Y (ed) Active perception. Erlbaum, Hillsdale NJ, pp 191-226
Vishton PM, Cutting JE (1995) Wayfinding, displacements and mental maps: velocity fields are not typically used to determine one's aimpoint. J Exp Psychol: Human Percep Perform 21: 978-995
Warren WH Jr, Mestre DR, Blackwell AW, Marris MW (1991) Perception of circular heading from optical flow. J Exp Psychol: Human Percep Perform 17: 28-43

Part V

Neural Coding of Motion

Neural Encoding of Visual Motion in Real-Time
Anne-Kathrin Warzecha and Martin Egelhaaf

Real-Time Encoding of Motion: Answerable Questions and Questionable
Answers from the Fly's Visual System
Rob de Ruyter van Steveninck, Alexander Borst and William Bialek

A Comparison of Spiking Statistics in Motion Sensing Neurones of Flies
and Monkeys
Crista L. Barberini, Gregory D. Horwitz and William T. Newsome

Dynamic Effects in Real-Time Responses of Motion Sensitive Neurones
Ted Maddess

Neuronal Encoding of Visual Motion in Real-Time

Anne-Kathrin Warzecha and Martin Egelhaaf

Lehrstuhl für Neurobiologie, Universität Bielefeld, Bielefeld, Germany

Contents

1. Abstract

Changes in the activity of sensory neurones carry information about a given stimulus. However, neuronal activity changes may also arise from noise sources within or outside the nervous system. Here, the reliability of encoding of visual motion information is analysed in the visual motion pathway of the fly and compared to the findings obtained in other animal species. Several constraints determine and limit the reliability of encoding of visual motion information: (i) the biophysical mechanisms underlying the generation of action potentials; (ii) the computations performed in the motion vision pathway; and (iii) the dynamical properties of motion stimuli an animal encounters when moving around in its natural environment. The responses of fly motion-sensitive neurones are coupled to visual motion on a timescale of milliseconds up to several tens of milliseconds, depending on the dynamics of the motion stimuli. Only rapid velocity changes lead to a precise time-locking of spikes to the motion stimuli on a millisecond scale. Otherwise, the exact timing of spikes is mainly determined by fast stochastic membrane-potential fluctuations. It is discussed on what timescale behaviourally relevant motion information may be encoded.

2. Introduction

We usually take it for granted that we are able to react appropriately in most situations – even when navigating through hectic everyday traffic – and thus to arrive safely at our destination. This capability requires fast and reliable decisions. Such behavioural decisions are inevitably preceded within the organism by a series of neuronal processing steps. First the outside world has to be captured by the sensory system. Then the relevant features of the stimuli have to be extracted in the nervous system from the activity profile of the sensory cells before the motor programs can be initiated to execute the appropriate behavioural reactions. Visual systems of many animals including humans often exploit motion cues to control their path of locomotion. Retinal motion experienced by an organism moving through its environment, the so-called optic flow, provides a wealth of information about the self-motion of the organism and the three-dimensional structure of its surroundings (e.g. Koenderink 1986). The nervous systems of many mobile animals are able to evaluate the optic flow and to exploit this information to mediate appropriate steering manoeuvres (reviews: Miles and Wallman 1993). The activity profile of neurones in the visual motion pathway, thus, provides crucial information about the animal's self-motion and its environment.

Despite the fact that many animals appear to be able to extract without much effort the relevant information about optic flow, the underlying computations are far from being trivial (see also Dahmen et al. this volume). In particular, the representation of the dynamical features of optic flow in real time is constrained by the underlying neuronal mechanisms:

- The activity level of a motion-sensitive neurone does not unambiguously signal the time course of the retinal velocity within its receptive field. Various combinations of different stimulus parameters such as the velocity of a moving object, its acceleration, texture and contrast, may lead to the same activity level of a given neurone (for review, see Egelhaaf and Borst 1993b).

- The temporal precision with which sensory stimuli can be encoded by neurones is constrained by the biophysical mechanisms underlying neuronal signal processing.

Many events proceeding in nerve cells are not completely deterministic and thus lead to responses that are unpredictable to a certain degree even if they are elicited by identical stimuli. This uncertainty can have several causes such as the stochastic opening and closing of ion channels (White et al. 2000), or probabilistic transmitter release at chemical synapses (e.g. Allen and Stevens 1994). Moreover, even the sensory input itself may be noisy, as is the case for the visual system due to the stochastic nature of light. Thus, when a given stimulus is repeatedly presented to a neurone, the response exhibits a considerable amount of variability. It has been reported that the variance of the spike count elicited by repeated presentation of a visual stimulus may be as large as the average spike count (e.g. Tolhurst et al. 1983; Vogels et al. 1989; Britten et al. 1993; see also Barberini et al., this volume). Hence, the problem arises how to infer from changes in the neuronal responses whether these changes carry information about the actual stimulus or whether they are just due to noise.

To estimate the response component induced by the stimulus, experimenters usually average over the individual responses to many stimulus presentations. This procedure is intended to eliminate the stochastic component from the neuronal responses. It needs to be mentioned that not all variability observed in responses to identical stimulation is due to stochastic sources. Neuronal variability may also be due to deterministic processes that do not occur time-coupled to the stimulus and thus are not controlled by the experimenter. Because these two sources of neuronal variability cannot easily be separated experimentally, they will be referred to as noise in the following without further distinction. In real life an animal can hardly ever average over the responses to many repetitions of the same situation. It often has only a single and brief chance to react appropriately, for instance in order to evade a predator. A neuronal strategy to increase the signal-to-noise ratio of the neuronal response could be to pool the activity of many equivalent neurones. Indeed, this strategy has been proposed to be exploited by nervous systems (e.g. by neurones in area MT of the monkey, see Zohary et al. 1994; Shadlen et al. 1996). However, it is useful only if the stochastic fluctuations in the signals of the neurones that are being pooled do not covary too much and the pooling neurone itself does not introduce a large amount of noise from its own biophysical machinery. Another strategy to increase the neuronal reliability could be to use the average neuronal activity within a certain time interval. However, by temporally averaging, fast changes in the stimulus cannot be detected in the neuronal activity. Hence, the extent of temporal averaging that helps to increase the signal-to-noise ratio largely depends on the timescale on which the stimulus-induced response component varies.

This paper will focus on the constraints imposed by the neuronal mechanisms underlying visual motion computation on the evaluation of optic flow in real time. An understanding of the real-time performance of motion vision systems is essential if we want to understand how optic flow can be exploited by an animal to safely navigate through its environment. In particular, we will discuss the timescale on which visual motion is processed and the temporal precision with which motion stimuli are represented in the visual system. We will concentrate on the motion pathway of the fly and the behavioural responses controlled by this pathway, because a wide range of aspects of visual motion computation have been analysed in this animal in great detail. The capabilities and limitations of the fly visual motion system will be related to other animals and the computational tasks they have to solve.

3. The fly as a model system

Flies of the families Muscidae, Calliphoridae and Sarcophagidae have provided us with a well established model system for the analysis of motion information processing because both the behavioural and the neuronal level are well amenable to experimental analysis (for reviews see e.g. Reichardt and Poggio 1976; Egelhaaf et al. 1988; Hausen and Egelhaaf 1989; Bialek and Rieke 1992; Egelhaaf and Borst 1993a; Egelhaaf and Warzecha 1999). Visually guided flight manoeuvres of these insects can be investigated in free flight or with a flight simulator under the well controlled laboratory conditions of tethered flight. Some of these flight manoeuvres are impressively aerobatic and outperform those of most other animals as well as those of any man-made machine. One of the most virtuosic visually controlled manoeuvres of flies is the chasing behaviour during which male flies chase potential mates in order to catch and finally mate with them (Land and Collett 1974; Wagner 1986a). Other well studied behavioural routines that are controlled visually include the detection of objects on the basis of relative motion between the object and its background (Virsik and Reichardt 1976; Reichardt et al. 1983; Heisenberg and Wolf 1984; Egelhaaf et al. 1988; Götz 1991; Egelhaaf and Borst 1993a; Kimmerle et al. 1996; Kimmerle et al. 1997) and the ability to stabilize the flight path against disturbances by compensatory optomotor reactions (Götz 1968; Götz 1975; Reichardt and Poggio 1976; Heisenberg and Wolf 1984; Egelhaaf et al. 1988; Egelhaaf and Borst 1993a; Warzecha and Egelhaaf 1996).

The details of the neuronal circuits controlling components of the visually guided behaviour of the fly are in some instances known down to the level of individual cells (for review see Egelhaaf et al. 1988; Egelhaaf and Borst 1993a). All these neuronal circuits have in common that the retinal images are initially processed in the first and second visual neuropile by successive layers of retinotopically arranged columnar neurones (e.g. Järvilehto et al. 1989; Strausfeld 1989; Laughlin 1994). In the third visual neuropile, the main centre for optic flow analysis, the retinotopic organization of the columnar elements is abandoned. In this brain region, about 40 to 60 so-called tangential cells pool with their extended dendrites the output signals of many excitatory and inhibitory motion-sensitve columnar neurones that have opposite preferred directions (see Douglas and Strausfeld, this volume). As a consequence of this input the tangential cells have very large receptive fields (Hausen and Egelhaaf 1989; Egelhaaf and Borst 1993a; Egelhaaf and Warzecha 1999). Owing to this input and to synaptic interactions with other tangential cells they obtain their characteristic sensitivity to optic flow patterns (Hausen 1981; Egelhaaf and Borst 1993a; Egelhaaf and Warzecha 1999; Krapp 1999; Horstmann et al. 2000; see also Dahmen et al., this volume). The various tangential cells differ in their preferred direction of motion, in their receptive field properties and in their response mode. During visual motion stimulation some of them respond with a sequence of spikes, others with graded changes of their membrane potential and still others mainly in the graded mode, however, with additional spike-like depolarizations (Hengstenberg 1977; Hausen 1982a; Haag et al. 1997; Haag and Borst 1998). The tangential cells can be identified individually in each animal on the basis of their physiological and anatomical properties making them ideal for a quantitative analysis of their performance.

In particular, the spike activity of one of these neurones, the so-called H1-cell, can be recorded over extended periods of several hours. Hence, this neurone is well suited for an experimental analysis of the reliability of neural coding and has been used as a model system for this purpose in quite a number of studies (Mastebroek 1974; Gestri et al. 1980; de Ruyter van Steveninck and Bialek 1988; Bialek et al. 1991; de Ruyter van Steveninck and Bialek 1995; Haag and Borst 1997; de Ruyter van Steveninck et al. 1997; Warzecha and Egelhaaf 1997; Warzecha et al. 1998; Warzecha and Egelhaaf 1999).

4. How are motion stimuli represented by fly tangential cells?

Fly tangential cells respond directionally selective to visual motion. Spiking tangential cells increase their spike rate above the resting level during motion in their preferred direction and decrease the spike rate during motion in the opposite direction. Since the spontaneous activity of most spiking tangential cells is low, they usually stop firing when the pattern moves in the null direction. Tangential cells that encode their input signals by graded changes in their membrane potential depolarize during preferred direction motion and hyperpolarize during null direction motion (Hausen 1982a; Hengstenberg 1982; Hausen and Egelhaaf 1989). How spiking tangential neurones in the fly's third visual neuropil represent visual motion is shown for one example, the H1-cell, in figure 1.

The fly was stimulated in this example by a grating moving with either a constant velocity or randomly fluctuating velocities. At the onset of constant-velocity motion the spike activity of the H1-neurone increases and, after a transient phase, reaches a more or less constant steady-state level. Even during the steady-state, the interspike intervals are not of constant length but vary considerably. During dynamic stimulation the spike activity fluctuates strongly, following to some extent the pattern velocity. When comparing individual response traces elicited by identical motion stimuli, it becomes obvious that the spike activity is not entirely determined by the motion stimulus, but is also driven by some other process, possibly due to noise. Although the overall pattern of spike activity looks quite similar in each of the individual responses, there is some variability in their temporal fine structure across trials (de Ruyter van Steveninck et al. 1997; Warzecha et al. 1998; Warzecha and Egelhaaf 1999). Similar responses can be observed for other spiking tangential cells. Furthermore, tangential cells with graded changes of their membrane potential exhibit similar properties with respect to the motion-induced response component and show variability in the fine structure of their responses across trials (Fig. 6 and Warzecha 1994; Haag and Borst 1997, 1998; Warzecha et al. 1998).

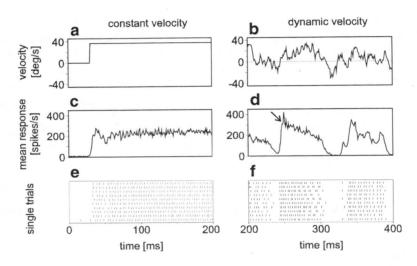

Fig. 1 Responses of the H1-cell to constant and dynamic velocity stimuli. **a, b** Velocity profile of the stimulus. Squarewave gratings were used as stimuli (wavelength: 18.2°, contrast: 0.99) and displayed on a monitor covering the central part of the cell's receptive field (horizontal extent: 91°, vertical extent: 76.6°). To obtain the dynamical motion stimuli, white-noise velocity fluctuations were generated according to a Gaussian distribution with a standard deviation of 0.57°/ms. The resulting velocity trace was lowpass-filtered with a cutoff at 80 Hz to avoid aliasing due to the frame rate limit of 183 Hz. **c, d** Spike frequency histogram obtained from the responses of 5 H1-cells to repetitive presentation of the motion traces shown in (a) and (b). For each of the five neurones between 140 and 300 responses were averaged. The responses were shifted by 30 ms to compensate for the latency between stimulus and response. Temporal resolution: 1ms. **e, f** Subsequent sample traces of individual responses to repetitive presentation of the same motion trace plotted underneath each other. Each vertical line denotes the occurrence of a spike. The section of the responses shown in (e) and (f) corresponds to the section of the spike frequency histogram shown in (c) and (d). (a,c,e modified from Warzecha and Egelhaaf 1999)

To understand what information about the time course of the continually changing retinal input is encoded by fly tangential neurones and how reliably this is accomplished the following topics will be discussed:

- The encoding of dynamic velocity stimuli by the motion-induced response component of the tangential neurones will be analysed.
- The variance between consecutive responses to identical stimuli will be determined in order to find out whether the variability of spike responses is affected by the stimulus dynamics.
- The temporal precision, i.e. the extent to which spikes are time-locked to visual motion, will be analysed by comparing the timing of spikes in consecutive response traces.

4.1. Representation of the time course of visual motion

By averaging over many individual responses to the identical motion stimulus we eliminate the noise component and derive the component of the response that is induced by the motion stimulus. For a spiking neurone the motion-induced response component corresponds to the spike frequency histogram. Interestingly, the motion-induced response components of spiking and graded potential neurones elicited by the same stimuli are quite similar. When the fly is stimulated with constant velocity, the motion-induced response is often more or less constant after a brief transient phase (Figs. 1c and e). Sometimes the response decreases slightly over several seconds (Maddess and Laughlin 1985). The response amplitude is modulated only slightly by the spatial structure of the moving pattern. When the velocity fluctuates randomly, the motion-induced response component is strongly modulated, occasionally reaching momentary peak frequencies of more than 300 spikes/s (Fig. 1d). Although the time course of the motion-induced response component is modulated by velocity changes, it is not proportional to the velocity of the stimulus pattern. Particularly pronounced velocity transients in the cell's preferred direction lead to considerably larger responses than expected for a cell which represents the pattern velocity. This is particularly obvious at the onset of the constant velocity stimulus but also during dynamical motion stimulation when the stimulus reverses its direction of motion (see e.g. arrow in Fig. 1d). Indeed, the time course of the motion-induced response of fly tangential neurones does not only depend on pattern velocity but also, in a non-linear way, on acceleration and higher-order temporal derivatives (Egelhaaf and Reichardt 1987; Egelhaaf and Borst 1989). These deviations of the responses from being proportional to pattern velocity are most prominent when the motion stimulus reverses its direction frequently and when the instantaneous velocity becomes large (Egelhaaf and Reichardt 1987, see Fig. 2).

Information about the time course of pattern motion cannot only be derived from the motion-induced response component, as obtained by averaging over many individual response traces, but also in real-time by temporal filtering the individual time-dependent responses. The linear filter which leads to an optimal estimation of the instantaneous velocity has been derived both for spiking tangential cells and for tangential cells which encode motion information mainly with graded potentials (Bialek et al. 1991; Haag and Borst 1997). The time course of pattern velocity can be estimated quite well in this way for low frequencies and sufficiently small stimulus amplitudes (Haag and Borst 1997), in accordance with the results displayed in figure 2. The deviations of the time-dependent responses of tangential cells at high frequencies and large amplitudes from the time course of stimulus velocity are to a large extent the consequence of nonlinearities in the mechanisms underlying motion computation (Egelhaaf and Reichardt 1987; Haag and Borst 1997).

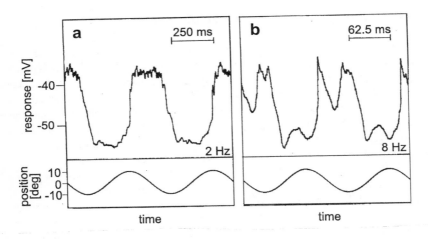

Fig. 2 Dynamical response properties of HS-cells. The cell was stimulated by a grating pattern oscillating sinusoidally with 2 Hz (**a**) or 8 Hz (**b**) at an amplitude of ±10° (see bottom traces; note the different time scales). HS-cells signal information about visual motion mainly with graded membrane potential changes that can be superimposed by small spike-like events. At 2 Hz the response follows the time course of the pattern velocity smoothly and the cell is depolarized during front-to-back motion and hyperpolarized during back-to-front motion; thus the response more or less follows the pattern velocity. At 8 Hz characteristic distortions become prominent, since hyperpolarizing deflections of the membrane potential occur during front-to-back motion and depolarizing deflections during back-to-front motion. Thus, the response profile is no longer proportional to the time course of pattern velocity if the velocity changes get too transient. Averages from 16 stimulus presentations. (Modified from Egelhaaf and Reichardt 1987).

There are indications that motion-sensitive neurones of other species resemble fly tangential neurones with respect to their representation of the time course of pattern velocity (Movshon et al. 1990; Ibbotson et al. 1994; Lisberger and Movshon 1999). For example, the responses of neurones in area MT of macaque monkeys, which may be involved in the control of pursuit eye movements, do not only depend on the velocity but also on the acceleration of pattern motion (Movshon et al. 1990; Lisberger and Movshon 1999) in a way reminiscent of fly tangential neurones.

The peculiar dynamical properties of directionally selective neurones, such as fly tangential cells, can partly be understood as the consequence of the non-linearities and the time constants that are necessary constituents of any motion detection mechanism (for review see Borst and Egelhaaf 1989; Egelhaaf and Borst 1993b). The time constants involved in motion computation have been reported to cover a range of ten to several tens of milliseconds in a variety of animal species that are phylogenetically rather far apart (arthropods: for review see Egelhaaf and Borst 1993b; monkeys: Mikami et al. 1986). There are three further consequences of the mechanism underlying motion detection which need to be considered here:

- Even the steady-state response amplitude of motion-sensitive neurones increases with increasing velocity only within a limited velocity range (for review, see Buchner 1984; Egelhaaf and Borst 1993b). At a certain velocity the responses reach a maximum and then decrease when the velocity further increases. It should be noted that this kind of velocity dependence is not specific to fly tangential neurones but seems to be a general feature of motion-sensitive neurones.

- As a consequence of time-constants inherent to the processes underlying motion computation, the power spectrum of the motion-induced response component in fly tangential neurones obtained from velocity fluctuations containing a wide range of frequencies falls off earlier than that of the motion stimulus (Fig. 3). Through such a lowpass characteristic high-frequency velocity changes are greatly attenuated by the fly motion vision system (Haag and Borst 1997; Warzecha et al. 1998). As might be expected from the time constants described for the processes underlying motion computation in other animal species, this property appears to be a general feature of motion vision systems. For instance, the mean responses of neurones in area MT of the monkey exhibit a similar lowpass characteristic (Bair and Koch 1996).

- The time-constants of the motion detection system in the fly and other insects have been concluded not to be constant, but to adapt to some extent (Maddess and Laughlin 1985; de Ruyter van Steveninck et al. 1986; Borst and Egelhaaf 1987; Harris et al. 1999; see also Maddess this volume). It is assumed that in this way the operating range of the activated neurones is extended so that they are most sensitive under the prevailing stimulus conditions.

In conclusion, the motion-induced response component of fly tangential neurones as well as neurones in other visual systems does not faithfully represent the time course of the retinal velocity. The representation of velocity is highly ambiguous, since the responses do not only depend on pattern velocity but also on the acceleration and higher temporal derivatives of the velocity as well as on the adaptational state of the motion detection system. Moreover, it has long been known that, the responses also depend on the texture and contrast of the stimulus pattern (for review, see e.g. Egelhaaf and Borst 1993b). These characteristic features of motion-induced responses need to be taken into account if one wants to understand how optic flow is encoded in real-time, in situations as they are encountered by an animal moving around in its environment.

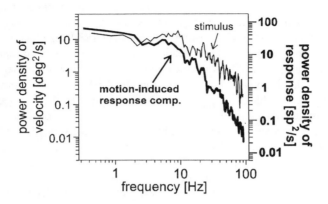

Fig. 3 Frequency response characteristic of the fly motion detection system. The thin line corresponds to the power spectrum of random velocity fluctuations which were used as motion stimuli (left scale). The thick line represents the power spectrum of the stimulus-induced response component of four H1-cells (right scale). The high frequency velocity components in the motion stimulus are attenuated by the motion detection system. For illustration the power spectra were smoothed by a 5 point boxcar filter. (Modified from Warzecha et al. 1998).

4.2 Variability of motion-induced responses in a spiking neurone

Obviously, an animal does not have the motion-induced response component at its disposal for controlling behavioural reactions. Instead, the motion-induced response component is contaminated by stochastic fluctuations in the individual responses of a neurone (see Figs. 1 and 6). This stochastic component constrains how reliably motion-induced information can be conveyed by neurones. For motion-sensitive tangential cells of the fly the reliability of coding has been suggested to be especially adapted to dynamical stimuli as they are encountered by an animal in its natural behavioural context (de Ruyter van Steveninck et al. 1997). The variance across trials has been used to quantify how variable individual responses to the same stimulus are within a given time interval relative to the onset of stimulation. Whereas during constant velocity stimulation the variance has been concluded to be in the range of the mean activity, the variance during dynamic velocity stimulation was found to be considerably smaller. On this basis, the H1-cell has been interpreted to represent dynamic stimuli more reliably than constant ones (de Ruyter van Steveninck et al. 1997). A different conclusion however is drawn from the results that are presented in the following.

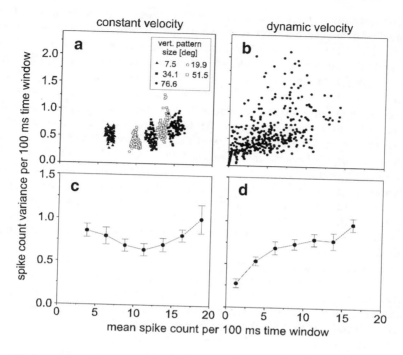

Fig. 4 Variance of the spike count of the H1-cell as a function of the mean count. Spikes were counted in consecutive trials within time windows of 100 ms time-locked to the onset of motion. Responses were obtained during constant (**a, c**) and dynamic (**b, d**) velocity stimulation (stimuli as in Fig. 1). The six stimulus conditions (5 constant velocity stimuli and 1 dynamical motion stimulus) were presented in a pseudo-random order before the next sequence started. a, c For constant velocity stimulation different response levels were obtained by presenting 5 stimuli that differed in their vertical extent (different symbols), centred at the cell's receptive field. The motion stimulus lasted for 2.5 s. Mean spike counts and variances were evaluated in 91 consecutive time windows starting 1.5 s after the motion onset. Consecutive time windows overlapped by 90 ms. b, d Dynamic velocity stimulation with the largest stimulus pattern used in (a) leads to pronounced response modulations (see Fig. 1d), which cover the whole activity range of the H1-cell. The motion stimulus lasted for 5 s. Mean spike counts and variances were evaluated in 481 consecutive time windows starting 100 ms after the motion onset. Consecutive time windows overlapped by 90 ms. a, b Data from one cell. Each symbol indicates the mean and corresponding variance within a given 100 ms interval time-locked to the onset of motion. c, d Data combined from 8 cells: The mean spike count was subdivided into activity classes with a width of 2.5 spikes per time window. Variances of each cell were averaged, if the corresponding mean spike count fell into the same activity class. It occurred for the analysis done for (c), that the variances of responses elicited by stimuli of different sizes fell into the same activity class and were therefore combined for the mean variance in that activity class. Then the mean variances of up to 8 cells were averaged for each activity class. Only those variances and corresponding activity classes are illustrated to which at least 4 cells contributed. Error bars denote SEMs. For each cell 60 individual responses were evaluated. (c, d modified from Warzecha and Egelhaaf 1999).

For constant as well as dynamic velocity stimulation the variance of the individual responses across trials is shown in the figure 4. The mean spike count was determined across trials within time windows time-locked to the onset of motion, and the variance of the spike count within these same time windows was calculated (Warzecha and Egelhaaf 1999). The variances obtained for the two different stimulus dynamics are very similar. The variance does not equal the mean spike count but is markedly smaller than the latter irrespective of stimulus dynamics. Moreover, the variances for the different stimulus dynamics are in the same range. De Ruyter van Steveninck et al. (1997) might have been led to a divergent conclusion because they evaluated the relationship between variance and mean spike count of responses to constant and dynamical motion stimulation in different ways (for details, see Warzecha and Egelhaaf 1999). It should be noted that the response variance across trials is a measure to quantify neuronal variability. Without further qualification, this measure does not allow one to answer the question to what extent information is represented by the exact timing of individual action potentials.

It is interesting to note that in accordance with our data on the fly H1-neurone, there is evidence that in motion-sensitive neurones in area MT the variances of responses to stimuli with different dynamical properties are fairly similar (Buracas et al. 1998). However, in contrast to the fly H1-neurone, in directionally selective neurones in the visual cortex of cats and monkeys (e.g. Britten et al. 1993; Geisler and Albrecht 1997; Buracas et al. 1998; Gershon et al. 1998; see however Gur et al. 1998) the variance was found to be at least as large as the mean spike rate (see also Barberini et al., this volume). The difference between the variability of responses of motion-sensitive cells in flies and "higher" vertebrates most likely does not have its cause in different reliabilities of the spike generating mechanisms. As is suggested by the reliable responses of cortical neurones to current injection (Mainen and Sejnowski 1995; Stevens and Zador 1998), the large variability in cortical neurones may originate from input fluctuations rather than from noise sources intrinsic to the neurones themselves. In general, cortical neurones do not only receive their input signals from peripheral elements along a given sensory pathway but also to a large extent from cortical feedback loops. Therefore, the large variability of responses of cortical neurones might be due to fluctuations of the latter type of input signals, which cannot be expected to time-lock to a given sensory stimulus. In contrast to cortical feedback circuitry, fly motion-sensitive neurones are likely to receive their input signals from neurones in the visual motion pathway which reside at the preceding or the same level of motion processing. There is no evidence for recurrent input from subsequent levels of processing. Hence, the different input organization of cortical and fly motion-sensitive neurones may be the reason for their different levels of variability (see also Barberini et al. this volume).

4.3 Temporal precision of encoding of visual motion information by spiking neurones

In order to quantify the temporal precision with which neuronal responses are coupled to the stimulus, we calculated the crosscorrelation between the different spike trains elicited by the same dynamic velocity motion. The power spectrum of the stimulus velocity covered large parts of the dynamic range of motion stimuli that are likely to be encountered by a fly in a natural situation (see Section 6). If there were no jitter in the timing of action potentials across trials, the spikes would be precisely locked to the stimulus on a millisecond timescale and the mean cross-correlogram (CCG) would be identical to the autocorrelogram of the spike train. It should then exhibit a sharp peak at time zero.

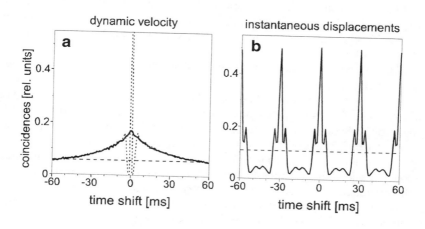

Fig. 5 Time-locking of action potentials to visual stimuli. Average crosscorrelation between pairs of individual spike responses of the H1-cell to identical stimulation. An ordinate value of 1 indicates that individual responses are identical at a temporal resolution of 1 ms. The dashed lines indicate the level of random coincidences expected on the basis of the mean activity. Temporal resolution: 1ms. **a** Crosscorrelogram (thick line, CCG) and autocorrelogram (dotted line) between 105 pairs of individual spike trains in response to dynamical motion. For the CCG, pairs were chosen pseudorandomly ensuring that every individual response trace was used and that no trace was correlated with itself. The autocorrelogram is shown only partly; at time zero it reaches a value of 1. Stimulus conditions as in figures 1b and d. The time interval evaluated for the correlograms started 100 ms after the motion onset and lasted for 4.9 s. The mean spike activity within this interval was 57 spikes/s. **b** CCG between 20 pairs of spike trains in response to instantaneous displacements of a squarewave grating in the cell's preferred direction that occurred every 30 ms. Stimuli were displayed via an LED-array that allowed to present a bright high-contrast wide-field stripe pattern and to displace this pattern within microseconds (mean luminance: 400 cd/m2; contrast: 1.0; extent of stimulus pattern: horizontal: 252°, vertical 46°; pattern wavelength: 36°; displacement: 2.25°). The time interval evaluated for the CCG started 1 s after the onset of the pattern displacements and lasted for 3 s. The mean spike activity within this interval was 108 spikes/s. (Data shown in b provided by B. Kimmerle).

In figure 5a an example of a mean CCG of responses to dynamic velocity motion is shown normalized to the peak amplitude of the corresponding mean autocorrelogram. The CCG is characterized by a rather broad and flat peak around time zero as compared to the width of the mean autocorrelogram. Hence, most action potentials of the H1-neurone are temporally coupled to the transient velocity fluctuations on a timescale of tens of milliseconds but not with a millisecond precision. The temporal precision with which spikes couple to visual motion strongly depends on the dynamics of the motion stimulus. Increasingly less action potentials are time-coupled to the motion stimulus when the velocity changes get smaller and less transient. The CCG between the different spike trains elicited by a constant velocity stimulus is basically flat (for details, see Warzecha et al. 1998).

These conclusions are in accordance with previous findings. The performance in discriminating between brief sections of dynamic motion sequences on the basis of the corresponding spike patterns as well as the reconstruction of pattern velocity from individual spike trains has been concluded to be robust to errors of several milliseconds in spike timing (de Ruyter van Steveninck and Bialek 1988; Bialek et al. 1991).

All these findings are not meant to imply that spikes in motion sensitive neurones of the fly cannot precisely time-lock to visual motion stimuli. CCGs only reflect a kind of average time-locking of spikes to the stimulus and, thus, do not tell us much about neuronal precision during special episodes of the time-dependent spike trains. Precise time-locking of spikes on a millisecond scale may occur when the velocity changes very rapidly. For instance, large instantaneous displacements of a spatially extended, high-contrast pattern in the cell's preferred direction leads to a precise time-locking of action potentials as can be inferred from the pronounced and narrow peaks in the corresponding CCG (Fig. 5b). Even less transient stimuli than instantaneous pattern displacements may lead to a precise time-locking of spikes. For instance, when the pattern suddenly reverses its direction of motion from null-direction to preferred direction motion, the initial spikes may be very precisely timed. This precision is reflected by a small temporal jitter of the first spike at the onset of a period of elevated firing (see arrow in Fig. 1d). The standard deviation of the timing of the first spike can lie below 1-2 ms at such instances for the H1-cell (see also Fig. 2 in de Ruyter van Steveninck et al., this volume). This precision is in accordance with that of neurones in cortical area MT of monkeys (Bair and Koch 1996) as well as of salamander and rabbit retinal ganglion cells (Berry et al. 1997).

It might be desirable to determine the standard deviation of spike jitter not only at the onset of motion transients when no spike has been generated for some time, but also during continuous motion. Although, the spike jitter relative to the time-dependent stimulus can be determined everywhere along a response trace, the result needs to be interpreted with caution. A standard deviation of the spike jitter as small as a few milliseconds indicates precise time-locking to the stimulus only if it is significantly smaller than, roughly speaking, half the interspike interval, which means, for instance, 2.5 ms for a firing rate of 200 Hz. In this case the small spike jitter contingent on the stimulus will also show up in distinct peaks in the spike-frequency histograms, as can be seen, for instance, at the onset of rapid motion (see Fig. 1d of the present account and Fig. 2 in de Ruyter van Steveninck this volume).

In conclusion, spikes can time-lock to visual motion on a wide range of timescales, depending on the dynamics of velocity changes. Which of these time-scales is of functional significance for the animal can only be assessed if motion stimuli as they occur in behavioural situations are taken into account (see Section 7).

5. What determines the timing of action potentials in fly tangential neurones?

The timing of spikes is mainly determined by the membrane potential and its temporal changes at the spike initiation zone. Therefore we analysed the dynamical properties of the motion-induced and stochastic components of the intracellular membrane potential of tangential neurones. In spiking tangential neurones, like the H1-cell, the postsynaptic potentials cannot easily be derived from intracellular recordings because spikes are superimposed on the postsynaptic potentials. Fortunately there are other tangential neurones such as the HS-cells (Hausen 1982a,b), in which the postsynaptic membrane potential is much less affected by active membrane properties. These cells respond mainly with graded membrane-potential changes which may be superimposed by small spike-like depolarizations. Their receptive fields largely overlap with that of the H1-cell. Since the input organization of these two cell-types is thought to be in principle the same (Hausen 1981), we proceed on the assumption that the graded potential changes recorded in HS-cells can be regarded as a "model" of the synaptically induced membrane-potential fluctuations that underly the spike activity in the H1-cell.

White-noise velocity fluctuations were used to determine the dynamic properties of the postsynaptic potentials elicited in one of the HS-cells (for details, see Warzecha et al. 1998). In response to these velocity fluctuations, the membrane potential de- and hyperpolarizes by several millivolts (Fig. 6a). The motion-induced response component obtained from averaging the responses to many presentations of the same motion stimulus (Fig. 6b), is similar to but somewhat smoother than the individual responses. The difference between a single response trace and the motion-induced response component yields the stochastic response component of the respective response trace (Fig. 6a-c). The power spectra of the stochastic and motion-induced response component are shown in figure 6d. The motion-induced response component contains most power below 20 Hz although the stimulus contained much higher frequencies. In the low frequency range, the power of the motion-induced response component is larger than that of the stochastic response component. Towards higher frequencies, the power of the motion-induced response component decreases steeply, similar to what has been described above for the motion-induced response component of the H1-cell (Fig. 6d). At frequencies above 30-40 Hz, the stochastic response component contains more power than the motion-induced component. Similar results were obtained by Haag and Borst (1997, 1998).

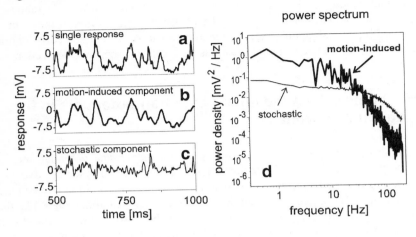

Fig. 6 Dynamic properties of the motion-induced and the stochastic response component. White noise velocity fluctuations were used to determine the dynamic response properties of the post-synaptic potentials elicited in one of the HS-cells, the HSN-cell. The resting potential of the HSN-cell (-53.4 mV) was set to 0 mV for illustration purposes. **a** Section of a sample record of an individual response trace. **b** Motion-induced response component as obtained from averaging 101 responses to the same motion trace. The same time interval relative to motion onset is shown as in (a). **c** Sample trace of the stochastic response component as obtained from the difference between the motion-induced response shown in (b) and the single response trace shown in (a). **d** Power spectrum of both the motion-induced response component (thick line) and of the stochastic response component averaged over 101 individual power spectra (thin line). (Modified from Warzecha et al. 1998)

How do these results help us to assess what factors determine the exact timing of action potentials in tangential cells? We have shown that spikes of the H1-cell are precisely coupled only to rapid velocity changes. At less transient episodes of time-dependent motion stimuli, spikes are time-locked to motion with a smaller temporal precision. As a consequence of inevitable temporal lowpass filters involved in motion detection (e.g. Reichardt 1961; Borst and Egelhaaf 1989; Egelhaaf and Borst 1993b), the response fluctuations induced even by velocity fluctuations with broad frequency spectrum contain most power below 20 Hz. (Haag and Borst 1997, 1998; Warzecha et al. 1998). In general, slow membrane potential fluctuations are less effective in evoking an action potential than fast rises of the membrane potential (Johnston and Wu 1995). Indeed, fluctuations of the membrane potential above approximately 30 Hz have been reported in various systems, including the fly, to be more effective in eliciting spikes time-locked to a stimulus than less transient fluctuations (Mainen and Sejnowski 1995; Haag and

Borst 1996; Nowak et al. 1997). This suggests that only very fast velocity changes lead to transient depolarizations that are sufficient to elicit precisely timed action potentials. Otherwise, the precise timing of spikes seems to be primarily governed by stochastic fluctuations of the membrane potential and the neuronal response is coupled to the motion stimuli on a coarser timescale.

Where do these stochastic fluctuations originate? In principle, the stochastic component observed in the responses of tangential cells could arise anywhere in the visual system starting from photon noise and the stochastic nature of photon absorption down to the level of the tangential cells themselves. It has been proposed that the performance of fly motion sensitive neurones in the third visual neuropil is limited by the photoreceptor noise (Bialek et al. 1991; de Ruyter van Steveninck and Bialek 1995). On the other hand, there is evidence that, depending on the light level, the synapse between photoreceptors and the first-order visual interneurones may also contribute a considerable amount of noise to the signal (Laughlin et al. 1987; Juusola et al. 1996; de Ruyter van Steveninck and Laughlin 1996). Apart from these noise sources in the periphery of the visual system, subsequent processing steps may also be noisy, although their significance in this regard has not yet been analysed. In addition to all these noise sources, the reliability of encoding of motion stimuli may also be affected by noise originating in the tangential cells themselves.

Whether noise intrinsic to the motion-sensitive tangential neurones mainly determines the exact timing of action potentials can be estimated by recording simultaneously the activity of two tangential cells which receive their input to a large extent from common retinotopically organized input elements. If the temporal jitter in the occurrence of spikes was primarily caused by noise originating in the common motion pathway peripheral to the tangential cells rather than by noise intrinsic to the tangential cells themselves or their input synapses, most spikes of the two cells should coincide. In contrast, when noise sources intrinsic to the tangential cells are most decisive, the spike activity of two tangential cells should not be significantly correlated.

The H1- and the H2- cell, two spiking tangential cells in the lobula plate of the fly, are thought to share large parts of their motion-sensitive input elements and not to be synaptically coupled to each other. Both neurones have largely overlapping receptive fields and the same preferred direction of motion (Hausen 1981). The mean activity of the H2-cell is lower than that of the H1-cell. The crosscorrelation between simultaneously recorded responses of the two neurones to white-noise velocity fluctuations, reveals a narrow peak (Fig. 7a). Both neurones are able to generate spikes with a much higher temporal precision than is expected on the basis of their time-coupling to the motion stimulus. This becomes obvious when we compare the CCGs of the simultaneously recorded and the randomly shuffled spike trains of both cells (Fig. 7). This conclusion is further corroborated by recordings from the H1- and H2-cell during stimulation with constant-velocity motion. The CCG of the simultaneously recorded activity reveals a similar peak as obtained for transient motion stimulation. Since the synchronicity of spikes is not elicited by the motion stimulus, we conclude that it has its origin in a common noise source in the peripheral motion pathway or in the stochastic nature of light (for details, see Warzecha et al. 1998). Obviously, the spike generating mechanism does not introduce much jitter in the timing of spikes. This conclusion is in accordance with previous results obtained in other systems (Calvin and Stevens 1968; Mainen and Sejnowski 1995).

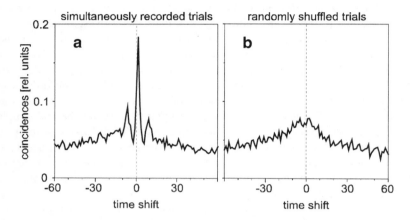

Fig. 7 Synchronization of spikes in neurones with common synaptic input. Crosscorrelograms between the simultaneously recorded responses of the H1- and H2-neurone (**a**) and shuffled CCGs of the same number of pseudorandomly chosen H1- and H2-responses that were not recorded simultaneously but obtained from repetitive stimulation with the same dynamic velocity stimulus (**b**). CCGs were normalized to the square root of the product of the peak values in the autocorrelograms of the H1- and H2-cell. An ordinate value of 1 indicates that the responses of both cells are identical at a temporal resolution of 1.1 ms. This value cannot be reached, even if all spikes of the H2-cell coincide with a spike of the H1-cell, because the H2-cell generates action potentials less frequently than the H1-cell. (**a**) The narrow peak illustrates that the H1- and the H2-cell are able to time-lock spikes to input fluctuations with a millisecond precision. The peak in the CCG is slightly shifted indicating that the H1 spikes tend to precede spikes of the H2-neurone by 1.1 ms. (**b**) The broad and flat peak in the randomly shuffled CCG shows that both neurones do not generate spikes in response to motion with a millisecond precision (for details, see Warzecha et al. 1998).

Taken together these results indicate that most spikes elicited by white-noise velocity fluctuations are not precisely time-coupled to stimulus-induced membrane potential fluctuations. Instead, spikes usually time-lock to fast fluctuations in the membrane potential that are not induced by the stimulus but are stochastic. Only, when the stimulus-induced response component is sufficiently transient may precise time-locking to the stimulus occur, as can be elicited by sudden velocity changes. These conclusions have been further corroborated by model simulations (Kretzberg et al., personal communication). Similar conclusions have been drawn for directionally selective neurones in area MT of monkeys. Here the visual stimulus also causes the neurone to modulate its spike rate on a coarse time scale consistently from trial to trial, whereas the actual timing of individual spikes has been concluded to be effectively random (for review, see Shadlen and Newsome 1998).

6. Performance of spiking and graded potential neurones in the encoding of visual motion

Fly tangential cells encode visual motion either by sequences of action potentials (e.g., H1-cell) or by graded changes in the membrane potential that may be superimposed by spike-like events (e.g., HS-cells; see also Hengstenberg 1977; Hausen 1982a; Haag et al. 1997; Haag and Borst 1998). These two response modes are not only characteristic of fly tangential neurones but are commonly employed at various processing stages of different sensory modalities in both vertebrates and invertebrates (for review, see Roberts and Bush 1981). Are there any differences between these coding strategies with respect to the reliability with which the respective neurones represent and process motion information?

The spiking and the graded response mode were compared on the basis of responses of the H1- and HS-cell by using several different criteria which address the following questions:

- How reliably do the two types of neurones signal the presence of a motion stimulus?
- How fast can they signal a motion onset?
- How many different stimulus states can be discriminated on the basis of the activity of both types of neurones?
- How much information about stimulus velocity is represented by spiking and graded potential tangential cells?

How reliably is a motion stimulus detected? As a criterion for comparing the performance of both response modes, the reliability was determined with which one can detect a constant velocity stimulus on the basis of the responses of the H1- and the HS-cell. The neuronal activity recorded before and during motion stimulation was evaluated by a statistical procedure derived from signal-detection theory (Green and Swets 1974). This procedure assumes a hypothetical ideal observer who looks through a pair of windows at the neuronal activity before and during motion stimulation. The ideal observer does not know which window belongs to which stimulus condition. Rather it is the observer's task to carry out this assignment on the basis of the integrated activity within each of the windows. The ideal observer bases his/her decision on the knowledge that, on average, the stimulus leads to an increased activity. If this procedure is applied to many responses elicited by identical stimulation, the proportion of correct decisions can be determined as a statistical measure of reliability. A value of 0.5 indicates that false and correct decisions occur equally often, i.e. the activity during motion stimulation cannot be

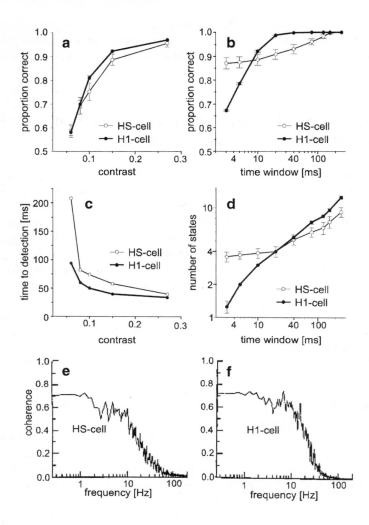

Fig. 8 Performance of HS-cell which mainly responds with graded changes in its membrane potential, and H1-cell that generates regular action potentials. **a, b** Proportion of correct decisions with which a motion stimulus can be detected on the basis of the steady state responses of the H1-cell (filled circles) and the HS-cell (open circles), plotted as a function of pattern contrast (a, time window 10 ms) and the time window within which the neuronal activity was integrated (b, contrast 0.15). **c** Time to detection as a function of pattern contrast was determined within an 80 ms time window as the interval between the motion onset and the first time instant 75% of correct decisions was reached. The time to detection was not calculated for each cell separately because at low contrasts a reliability level of 75% was only reached for some cells. **d** Number of discriminable states as a function of the width of the time window (methods described in text). (a-d): The responses of 8 H1-cells and 9 HS-cells were evaluated. For each H1-cell (HS-cell) responses to 20 (5-10) presentations of the same stimulus were taken into account. Error bars denote SEMs. (Modified from Warzecha 1994). **e, f** Average coherence as a function of stimulus frequency, calculated from the responses of 6 HS-cells (e) and 10 H1-cells (f) with 5 to 20

discriminated from the resting activity. A value of 1 is obtained if there are only correct decisions (for details of the method, see Warzecha and Egelhaaf 1998). For the comparison of the performances of the H1- and HS-cell, pattern contrast was varied to cover large parts of the neurones' activity range.

For both the H1- and the HS-cell the proportion of correct decisions increases in a similar way with increasing pattern contrast and thus an increasing mean response amplitude (Fig. 8a). The proportion of correct decisions also increases with increasing size of the time window within which the neuronal activity is integrated, because large time windows average out the stochastic component to a larger extent than do small time windows (Fig. 8b). For small time windows the responses of the HS-cell are more reliable than those of the H1-neurone. For intermediate window sizes the responses of the H1-cell lead to higher proportions of correct decisions. Large time windows (above 100 ms) result in a very reliable detection of the motion stimulus for both cell types.

How fast is a motion onset detected? The time it takes to detect the onset of a motion stimulus is used as a further criterion to compare the performance of the two types of tangential cells. To evaluate how long it takes until the motion onset is reliably signalled the proportion of correct decisions is determined as a function of time. The "time-to detection" is defined as the time interval between the motion onset and the moment when a reliability of 75% correct decisions is reached (for details of the method, see Warzecha and Egelhaaf 1998). For both the H1- and the HS-cell the time-to-detection decreases considerably with increasing pattern contrast reaching reliable responses for high-contrast motion stimuli already after about 30 ms (Fig. 8c). The graded potential neurone does not signal the motion onset any faster than the spiking neurone. Within limits, this result does not much depend on the size of the time window with which the performance of the neurones was assessed (for details, see Warzecha 1994).

How many stimulus states can be discriminated? As another criterion to assess the performance of spiking and graded potential neurones, the number of stimulus states has been determined that can be discriminated with a reliability of 75% on the basis of the neuronal activity. Again the concept of an ideal observer assigning responses to one of two stimuli was used. The stimuli which had to be discriminated were constant velocity stimuli which differed in their contrast. For each contrast the distribution of the responses to repeated stimulation was determined by integrating for each individual response the steady-state neuronal activity

---

individual responses per cell. The coherence indicates how much information about the stimulus velocity is preserved in the neuronal responses. (Modified from Fig. 5, Haag and Borst 1997). Due to the low spontaneous activity of the H1-cell, the activity is modulated over a much larger range by motion in the preferred than by motion in the null direction, whereas the HS-cell is not restricted in this way because it hyperpolarizes during motion in null direction. To enable a direct comparison between the performance of a spiking and a graded response cell, the original velocity trajectory was replayed to the H1-cell also in a mirror-symmetrical version so that every displacement of the stimulus pattern occurred in the direction opposite to the original one. The spike trains of the H1-cell to both versions of the stimulus trace were combined into a single response with positive and negative spikes for calculation of the coherence function.

within a given time window. This was also done for the resting activity. The distribution obtained for the resting activity is used as the first reference distribution. A response distribution is then sought such that a randomly drawn sample of this distribution can be discriminated with a reliability of 75% from a randomly drawn sample of the reference distribution. Since the response distributions could only be determined for a limited number of contrasts, samples of adjacent response distributions are unlikely to be discriminable with a reliability of exactly 75%. Hence, intermediate distributions were interpolated from the experimentally determined distributions obtained with the next higher and next lower contrast. Once the distribution was determined that can be discriminated with a reliability of 75% from the resting distribution, this distribution was used as the next reference distribution. The whole procedure was repeated until the response distribution was reached that is associated with the highest contrast (for details, see Warzecha 1994). For small time windows more stimulus states can be discriminated on the basis of the HS-cell responses than on the basis of the responses of the H1-cell. For larger time windows the performance of both cells does not differ much, although the spiking neurone may be able to encode slightly more states than the graded potential cell (Fig. 8d). Hence, with respect to the number of stimulus states which can be discriminated on the basis of the neuronal activity, graded potentials have advantages over spikes only on a short time scale.

How much information about stimulus velocity is preserved in spiking and graded potential neurones? The performance of spiking and graded potential tangential neurones was further compared by determining how well the velocity of a randomly fluctuating motion stimulus is represented by the two neurones with different response modes (Haag and Borst 1997). The reverse reconstruction method (Eggermont et al. 1983; Bialek et al. 1991; Theunissen et al. 1996; Borst and Theunissen 1999) was used to reconstruct the stimulus velocity from the neuronal responses. The analysis was done in the frequency domain yielding the best linear filter that transforms the neuronal responses into the stimulus velocity. The coherence between the real and the reconstructed stimulus calculated as a function of frequency served as a measure to quantify how well either cell can represent the velocity. Up to oscillation frequencies of approximately 10 Hz the stimulus velocity is well preserved in the responses of both types of neurones as long as the pattern moves into the preferred direction of motion. However, the spike activity of the H1-cell does not very well encode the velocity of a pattern moving in the cell's null direction. Apart from this difference that results from the low spontaneous activity of the H1-cell and the accordingly limited dynamic range for motion in the null direction, both the HS-cell and the H1-cell perform almost indistinguishably (Figs. 8e and f). At higher oscillation frequencies and larger stimulus amplitudes the coherence between stimulus velocity and the time course of the neuronal responses becomes much smaller (Haag and Borst 1997). This deviation from representing stimulus velocity is only partly due to stochastic fluctuations in the neural responses. Since the stimulus-induced responses of fly tangential neurones do not only depend on the velocity but also on its temporal derivatives (Egelhaaf and Reichardt 1987; see also Section 4.1), it is to be expected that the coherence between stimulus velocity and the neuronal responses should decrease for more transient motion stimuli. Irrespective of these complications, it is important to note in the present context, that the spiking and the graded potential neurone perform in basically the same way in representing dynamical motion stimuli in the cells' preferred direction.

In conclusion, as judged by a wide range of criteria, a spiking and a graded potential neurone that are located at the same level of information processing in the fly's motion pathway resemble each other closely with respect to their reliability in representing motion information. Only at a fine timescale are there pronounced differences in the performances of both response modes, where the graded potential cell may be superior over the spiking one (Figs. 8a-d). Of course, the possibility that other graded and spiking neurones perform differently to the HS- and H1-cell, respectively, cannot be excluded. Moreover, by using other criteria to assess the performance of spiking and graded potential neurones more significant differences between both response modes may emerge. A final assessment with respect to the still open problem why some neurones convey information by graded potentials whereas others do so by generating trains of spikes is only possible, if it is known which parameters of neuronal activity carry behaviourally relevant information (e.g. Liebenthal et al. 1994).

7. Ecological constraints for neuronal representation of motion

The temporal precision required to represent visual motion mainly depends on the dynamical properties of the motion stimuli that an animal encounters when solving a particular task. An animal is confronted with visual motion in two kinds of situations, either, when it views moving objects or when it moves through its eviron ment (see Eckert and Zeil, this volume). In the latter case the images of objects in the surroundings move across the retina of the animal even if these objects are stationary. The dynamical properties of retinal motion do not only depend on the time-varying velocity of the moving animal or of objects moving in its visual field, but also on the three-dimensional layout of the environment. Changes in the retinal velocity are thus not necessarily due to velocity changes of the moving animal or of objects moving in its visual field. During translation of the animal, they may also have their cause in changing distances between the eye and stationary or moving objects in the surround. Changes in the direction and speed of an animal's self-motion are limited by inertia and friction. The relative contribution of these two factors are influenced by the size of the animal and the substrate on or in which it moves. As a consequence of these physical constraints, retinal motion stimuli will not change their direction at arbitrarily high frequencies under natural conditions.

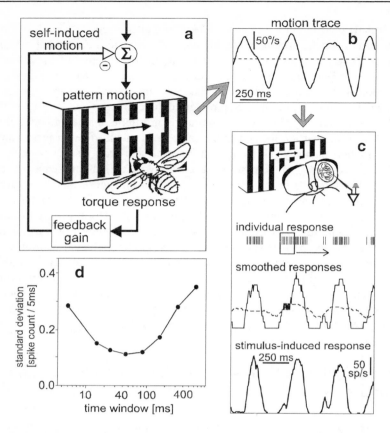

Fig. 9 Behaviourally generated motion stimuli and their representation by the H1-cell. **a** Generation of the motion traces in a behavioural closed-loop situation in a flight simulator. The fly is
tethered to a torque compensator (not shown) which allows to determine the fly's instantaneous
yaw torque. The visual consequences of self-motion are simulated by transforming the torque
signals into image displacements on a CRT screen. **b** Short section of a motion trace generated
by a fly in the situation illustrated in (a). **c** In the upper part a fly looking at a monitor screen is
shown schematically from behind with a hole cut into its head capsule and an electrode inserted
into the brain. Behaviourally generated motion traces were replayed to the fly while recording
the spike activity of the H1-cell in the right half of the brain. A section of a spike train obtained
in this way is shown in its original form (with each vertical line indicating the time of occurrence
of a spike; upper trace) and in two temporally smoothed versions (middle traces), obtained by
integrating the number of spikes within a time window that was sliding across the spike trains.
The size of the time window was either 40 ms (middle trace, solid line) or 320 ms (middle trace,
dashed line). The bottom trace shows the corresponding section of the stimulus-induced response
component as obtained by averaging over 40 individual responses to the identical motion trace. **d**
Similarity between smoothed individual and stimulus-induced responses given by the standard
deviation: For each time window the squared difference between each smoothed individual
response and the stimulus-induced response was averaged across time and across trials. For each
cell the square root of this value was determined. The analysis was performed at a temporal
resolution of 5 ms. Values indicate means of three cells. The stimulus-induced response component
elicited by behaviourally generated motion stimuli can be estimated best for each instance of time
from noisy neuronal signals if these are smoothed by medium-sized time windows (40-100 ms).

Not much is known about the dynamical properties of behaviourally relevant motion stimuli. They can only be characterized by analysing the behaviour of the animal and by reconstructing the optic flow encountered by the animal in the real world. In the fly, a first modest attempt has been made in the context of optomotor course control to analyse neural coding of visual motion stimuli that were generated by the fly's own actions and reactions. The analysis was done with a flight simulator operated under closed-loop conditions where the tethered flying fly could control by its yaw torque the horizontal displacements of its retinal input (Fig. 9a).

When the visual consequences of a disturbance of the fly's flight course are simulated by displacing the stimulus pattern with a constant velocity in one direction, flies are able to compensate for this disturbance to a large extent. Nonetheless, the torque and thus the retinal motion is characterized by pronounced temporal fluctuations (Fig. 9b, Warzecha and Egelhaaf 1996). The power spectrum of these velocity fluctuations is characterized by low frequencies of up to 5 Hz (Heisenberg and Wolf 1988; Warzecha and Egelhaaf 1997). The motion stimuli generated by a behaving fly were replayed in electrophysiological experiments while the activity of the spiking H1-cell was recorded (Fig. 9c). The spike activity was modulated depending on the direction and speed of pattern motion, with some jitter in the timing of spikes. As a consequence, the spike rate as obtained by averaging over many presentations of the same behaviourally generated motion trace varied smoothly, similar to the responses elicited by white-noise velocity fluctuations (for details, see Warzecha and Egelhaaf 1997).

How well do these neuronal responses encode the behaviourally generated motion stimuli? This question relates to two aspects which should not be confounded, i.e. (i) the parameters of the motion stimulus that are encoded by the neuronal responses and (ii) the reliability with which this is done. The first of these aspects can be analysed by reconstructing a certain feature of the motion stimulus from the time-dependent spike trains. This approach has been employed for dynamical motion stimuli, though not behaviourally generated ones, by determining the linear filter that leads, on the basis of the individual response traces, to an optimal estimation of the time-dependent velocity of motion (see Section 4.1). In this way, it is possible to assess how well individual responses represent a particular stimulus parameter, such as pattern velocity. However, this approach does not easily allow us to assess how reliably motion information is processed because deviations of the estimated velocity from the real pattern velocity can have two reasons. One reason is the nonlinear relationship between the velocity of the motion stimulus and the neuronal responses (see Section 4.1.). The other reason is neuronal noise (Warzecha and Egelhaaf 1997; Haag and Borst 1997). The reliability with which motion stimuli are encoded, i.e. the second of the above-mentioned aspects, has been assessed for the behaviourally generated motion stimuli obtained in the context of optomotor course control by relating the individual response traces to the stimulus-induced response component, i.e. to the instantaneous spike rate. If there were no jitter in the timing of spikes and the responses were precisely coupled to the motion stimulus, the individual responses and the stimulus-induced response component should look alike. Obviously, this is not the case (Fig. 9c) and it is not possible to predict the stimulus-induced response component very well for each instant of time from the instantaneous activity of individual responses. A much better prediction of the stimulus-induced response component is possible for each instant of time, when the individual spike trains are temporally smoothed to some extent (compare the different traces in Fig. 9c). It is obvious that, if the time window within which spikes are temporally averaged is too small, the filtered individual responses and the stimulus-induced response component differ greatly. On the other hand, if the time window is too large, fluctuations of the response component that are elicited by the motion stimulus are much attenuated. Hence, there should be an optimal time window leading to the best representation of the stimulus-induced response component on the basis of individual responses or, in other words to the best prediction of the spike rate for each instant of time. The similarity between the filtered individual responses and the stimulus-induced response component was assessed in two ways (i) on the basis of an information theoretic approach (Warzecha and Egelhaaf 1997, see Appendix) and (ii) by determining for each instant of time the square root of the mean squared differences between the filtered individual response traces and the stimulus-induced response component ("difference measure"). Both approaches led to the same result: As is shown in figure 9d for the difference measure, the stimulus-induced response to motion generated by the fly during optomotor course stabilization can be estimated best for each time instant when individual spike trains are temporally smoothed within time windows of some tens of milliseconds, that is on a timescale on which the exact timing of individual spikes does not matter much. This finding may be not very surprising, as in this task the motion stimuli and, accordingly, the motion-induced response component did not contain much power for oscillation frequencies above 6-10 Hz. The optimal time window will be smaller when the motion stimuli and thus the stimulus-induced response component contain higher frequencies than the stimuli generated by the behaving fly in the context of optomotor course control and the corresponding responses used here for analysis (see Appendix).

Although the retinal image displacements are expected to be much smaller while the fly tries to stabilize its course against disturbances than during voluntary turns, the significance of the findings summarized in figure 9 will be qualified in the following. First of all, it is necessary to exclude the possibility that the relatively slow dynamics of the behaviourally generated motion traces is the consequence of the lowpass properties of the flight simulator (Warzecha and Egelhaaf 1997). Since in another behavioural context (object detection and fixation), that was also analysed using the flight simulator, much faster responses are generated by the fly than during optomotor course control (Egelhaaf 1987; Zanker et al. 1991; Kimmerle et al. 1997), the dynamics of the flight simulator are not the main determinant of the dynamics of the behaviourally generated optomotor responses. In addition, increasing or decreasing the cut-off frequency of the lowpass filter in the flight simulator by a factor of two does not change the dynamics of the opto-motor responses in any obvious manner (Warzecha and Egelhaaf 1997). More-over, there is evidence that the signals of the output elements of the third visual neuropile are temporally lowpass filtered somewhere between the third visual neuropile and the steering muscles mediating optomotor course stabilization (Egelhaaf 1987, 1989). The time constant estimated for this lowpass filter is much larger than that of the flight simulator. Hence, the relatively slow dynamics of image motion during optomotor course stabilization, as analysed in the flight simulator, can be concluded to be a genuine property of this control system. Of course, it is not easily possible to derive from the experiments done with the flight simulator the functional significance of the optomotor responses under free-flight conditions. This problem cannot be resolved at present, since no one has analysed under free-flight conditions optomotor course stabilization in blowflies (Calliphoridae) and houseflies (Muscidae) on which most behavioural experiments in the flight simulator were done. However, there is evidence that in several insect species optomotor course stabilization is comparatively slow also under free-flight conditions (Collett 1980; Farina et al. 1995; Kern and Varjú 1998). Although hoverflies, for instance, can execute tremendously virtuosic flight manoeuvres and can change their flight direction very rapidly (Collett and Land 1975), their compensatory turning responses – even under free-flight conditions – have a temporal lowpass characteristic which leads to attenuation of the optomotor response at frequencies above 1 Hz (Collett 1980) and thus in a similar range as found in the flight simulator with houseflies (Egelhaaf 1987). Hence, there might exist common computational reasons for optomotor course stabilization to operate on a relatively slow timescale (for discussion see Collett 1980; Egelhaaf 1987).

The temporal characteristics of the retinal image displacements may differ considerably for behaviour other than optomotor course stabilization. The relevant timescales of retinal image motion are thus likely to vary over a wide range in different behavioural contexts. Unfortunately, there are only few examples where information about the dynamics of behaviourally relevant motion stimuli is avail-able. One such example is optomotor position stabilization of the hummingbird hawkmoth which hovers almost stationarily in front of flowers while sucking nectar. Flowers, on which the hawkmoth is feeding, were found to wiggle in the wind at frequencies between about 1 and 2 Hz and thus change their direction of motion on a relatively slow timescale (Farina et al. 1994). Another example are the retinal image displacements experienced by solitary wasps while acquiring a visual representation of the environment of their nest in systematic orienting flights. During these flights the animals fly in ever increasing arcs around their nest, thereby changing their direction of motion and, thus, the overall direction of retinal image displacements at frequencies in the range of 0.5 Hz. During these movements the wasps do not change the orientation of their body axis smoothly but rapidly in a saccade-like manner. Nevertheless the prevailing frequencies in the angular velocity profiles hardly lie above 10 Hz (Zeil 1993; Voss and Zeil 1998). Similar saccade-like turns can be observed during free-flight manoeuvres of flies (Wagner 1986a,b; Land 1993). In recent experiments where the turning dynamics of freely flying flies could be analysed at very high spatial and, in particular, temporal resolution, even the most rapid turns appear to take at least 15 ms (van Hateren and Schilstra 1999; Schilstra and van Hateren 1999). So far, no one has recorded neuronal responses to these rapid velocity changes. However, as judged by the responses to more conventional laboratory stimuli, it is well conceivable that the initial spikes elicited by changing the direction of the body axis as rapidly may well be timed precisely on a millisecond timescale (see also de Ruyter van Steveninck et al., this volume).

In this context it needs to be reiterated that not all visual motion stimuli an animal is confronted with in real life are as transient. The rapid turns as described above are certainly just one extreme. Also chasing behaviour of male flies in the context of mating behaviour is extreme with respect to speed and virtuosity. Since in many free-flight studies – for technical reasons – the flies were restricted to fly within a relatively small space, requiring frequent changes of flight direction, we do not know anything about the dynamics of retinal motion in other behavioural contexts, for instance, during cruising flight over distances of some tens to hundreds of meters or when the animal compensates for a disturbance of its flight course. Hence, no firm conclusions concerning the relevant timescales of encoding of motion information in normal behavioural situations and in a variety of relevant behavioural contexts of a fly are possible at present.

8. Conclusions

All information about the outside world available to an animal is somehow encoded in the temporal activity patterns of its neurones. A variety of approaches has been employed to demonstrate that for directionally selective, motion sensitive neurones in the fly different aspects of visual motion stimuli can be derived even from individual responses. (i) The onset and presence of motion can be detected reliably (Section 6). Under certain conditions, even the pattern velocity can be reconstructed from individual spike trains (see Section 4.1; Bialek et al. 1991; Haag and Borst 1997). (ii) It is possible to discriminate between spatial displacements which differ by less than the spacing of two photoreceptors on the basis of the timing of the first spike or the first pair of spikes generated after stimulus presentation (de Ruyter van Steveninck and Bialek 1995). (iii) Pairs of spikes may carry considerably more information about the motion stimulus than the sum of the information contributed by each spike separately (de Ruyter van Steveninck and Bialek 1988).

Although these findings clearly reveal that temporal patterns of neural activity provide significant information about visual motion, they do not allow us, without further qualification, to make inferences about the timescale on which spikes are locked to the timecourse of motion. We have argued that the real-time performance of motion-sensitive neurones in the fly visual system is constrained by various properties of the underlying neuronal machinery. As a consequence of the biophysical properties of neurones, action potentials time-lock to fast membrane potential changes with a higher temporal precision than to slower changes. Hence, spikes are likely to time-lock on a millisecond scale to rapid stochastic membrane potential changes which dominate in the postsynaptic potentials of the cells, if the stimulus-induced membrane potential changes are not sufficiently fast (see Sections 4.3 and 5). Although fast velocity transients are much attenuated by the motion pathway as a consequence of temporal filters inherent to the mechanism of motion computation, it is possible to generate rapid velocity changes that lead to a precise time-locking of spikes on a timescale of 1-2 ms. Whether changes in the optic flow occurring in natural flight situations are transient enough to lead to precise time-locking of spikes needs to be analysed in future experiments. As mentioned above, there are indications that this might well be the case under special conditions (de Ruyter van Steveninck et al., this volume). Nonetheless, most spikes elicited by broad-band velocity fluctuations are coupled to the stimulus on a timescale of some tens of milliseconds. We thus conclude that the timing of spikes on a millisecond scale is governed to a large extent by stochastic membrane potential fluctuations that are more transient than the membrane potential fluctuations elicited by most episodes of visual motion traces. As a consequence, the stimulus-induced neuronal response component elicited by motion stimuli generated by the behaving fly while stabilizing its course against disturbances can be estimated best for each instant of time on the basis of individual response traces, if these are temporally smoothed on a timescale of several tens of milliseconds (see Section 7).

At first sight these findings appear to contradict recent results of an elaborate information theoretic analysis of responses of the H1-neurone by Strong et al. (1998). Their study has shown that spike trains of the H1-cell elicited by broadband velocity fluctuations transmit increasingly more information about the stimulus when the spike responses are evaluated at an increasingly finer temporal resolution down to millisecond precision. It is puzzling why seemingly contradictory results have been obtained on the basis of the activity of the same neurone with, at least partly, the same type of stimuli, and both studies using a similar measure for the neurone's reliability. A close look, however, reveals that the results of both studies do not necessarily contradict each other. Instead, different aspects of encoding of motion information have been analysed. Strong et al. (1998) were looking at a given signal, i.e. the spike trains of the H1-cell, with different temporal resolutions and demonstrated that the information transmitted by the individual responses about the stimulus increases with the temporal resolution. In contrast, we compared with a given temporal resolution different signals, i.e. the individual spike trains and their temporally smoothed versions and demonstrated that the stimulus-induced response component can be estimated best from noisy individual spike trains if these are smoothed to some extent (Figs. 9c and d, see Appendix).

The conclusions regarding the temporal precision with which spikes are locked to the stimulus raise a principal conceptual question. All inferences an animal can make about its outside world are necessarily based on the electrical activity evoked by its sensory input. Even if part of the incoming action potentials are time-locked very precisely to the stimulus, in the absence of additional information the animal has no chance to infer which of the action potentials can be relied on with respect to the timing of the stimulus and which are less reliable in this regard. It needs to be analysed whether there are means in the fly's motion pathway to make use of the information which is potentially carried by precisely timed spikes.

Although not much is known about the relevant timescales of visual motion an animal such as the fly encounters in normal life, it appears to be plausible to assume that the dynamics of retinal motion covers a wide range and may vary systematically and quite a lot with the behavioural context (see Section 7). Hence, it is hard to define the computational needs for a temporal coupling of spikes to visual motion stimuli on a millisecond scale. However, very precise timing of spikes to the sensory input is necessary, in other computational contexts, for instance when an object is to be localized on the basis of acoustic or electrical signals or when the activity of flight motor neurones is coupled to the temporal phase of the wingbeat as is the case in flies. Accordingly, very rapid stimulus-induced membrane potential fluctuations and, thus, a very precise time-locking of spikes to stimuli are found in neurones of the fish electrosensory system involved in object detection (Kawasaki 1993), in the system mediating acoustic sound localization of vertebrates (Carr 1993) and the mechanosensory system of flies involved in coupling the activity of flight motor neurones to the temporal phase of wing beat (Fayyazuddin and Dickinson 1996). Motion vision systems that do not need to represent such high-frequency fluctuations under most behaviourally relevant stimulus conditions appear to be adapted to the more slowly changing stimulus-induced fluctuations they encounter in the real world. In fact, it has been suggested on the basis of the temporal tuning of motion-sensitive neurones in a variety of insect species, that the temporal filtering properties of motion detection systems are adapted to the lifestyle of the respective animal (O'Carroll et al. 1996). However, the quantitative characterization of naturally occuring motion stimuli poses many technical problems and has not been done in any detail so far. We are just beginning to quantify behaviourally relevant visual motion signals by reconstructing the time-dependent retinal input from video films taken of freely moving flies in a variety of behavioural contexts (Kern et al. 1999, 2000). It will be one of the challenges in the near future to investigate how and on what time-scale motion stimuli that are encountered by the animal in natural situations are represented by its nervous system.

In conclusion, the relevant timescale needed for the encoding of motion information cannot be derived directly from information theoretic or system analytical approaches to responses elicited by artificial stimuli. Rather, it is essential also to take an ecological perspective to find out what stimuli an animal encounters in real life. Only then it is possible to tell what might be really relevant for an animal and thus to understand what neural codes are used by a particular nervous system in a given task.

Acknowledgements

We thank our co-workers R. Kern, B. Kimmerle, H. Krapp and J. Kretzberg, for reading and discussing the manuscript. B. Kimmerle performed the experiment shown in figure 5b which is gratefully acknowledged. J. Zanker and J. Zeil as well as two anonymous referees made many critical annotations to a previous version of the paper and, thus, helped to improve it considerably.

Appendix

On the estimation of the stimulus-induced response component from individual spike trains: Misconceptions and misunderstandings

Individual spike trains elicited by dynamical motion stimuli can be used to estimate the corresponding velocity trajectory or the stimulus-induced response component which is reflected in the time course of the mean spike rate (see Sections 4.1 and 7). The best estimates are obtained if the noisy individual spike trains are smoothed to some extent by appropriate temporal filtering (Bialek et al. 1991; Haag and Borst 1997; Warzecha and Egelhaaf 1997; see also Fig. 9c). Temporal filtering increases the similarity between the individual spike trains and the signal to be estimated only if the signals are not statistically independent at subsequent instants of time, but contain temporal correlations. If there are temporal correlations on a coarse timescale the time constant of the best filter is larger than when there are correlations mainly on a finer timescale. Accordingly, the time-constant of the filter can be expected to be smaller for rapidly varying stimuli than for slowly varying ones (see below). Although all this is already obvious from comparing the filtered spike trains with either the stimulus velocity (e.g., Fig. 2 in Bialek et al. 1991) or the stimulus-induced response component (see Fig. 9c), the quality of the estimation needs to be quantified, especially if the performance of different filters is to be compared. There are various ways how this comparison can be done. Perhaps the most straightforward way is to determine $(\chi^2 = \int [s(t) - s_{est}(t)]^2 dt$, with s(t) representing either the time-dependent stimulus velocity or the stimulus-induced response component and $s_{est}(t)$ corresponding to the estimated signal based on temporally filtering the individual responses (see difference measure in Section 7 and Bialek et al. 1991).

In a previous account we employed another measure of how well the stimulus-induced response component can be estimated for each instant of time on the basis of the original and filtered individual responses (Warzecha and Egelhaaf 1997). To obtain a measure for the similarity we first determined $p(s_i | r_j)$, the conditional probability with which a particular level of the stimulus-induced response component (s_i) occurs given a particular individual response (r_j), and $p(s_i)$, the probability of occurrence of (s_i). We then calculated the ratio of both terms $p(s_i | r_j)/p(s_i)$. This ratio will be unity and at its minimum, if the time-dependent individual responses and the stimulus-induced response are completely unrelated, i.e. knowledge of r_j does not help to predict s_i, (i.e. $p(s_i | r_j)=p(s_i)$). The ratio will be larger than 1, if r_j restricts the possible range of stimulus-induced responses (i.e. $p(s_i | r_j)>p(s_i)$). The more similar individual responses, r_j, are to the corresponding stimulus-induced response component, si, the better we can predict what si is likely to be when we know r_j and the larger the ratio $p(s_i | r_j)/p(s_i)$ will be. This ratio can be determined for each activity level of the stimulus-induced response component and each activity level of the individual responses and it can be used as a measure of the similarity between both signals. Using the logarithm of these ratios and taking into account that not all combinations of activity levels si and rj are equally likely to occur by weighting the different ratios by the probability of their occurrence, a measure of similarity is obtained which is formally equivalent to the so-called transinformation

$$T=\sum_{i,j} p(s_i,r_j) \log_2[p(s_i | r_j)/p(s_i)].$$

The transinformation is the information which is transmitted by a signal, r, about another signal,s (e.g. Shannon and Weaver 1949; Rieke et al. 1997). In our previous study we plotted this transinformation in relative units (Fig. 3 in Warzecha and Egelhaaf 1997) because we only wanted to use the transinformation as a measure of the overall similarity between the stimulus-induced response component and the filtered or unfiltered individual responses. It should be noted that the transinformation was always determined with the same temporal resolution (time bins of 5 ms) irrespective of how the individual spike trains were temporally smoothed. In accordance with the analysis shown in the present account (see Fig. 9), we obtained for the behaviourally generated motion traces the largest similarity, if the individual spike trains were smoothed by time windows of a width of 40 to 100 ms (Warzecha and Egelhaaf 1997). If the stimulus-induced response component contains higher frequency components than the one evoked by the behaviourally generated stimuli, the optimal time window is shifted to smaller values. This shift is illustrated in figure 10 for a model simulation. Two different motion-induced response traces were simulated by lowpass-filtering the same white noise sequence with a cut-off at either 80 Hz ("high-frequency response") or 20 Hz ("low frequency response"). These simulated time-dependent motion-induced responses were used to generate individual response traces (for details, see legend of Fig. 10). After smoothing individual spike trains to a variable extent, the transinformation was calculated separately for the high-frequency response and the corresponding individual spike trains on the one hand and the low frequency response and the corresponding spike trains on the other hand. For both the low- and the high-frequency responses the transinformation increases when individual responses are temporally smoothed before calculating the transinformation. The size of the optimal time window increases when the power of the motion-induced response is decreased in the high frequency range (compare triangles to circles in Fig. 10). This conclusion does not depend on the specific statistics of spike activity used for the simulations. Similar results are also obtained for experimental data. When instead of the motion stimuli generated by the behaving fly in the context of optomotor course control, white-noise stimuli as shown in figure 1 are presented to an H1-cell, the optimal time window for smoothing individual responses reduces from 80 to about 10 ms (Warzecha unpublished).

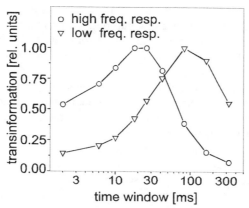

Fig. 10 Transinformation between artificially generated individual spike trains and the corresponding stimulus-induced response obtained by averaging 500 individual spike trains. In the first step of the simulation two different motion-induced responses and the corresponding individual spike trains were calculated. The responses differed with respect to their frequency content. A sequence of 5000 Gaussian distributed random numbers was generated corresponding to a response trace of 10 s with a temporal resolution of 2 ms. To obtain the stimulus-induced response, the sequence of random numbers was temporally filtered by a first-order lowpass with a cutoff at either 80 Hz ("high frequency response") or 20Hz ("low frequency response"). The high-frequency response was normalized to a range of values between 0 and 0.7. The low-frequency response was normalized by the same factor. These time-dependent responses determined the probability of spike generation as a function of time. The individual response traces were generated by comparing for each elementary time bin of 2 ms a uniformly distributed random number between 0 and 1 with the spike probability given by the respective stimulus-induced response in the respective time bin. If the spike probability exceeded the random number a spike was assigned to the corresponding time bin. 500 individual spike trains were generated in this way for each stimulus-induced response (80 Hz and 20 Hz). Prior to the calculation of the transinformation, the individual response traces were temporally smoothed by integrating the number of spikes within time windows of varying size (abscissa). For the calculation of the transinformation responses were subdivided into 11 equally sized activity classes (for details of the calculation, see Warzecha and Egelhaaf 1997). The transinformation was determined by taking into account the mean activity within 6 ms time bins. The transinformation was normalized to its maximum separately for the high frequency response (circles) and the low frequency response (triangles), respectively. The transinformation first increases with increasing temporal smoothing of the individual response traces, reaches an optimum and then decreases again.

Having in mind the basic information theoretic theorem, that the information transmitted by a signal cannot be increased by temporally filtering the signal (Shannon and Weaver 1949), our result may, at first sight, appear paradoxical (see de Ruyter van Steveninck et al., this volume). There is no doubt that the information transmitted by a time-dependent signal cannot increase by filtering it. It should be noted, that one cannot determine the information content of time-dependent signals in the way described above, unless it is ensured that the neuronal signals in subsequent time units that form the basis for calculating the information are statistically independent. This is a critical point which is usually not tested when calculating the information content of a neuronal signal. It is quite clear that the statistical independence is not given for subsequent time bins of 5 ms which formed the basis of our calculations. Moreover, it is also evident that temporal filtering the individual spike trains introduces correlations at a larger time-scale and therefore fewer statistically independent activity levels per time unit. It should be noted that these correlations are not just a consequence of the filtering procedure but reflect correlations which are intrinsic in the time-modulated stimulus-induced response component. As a consequence of these correlations the stimulus-induced response component can be predicted more reliably on the basis of the individual responses, when these are filtered appropriately. However as a consequence of these inevitable correlations, it would be a severe misconception, to interpret the measure of similarity we used in our previous account (Warzecha and Egelhaaf 1997) in terms of information rates. One reason for this misinterpretation may be that the measure we used for assessing the similarity is called "transinformation" and thus suggests that the outcome of the calculations may be interpreted as information rate. Certain formulations in our previous account (Warzecha and Egelhaaf 1997) were not sufficiently precise in this regard and, thus, might have facilitated misinterpretations of our results.

References

Allen C, Stevens CF (1994) An evaluation of causes for unreliability of synaptic transmission. Proc Natl Acad Sci USA 91: 10380-10383

Bair W, Koch C (1996) Temporal precision of spike trains in extrastriate cortex of the behaving macaque monkey. Neural Comput 8: 1185-1202

Berry MJ, Warland DK, Meister M (1997) The structure and precision of retinal spike trains. Proc Natl Acad Sci USA 94: 5411-5416

Bialek W, Rieke F (1992) Reliability and information transmission in spiking neurons. Trends Neurosci 15: 428-433

Bialek W, Rieke F, de Ruyter van Steveninck R, Warland D (1991) Reading a neural code. Science 252: 1854-1857

Borst A, Egelhaaf M (1987) Temporal modulation of luminance adapts time constant of fly movement detectors. Biol Cybern 56: 209-215

Borst A, Egelhaaf M (1989) Principles of visual motion detection. Trends Neurosci 12: 297-306

Borst A, Theunissen FE (1999) Information theory an neural coding. Nature Neurosci 2: 947-957

Britten KH, Shadlen MN, Newsome WT, Movshon JA (1993) Responses of neurons in macaque MT to stochastic motion signals. Vis Neurosci 10: 1157-1169

Buchner, E (1984) Behavioural analysis of spatial vision in insects. In: Ali MA (ed) Photoreception and vision in invertebrates. Plenum Press, New York, London, pp 561-621

Buracas GT, Zador AM, DeWeese MR, Albright TD. (1998) Efficient dicrimination of temporal patterns by motion-sensitive neurons in primate visual cortex. Neuron 20: 959-969

Calvin WH, Stevens CF (1968) Synaptic noise and other sources of randomness in motoneuron interspike intervals. J Neurophysiol 31: 574-587

Carr CE (1993) Processing of temporal information in the brain. Ann Rev Neurosci 16: 223-243.

Collett TS (1980) Angular tracking and the optomotor response. An analysis of visual reflex interaction in a hoverfly. J Comp Physiol 140: 145-158.

Collett TS, Land MF (1975) Visual control of flight behaviour in the hoverfly *Syritta pipiens* L. J Comp Physiol 99: 1-66.

Egelhaaf M (1987) Dynamic properties of two control systems underlying visually guided turning in house-flies. J Comp Physiol A 161: 777-783

Egelhaaf M (1989) Visual afferences to flight steering muscles controlling optomotor response of the fly. J Comp Physiol A 165: 719-730

Egelhaaf M, Borst A (1989) Transient and steady-state response properties of movement detectors. J Opt Soc Am A 6: 116-127

Egelhaaf M, Borst A (1993a) A look into the cockpit of the fly: Visual orientation, algorithms, and identified neurons. J Neurosci 13: 4563-4574

Egelhaaf M, Borst A (1993b) Movement detection in arthropods. In: Wallman J, Miles FA (eds) Visual motion and its role in the stabilization of gaze, Elsevier, Amsterdam, London, New York, pp 53-77

Egelhaaf M, Reichardt W (1987) Dynamic response properties of movement detectors: Theoretical analysis and electrophysiological investigation in the visual system of the fly. Biol Cybern 56: 69-87

Egelhaaf M, Warzecha A-K (1999) Encoding of motion in real time by the fly visual system. Curr Opinion Neurobiol 9: 454-460

Egelhaaf M, Hausen K, Reichardt W, Wehrhahn C (1988) Visual course control in flies relies on neuronal computation of object and background motion. Trends Neurosci 11: 351-358

Eggermont JJ, Johannesma PIM, Aertsen AMHJ (1983) Reverse-correlation methods in auditory research. Quart Rev Biophys 16: 341-414

Farina WM, Kramer D, Varjú D (1995) The response of the hovering hawk moth *Macroglossum stellatarum* to translatory pattern motion. J Comp Physiol A 176: 551-562

Farina WM, Varjú D, Zhou Y (1994) The regulation of distance to dummy flowers during hovering flight in the hawk moth *Macroglossum stellatarum*. J Comp Physiol 174: 239-247

Fayyazuddin A, Dickinson MH (1996) Haltere afferents provide direct, electrotonic input to a steering motor neuron in the blowfly, *Calliphora*. J Neurosci 16: 5225-5232

Geisler WS, Albrecht DG (1997) Visual cortex neurons in monkeys and cats: detection, discrimination, and identification. Vis Neurosci 14: 897-919

Gershon ED, Wiener MC, Latham PE, Richmond BJ (1998) Coding strategies in monkey V1 and inferior temporal cortices. J Neurophysiol 79: 1135-1144

Gestri G, Mastebroek HAK, Zaagman WH (1980) Stochastic constancy, variability and adaptation of spike generation: Performance of a giant neuron in the visual system of the fly. Biol Cybern 38: 31-40

Götz KG (1968) Flight control in *Drosophila* by visual perception of motion. Kybernetik 4: 199-208

Götz KG (1975) The optomotor equilibrium of the *Drosophila* navigation system. J Comp Physiol 99: 187-210

Götz KG (1991) Bewertung und Auswertung visueller Zielobjekte bei der Fliege *Drosophila*. Zool Jb Physiol 95: 279-286

Green DM, Swets JA (1974) Signal detection theory and psychophysics. Robert Krieger Publ Comp, Huntington, New York

Gur M, Beylin A, Snodderly DM (1998) Response variability of neurons in primary visual cortex (V1) of alert monkeys. J Neurosci 17: 2914-2920

Haag J, Borst A (1996) Amplification of high frequency synaptic inputs by active dendritic membrane processes. Nature 379: 639-641

Haag J, Borst A (1997) Encoding of visual motion information and reliability in spiking and graded potential neurons. J Neurosci 17: 4809-4819

Haag J, Borst A (1998) Active membrane properties and signal encoding in graded potential neurons. J Neurosci 18: 7972-7986

Haag J, Theunissen F, Borst A (1997) The intrinsic electrophysiological characteristics of fly lobula plate tangential cells: II. Active membrane properties. J Comput Neurosci 4: 349-369

Harris RA, O'Carroll DC, Laughlin SB (1999) Adaptation and the temporal delay filter of fly motion detectors. Vision Res 39: 2603-2613

van Hateren JH, Schilstra C (1999) Blowfly flight and optic flow. II. Head movements during flight. J Exp Biol 202: 1491-1500

Hausen K (1981) Monocular and binocular computation of motion in the lobula plate of the fly. Verh Dtsch Zool Ges 74: 49-70

Hausen K (1982a) Motion sensitive interneurons in the optomotor system of the fly. I. The Horizontal Cells: Structure and signals. Biol Cybern 45: 143-156

Hausen K (1982b) Motion sensitive interneurons in the optomotor system of the fly. II. The Horizontal Cells: Receptive field organization and response characteristics. Biol Cybern 46: 67-79

Hausen K, Egelhaaf M (1989) Neural mechanisms of visual course control in insects. In: Stavenga D, Hardie R (eds) Facets of vision. Springer, Berlin, Heidelberg, New York, pp 391-424

Heisenberg M, Wolf R (1984) Vision in *Drosophila*. Springer, Berlin, Heidelberg, New York

Heisenberg M, Wolf R (1988) Reafferent control of optomotor yaw torque in *Drosophila melanogaster*. J Comp Physiol A 163: 373-388

Hengstenberg R (1977) Spike responses of 'non-spiking' visual interneurone. Nature 270: 338-340

Hengstenberg R (1982) Common visual response properties of giant vertical cells in the lobula plate of the blowfly *Calliphora*. J Comp Physiol 149: 179-193

Horstmann W, Egelhaaf M, Warzecha A-K (2000) Synaptic interactions increase optic flow specificity. Europ J Neurosci: in press

Ibbotson MR, Mark RF, Maddess T (1994) Spatiotemporal response properties of direction-selective neurons in the nucleus of the optic tract and dorsal terminal nucleus of the wallaby, *Macropus eugenii*. J Neurophysiol 72: 2927-2943

Järvilehto M, Weckström M, Kouvalainen E (1989) Signal coding and sensory processing in the peripheral retina of the compound eye. In: Singh RN, Strausfeld NJ (eds) Neurobiology of sensory systems. Plenum Press, New York, London, pp 53-70

Johnston D, Wu M-S (1995) Foundations of cellular neurophysiology. MIT Press, Cambridge, MA

Juusola M, French AS, Uusitalo RO, Weckström M (1996) Information processing by graded-potential transmission through tonically active synapses. Trends Neurosci 19: 292-297

Kawasaki M (1993) Temporal hyperacuity in the gymnotiform electric fish, *Eigenmannia*. Amer Zool 33: 86-93

Kern R, Varjú D (1998) Visual position stabilization in the hummingbird hawk moth, *Macroglossum stellatarum* L. I. Behavioural analysis. J Comp Physiol A 182: 225-237

Kern R, Lorenz S, Lutterklas M, Egelhaaf M (1999) How do fly interneurons respond to optic flow experienced in 3D-environments? In: Elsner N, Eysel U (eds) Proceedings of the 27th Göttingen Neurobiol Conf 1999. Thieme, Stuttgart, p 438

Kern R, Lutterklas M, Egelhaaf M (2000) Neural representation of optic flow experienced by walking flies with largely asymmetric visual input. J Comp Physiol A 186: 467-479

Kimmerle B, Srinivasan MV, Egelhaaf M (1996) Object detection by relative motion in freely flying flies. Naturwiss. 83: 380-381

Kimmerle B, Warzecha A-K, Egelhaaf M (1997) Object detection in the fly during simulated translatory flight. J Comp Physiol A 181: 247-255

Koenderink JJ (1986) Optic Flow. Vision Res 26:161-180

Krapp H (1999) Neuronal matched filters for optic flow processing in flying insects. In: Lappe M (ed) Neuronal processing of optic flow. Academic Press, San Diego, San Francisco, New York, pp 93-120

Land MF (1993) Chasing and pursuit in the dolichopodid fly *Poecilobothrus nobilitatus*. J Comp Physiol A 173: 605-613

Land MF, Collett TS (1974) Chasing behaviour of houseflies (*Fannia canicularis*). A description and analysis. J Comp Physiol 89: 331-357

Laughlin SB (1994) Matching coding, circuits, cells, and molecules to signals: general principles of retinal design in the fly's eye. Prog Retinal Eye Research 13: 165-196

Laughlin SB, Howard J, Blakeslee B (1987) Synaptic limitations to contrast coding in the retina of the blowfly *Calliphora*. Proc Roy Soc Lond B 231: 437-467

Liebenthal E, Uhlmann O, Camhi JM (1994) Critical parameters of the spike trains in a cell assembly: coding of turn direction by giant interneurons of the cockroach. J Comp Physiol A 174: 281-296

Lisberger SG, Movshon JA (1999) Visual motion analysis for pursuit eye movements in area MT of macaque monkeys. J Neurosci 19: 2224-2246

Maddess T, Laughlin SB (1985) Adaptation of the motion-sensitive neuron H1 is generated locally and governed by contrast frequency. Proc Roy Soc Lond B 225:251-275

Mainen ZF, Sejnowski TJ (1995) Reliability of spike timing in neocortical neurons. Science 268: 1503-1506

Mastebroek HAK (1974) Stochastic structure of neural activity in the visual system of the blow-fly. Doctoral Dissertation, Rijksuniversiteit te Groningen

Mikami A, Newsome WT, Wurtz RH (1986) Motion selectivity in macaque visual cortex. II Spatiotemporal range of directional interactions in MT and V1. J Neurophysiol 55: 1328-1339

Miles FA, Wallman J (1993) Visual motion and its role in the stabilization of gaze. Elsevier, Amsterdam, London, New York

Movshon JA, Lisberger SG, Krauzlis RJ (1990) Visual cortical signals supporting smooth pursuit eye movements. Cold Spring Harb Symp Quant Biol 55: 707-716

Nowak LG, Sanchez-Vives MV, McCormick DA (1997) Influence of low and high frequency inputs on spike timing in visual cortical neurons. Cerebral Cortex 7: 487-501

O'Carroll DC, Bidwell NJ, Laughlin SB, Warrant EJ (1996) Insect motion detectors matched to visual ecology. Nature 382: 63-66

Reichardt W (1961) Autocorrelation, a principle for the evaluation of sensory information by the central nervous system. In: Rosenblith WA (ed) Sensory communication. MIT Press and John Wiley and Sons, New York, London, pp 303-317.

Reichardt W, Poggio T, Hausen K (1983) Figure-ground discrimination by relative movement in the visual system of the fly. Part II: Towards the neural circuitry. Biol Cybern 46 (Suppl): 1-30

Reichardt W, Poggio T (1976) Visual control of orientation behaviour in the fly. Part I. A quantitative analysis. Quart Rev Biophys 9: 311-375

Rieke F, Warland D, de Ruyter van Steveninck R, Bialek W (1997) Spikes. MIT Press, Cambridge, MA

Roberts A, Bush BMH (1981) Neurones without impulses. Cambridge University Press, Cambridge, London, New York

de Ruyter van Steveninck R, Bialek W (1988) Real-time performance of a movement-sensitive neuron in the blowfly visual system: Coding and information transfer in short spike sequences. Proc Roy Soc Lond B 234: 379-414

de Ruyter van Steveninck R, Bialek W (1995) Reliability and statistical efficiency of a blowfly movement-sensitive neuron. Phil Trans Roy Soc Lond B 348: 321-340

de Ruyter van Steveninck R, Laughlin SB (1996) The rate of information transfer at graded-potential synapses. Nature 379: 642-645

de Ruyter van Steveninck R, Lewen GD, Strong SP, Koberle R, Bialek W (1997) Reproducibility and variability in neural spike trains. Science 275: 1805-1808

de Ruyter van Steveninck R, Zaagman WH, Mastebroek HAK (1986) Adaptation of transient responses of a movement-sensitive neuron in the visual system of the blowfly, *Calliphora erythrocephala*. Biol Cybern 54: 223-236

Schilstra C, van Hateren JH (1999) Blowfly flight and optic flow. I. Thorax kinematics and flight dynamics. J Exp Bio 202: 1481-1490

Shadlen MN, Britten KH, Newsome WT, Movshon JA (1996) A computational analysis of the relationship between neuronal and behavioral responses to visual motion. J Neurosci 16: 1486-1510

Shadlen MN, Newsome WT (1998) The variable discharge of cortical neurons: implications for connectivity, computation, and information coding. J Neurosci 18: 3870-3896

Shannon CE, Weaver W (1949) The mathematical theory of communication. The University of Illinois Press, Urbana

Stevens CF, Zador AM (1998) Input synchrony and the irregular firing of cortical neurons. Nature Neurosci 1: 210-217

Strausfeld NJ (1989) Beneath the compound eye: neuroanatomical analysis and physiological correlates in the study of insect vision. In: Stavenga DG, Hardie RC (eds) Facets of vision. Springer, Berlin, Heidelberg, New York, pp 317-359

Strong SP, Koberle R, de Ruyter van Steveninck R, Bialek W (1998) Entropy and information in neural spike trains. Physical Review Letters 80: 197-200

Theunissen F, Roddey JC, Stufflebeam S, Clague H, Miller JP (1996) Information theoretic analysis of dynamical encoding by four identified primary sensory interneurons in the cricket cercal system. J Neurophysiol 75: 1345-1364

Tolhurst DJ, Movshon JA, Dean AF (1983) The statistical reliability of signals in single neurons in cat and monkey visual cortex. Vis Res 23: 775-785

Virsik R, Reichardt W (1976) Detection and tracking of moving objects by the fly *Musca domestica*. Biol Cybern 23: 83-98

Vogels R, Spileers W, Orban GA (1989) The response variability of striate cortical neurons in the behaving monkey. Exp Brain Res 77: 432-436

Voss R, Zeil J (1998) Active vision in insects: An analysis of object-directed zig-zag flights in wasps (*Odynerus spinipes*, Eumenidae). J Comp Physiol A 182: 373-387

Wagner H (1986a) Flight performance and visual control of the flight of the free-flying housefly (*Musca domestica*). II. Pursuit of targets. Phil Trans Roy Soc Lond B 312: 553-579

Wagner H (1986b) Flight performance and visual control of the flight of the free-flying housefly (*Musca domestica*). III. Interactions between angular movement induced by wide- and small-field stimuli. Phil Trans Roy Soc Lond B 312: 581-595

Warzecha A-K (1994) Reliability of neuronal information processing in the motion pathway of the blowflies *Calliphora erythrocephala* and *Lucilia cuprina*. Doctoral Disseration, Universität Tübingen

Warzecha A-K, Egelhaaf M (1996) Intrinsic properties of biological motion detectors prevent the optomotor control system from getting unstable. Phil Trans Roy Soc Lond B 351: 1579-1591

Warzecha A-K, Egelhaaf M (1997) How reliably does a neuron in the visual motion pathway of the fly encode behaviourally relevant information? Europ J Neurosci 9: 1365-1374

Warzecha A-K, Egelhaaf M (1998) On the performance of biological movement detectors and ideal velocity sensors in the context of optomotor course stabilization. Vis Neurosci 15: 113-122

Warzecha A-K, Egelhaaf M. (1999) Variability in spike trains during constant and dynamic stimulation. Science 283: 1927-1930

Warzecha A-K, Kretzberg J, Egelhaaf M (1998) Temporal precision of encoding of motion information by visual interneurons. Curr Biol 8: 359-368

White JA, Rubinstein JT, Kay AR (2000) Channel noise in neurons. Trends Neurosci. 23: 131-137

Zanker JM, Egelhaaf M, Warzecha A-K (1991) On the coordination of motor output during visual flight control of flies. J Comp Physiol A 169: 127-134

Zeil J (1993) Orientation flights of solitary wasps (*Cerceris*, Sphecidae, Hymenoptera). I. Description of flights. J Comp Physiol 172: 189-205

Zohary E, Shadlen MN, Newsome WT (1994) Correlated neuronal discharge rate and its implications for psychophysical performance. Nature 370: 140-143

Note added in proof

In this volume de Ruyter van Steveninck et al. argue that we reached "conclusions so nearly opposite" (p 303) from their own. They come to this view by attributing conclusions to us that represent caricatures of the conclusions we have drawn. Moreover, they claim that our analysis has severe theoretical shortcomings. This claim is unwarranted and results from misinterpreting our data analysis (compare their pp 290-293 with our pp 251-257 and their pp 293-298 with our pp 269-272). There are no principal differences in the experimental data obtained by both groups but merely discrepancies concerning interpretations. Whereas de Ruyter van Steveninck et al. conclude that "individual spikes are reproducible on a milli-second time scale" (p 303), we argue that in motion-sensitive neurones only part of the spikes elicited under natural conditions are timed as precisely: The precision of spike timing is determined by the membrane potential changes at the spike initiation zone and the dynamics of these membrane potential changes depends on the dynamics of the visual input. At least if we accept that under natural conditions visual motion is not only the consequence of saccade-like turns of the animal (see our pp 261-266), spikes inevitably lock to visual motion stimuli on a wide range of timescales.

Real-Time Encoding of Motion: Answerable Questions and Questionable Answers from the Fly's Visual System

Rob de Ruyter van Steveninck[1], Alexander Borst[2] and William Bialek[1]

[1]NEC Research Institute, Princeton, USA; [2]ESPM-Division of Insect Biology, University of California at Berkeley, Berkeley, USA

1. Introduction

Much of what we know about the neural processing of sensory information has been learned by studying the responses of single neurones to rather simplified stimuli. The ethologists, however, have argued that we can reveal the full richness of the nervous system only when we study the way in which the brain deals with the more complex stimuli that occur in nature. On the other hand it is possible that the processing of natural signals is decomposable into steps that can be understood from the analysis of simpler signals. But even then, to prove that this is the case one must do the experiment and use complex natural stimuli. In the past decade there has been renewed interest in moving beyond the simple sensory inputs that have been the workhorse of neurophysiology, and a key step in this program has been the development of more powerful tools for the analysis of neural responses to complex dynamic inputs. The motion sensitive neurones of the fly visual system have been an important testing ground for these ideas, and there have been several key results from this work:

1. The sequence of spikes from a motion sensitive neurone can be decoded to recover a continuous estimate of the dynamic velocity trajectory (Bialek et al. 1991; Haag and Borst 1997). In this decoding, individual spikes contribute significantly to the estimate of velocity at each point in time.
2. The precision of velocity estimates approaches the physical limits imposed by diffraction and noise in the photoreceptor array (Bialek et al. 1991).
3. One or two spikes are sufficient to discriminate between motions which differ by displacements in the "hyperacuity" range, an order of magnitude smaller than the spacing between photoreceptors in the retina (de Ruyter van

Steveninck and Bialek 1995). Again this performance approaches the limits set by diffraction and receptor noise.

4. Patterns of spikes which differ by millisecond shifts of the individual spikes can stand for distinguishable velocity waveforms (de Ruyter van Steveninck and Bialek 1988), and these patterns can carry much more information than expected by adding up the contributions of individual spikes (de Ruyter van Steveninck and Bialek 1988; Brenner et al. 2000a).

5. The total information that we (or the fly) can extract from the spike train continues to increase as we observe the spikes with greater temporal resolution, down to millisecond precision (de Ruyter van Steveninck et al. 1997; Strong et al. 1998).

6. These facts about the encoding of naturalistic, dynamic stimuli cannot be extrapolated simply from studies of the neural response to simpler signals. The system exhibits profound adaptation (Maddess and Laughlin 1985; de Ruyter van Steveninck et al. 1986; Borst and Egelhaaf 1987; de Ruyter van Steveninck et al. 1996; Brenner et al. 2000b), so that the encoding of signals depends strongly on context, and the statistical structure of responses to dynamic stimuli can be very different from that found with simpler static or steady state stimuli (de Ruyter van Steveninck et al. 1997).

We emphasize that many of these results from the fly's visual system have direct analogs in other systems, from insects to amphibians to primates (Rieke et al. 1997).

In a series of recent papers, Egelhaaf and coworkers have called all of these results into question (Warzecha and Egelhaaf 1997, 1998, 1999; Warzecha et al. 1998. However, for a different assessment of some of these points, see their Chapter, Sect. 8). Several of these papers are built around a choice of a stimulus very different from that used in previous work. Rather than synthesize a stimulus with known statistical properties, they sample the time dependent motion signals generated by a fly tethered in a flight simulator. The simulator is operated in closed loop so that the fly, by producing a yaw torque which is measured electronically, moves a pattern on a CRT monitor, while the animal itself stays stationary. For experiments on the responses of the motion sensitive neurones these patterns and motions are replayed to another fly, again through a monitor. In their judgement these stimuli "are characteristic of a normal behavioural situation in which the actions and reactions of the animal directly affect its visual input" (Warzecha and Egelhaaf 1998).

For these stimuli, Warzecha and Egelhaaf claim that the timing of individual spikes has no significance in representing motion signals in the fly's motion sensitive neurones. Instead they suggest that the neurone's response should be averaged over time scales of the order of 40-100 ms to recover the essential information, and that timing of spikes within this averaging window is irrelevant. These claims are in conflict with points [1], [4], and [5] above. As part of their discussion of these points Warzecha and Egelhaaf make repeated references to the noisiness of the neural response, in apparent contradiction of points [2] and [3], although they do not address specifically the quantitative results of the earlier work. Finally, they suggest that the spike count variance of H1 when stimulated with constant velocity is similar to the variance obtained with dynamical stimulation, when compared at equal mean spike rates. This is in contradiction of point [6].

Obviously the recent work of Egelhaaf and colleagues raises many different issues. In this contribution we try to focus on three problems of general interest. First, how do we define a meaningful "naturalistic stimulus", and does their "behaviourally generated" stimulus fall into this category? In particular, how do we reach an effective compromise between stimuli that occur in nature and stimuli that we can control and reproduce reliably in the laboratory? Second, how do we characterize the neural response to complex dynamic inputs? In particular, how do we evaluate all the relevant time scales in the sensory signal itself and in the spike train? Again, these are issues that we must face in the analysis of *any* neural system for processing of sensory information; indeed there are even analogous issues in motor systems. Thus the fly's visual system serves here as an example, rather than as an end in itself.

Before we begin our discussion of these two points, we must be clear that the first question – What is a natural stimulus? – is a question about the biology and ecology of the animal we are studying, as well as a question about the design and constraints of a particular experimental setup. One might well disagree about the best strategy for generating naturalistic stimuli in the laboratory. On the other hand, our second question – How do we characterize the response to complex signals? – is a theoretical issue which is not tied to the particulars of biology. On this issue there are precise mathematical statements to be made, and we hope to make clear how these mathematical results can be used as a rigorous guide to the analysis of experiments.

The third and final question we address concerns the comparison between static and dynamic stimuli. Although we believe that the most interesting problems concern the way in which the brain deals with the complex, dynamic stimuli that occur in nature, much has been learned from simpler static stimuli and there are nagging questions about whether it really is "necessary" to design new experiments that need more sophisticated methods of analysis. For reasons that will become clear below, the comparison of static and dynamic stimuli also is crucial for understanding whether many of the lessons learnt from the analysis of the fly's motion sensitive neurones will be applicable to other systems, especially the mammalian cortex.

2. What is a natural stimulus?

The fly's motion sensitive neurone H1 offers a relatively simple testing ground for ideas about the neural representation of natural signals. This cell is a wide field neurone, so rather than coding the motion of small objects or a component of the local velocity flow field, H1 is responsible primarily for coding the rigid body horizontal (yaw) motion of the fly relative to the rest of the world. Thus there is a limit in which we can think of "the stimulus" as being a single function of time, $v(t)$, which describes this angular velocity trajectory. It should be clear that this description is incomplete: the neural response is affected also by the mean light intensity, the spatial structure of the visual stimulus, and the area of the compound eye that is stimulated. Further, the system is highly adaptive, so that the encoding of a short segment of the trajectory $v(t)$ will depend strongly on the statistics of this trajectory over the past several seconds.

Traditional experiments on motion sensitive neurones (as on other sensory cells) have used constant stimuli (motion at fixed velocity), pulsed stimuli (step-wise motion), or have analysed the steady state behaviour in response to sinusoidal motion at different frequencies. In nature, trajectories are not so simple. Instead one can think of trajectories as being drawn from a distribution $P[v(t)]$ or "stimulus ensemble". A widely used example of stimulus ensembles is the Gaussian ensemble, in which the distribution of trajectories is described completely by the spectrum or correlation function. We can construct spectra and correlation functions so that there is a single characteristic stimulus amplitude – the dynamic range $v_{rms}(t)$ of velocity signals – and a single characteristic time τ_c in the dynamics of these signals. A reasonable approach to the study of naturalistic stimuli might then be to explore the coding of signals in H1 using stimulus ensembles parametrized by $v_{rms}(t)$ and τ_c. Most of the results enumerated above have been obtained in this way.

In their recent papers (Warzecha and Egelhaaf 1997; Warzecha et al. 1998), as well as in their contribution to this volume, Warzecha and Egelhaaf argue that the stimulus ensembles used in experiments on H1 have been restricted unfairly to short correlation times. Put another way, the stimuli used in these experiments have included high temporal frequency components. Warzecha and Egelhaaf suggest that these high frequency components bias the response of the motion sensitive cells to artificially high temporal precision which is not relevant for the behaviourally generated stimuli that they use[1]. The question of whether timing precision is important under truly natural conditions is left open.

---

[1] In fact Warzecha and Egelhaaf make two different arguments about high frequency stimuli. They make repeated references to the integration times and noise in the fly's visual system, all of which limit the reliability of responses to high frequency components in the input. These arguments generally are presented in qualitative terms, but Warzecha and Egelhaaf (1999) state explicitly that signals above 30 Hz are undetectable above the noise and hence can have no impact on the statistics of the spike train. On the other hand, Warzecha and Egelhaaf (1997) argue that the inclusion of high frequency components in the input causes an unnaturally tight locking of spikes to stimulus events, causing us to overestimate the significance of spike timing for the coding of behaviourally relevant stimuli. It should be clear that these two arguments cannot both be correct.

Independent of what is truly natural, one can argue that experiments with short correlation times have provided evidence on what the fly's visual system *can* do. Although we seldom sit in dark rooms and wait for dim flashes of light, such experiments led to the demonstration that the human visual system can count single photons (Hecht et al. 1942). In this spirit, studies of H1 using stimuli with short correlation times have revealed that the fly's nervous system can estimate velocity with a precision limited by noise in the photoreceptor array and that timing relations between neural responses and stimulus events can be preserved with millisecond precision, even as the signals pass through four stages of neural circuitry. It would seem strange that such impressive performance would evolve if it were irrelevant for fly behaviour.

Instead of choosing trajectories $v(t)$ from a known probability distribution, we could try to sample the trajectories that actually occur in nature. Here we have to make choices, and these will always be somewhat subjective: Dethier (1976) reports that female flies spend 12.7%, and male flies 24.3% of their time walking or flying. The other activities on Dethier's list are feeding, regurgitating, grooming and resting, during which information from the fly's motion sensitive cells presumably is not too relevant. So it seems the fly could live quite happily without its tangential cells most of its time. On the other hand, during periods of flight, the responses of its motion sensitive cells are strongly modulated. On top of that, the depth and speed of modulation may vary as the fly switches from periods of relatively quiet cruising to episodes of fast and acrobatic pursuit or escape, and back (Land and Collett 1974). Although it is not clear at the outset what portion of the total behavioural repertoire we should analyse, the thing that presumably tells us most about the "design" of the fly is the dynamics of neural signal processing during top performance. Correspondingly, Warzecha and Egelhaaf propose to use stimuli that are representative of the trajectories experienced by a fly in flight, and we agree that this is an excellent choice. There are still some difficulties, however. Warzecha and Egelhaaf propose that meaningful data can be obtained from "behaviourally generated" trajectories $v(t)$ recorded from flies that are tethered in a flight simulator apparatus in which the fly's measured torque is to move a pattern on a CRT monitor in the visual field of the fly. The combination of fly, torque meter, and moving pattern thus acts as a closed loop feedback system whose dynamical properties are determined both by the fly and by the gain and bandwidth of the mechanical and electronic components involved. The data presented by Warzecha and Egelhaaf (1997, 1998, and this volume) strongly suggest that the dynamics of the feedback system are dominated by the electromechanical properties of their setup, and not by the fly itself. This is most clearly seen from direct comparisons between the trajectories in the flight simulator and those observed in nature.

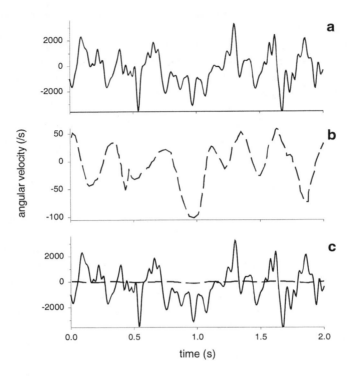

Fig. 1 Comparison of the rotational velocity traces reported from free flying and tethered flies. **a** Rotation velocity of a fly (*Fannia canicularis*) in free flight, derived from video recordings by Land and Collett (1974). **b** Rotation velocity of a pattern in a flight simulator, derived from torque signals measured from a tethered fly, as reported by Warzecha and Egelhaaf (1997). **c** The data from (a) and (b) plotted on the same scale.

Trajectories during free flight were recorded in the classic work of Land and Collett (1974), who studied chasing behaviour in *Fannia canicularis* and found turning speeds of several thousand °/s. Wehrhahn (1979), Wehrhahn et al. (1982) and Wagner (1986a,b,c) report very similar results for the housefly *Musca*, and recent publications (Schilstra and van Hateren 1998; van Hateren and Schilstra 1999) report flight measurements at high temporal and spatial resolution, from *Calliphora* flying almost free. In their published dataset flies made about 10 turns per second, during which head velocities easily exceeded 1000°/s, while maximum head turning velocities were well over 3000°/s. If we compare the results of these studies to the motion traces used in the experiments by Warzecha and Egelhaaf (1997, 1998) we see that their traces are considerably smoother, and do not go beyond 100°/s. These differences are illustrated in figure 1, where we make an explicit comparison between free flight data obtained by Land and Collett (1974) and the motion traces data presented in figure 1 of Warzecha and Egelhaaf (1997).

It is clear that there are dramatic differences in the frequency of alternation, and especially, in the amplitude of the motion signals. Perhaps surprisingly, Warzecha and Egelhaaf do not seem to think that these differences are relevant. In Warzecha et al. (1998) they use synthetic motion stimuli with a velocity standard deviation of 22°/s, and point out that "Much larger stimulus amplitudes and a larger frequency range than those used here could not be tested with our present stimulation equipment" (page 362). However, given the literature on free flight behaviour, we are not sure how they can maintain their claim about these experimental conditions that "Nonetheless, there are likely to be few instances in the normal world where visual motion encompasses a wider dynamic range than that which could be tested here."

Simple theoretical arguments suggest that these differences between the flight simulator trajectories and true natural trajectories will have enormous consequences for the reliability of responses in the motion sensitive neurones. Warzecha and Egelhaaf (Fig. 6 of their contribution in this volume) report estimates of the signal and noise power spectra in the graded voltage response of a motion sensitive cell. If we scale the signal to noise power ratio they present in proportion to the ratio between the power spectrum of natural motion and the velocity power spectrum they used, then the signal to noise ratio will increase so much that the natural trajectories will produce signal resolvable against the noise at frequencies well above 200 Hz. This would mean that events in natural stimuli will be localizable with millisecond precision.

There are other differences between the stimulus conditions studied by Warzecha and Egelhaaf and the natural conditions of free flight. Outdoors, in the middle of the afternoon, light intensities typically are two orders of magnitude larger than are generated with standard laboratory displays (Land 1981). Further, the wide field motion sensitive cells gather inputs from large portions of the compound eye (Gauck and Borst 1999), which extends backward around the head to cover a large fraction of the available solid angle; rotation of the fly produces coherent signals across this whole area, and it is very difficult to reproduce this "full vision" in the laboratory with CRT displays. While it is difficult to predict quantitatively the consequences of these differences, the qualitative effect is clear: natural signals are much more powerful and "cleaner" than the stimuli which Warzecha and Egelhaaf have used.

We can take a substantial step toward natural stimulus conditions by recording from a fly that itself rotates in a natural environment along a trajectory representative of free flight. Preliminary results from such experiments will be analysed in more detail below, and a detailed account is forthcoming (Lewen et al. submitted). A female wild fly (*Calliphora*), caught outdoors, was placed in a plastic tube and immobilized with wax. A small incision was made in the back of the head, through which a microelectrode could be advanced to the lobula plate to record from H1. The fly holder, electrode holder and manipulator were assembled to be as light and compact, yet rigid, as possible. In this way the fly and the recording setup could be mounted on the axle of a stepper motor (Berger-Lahr,

RDM 564/50, driven by a Divi-Step D331.1 interface with 10,000 steps/revolution) and rotated at speeds of up to several thousand °/s. The motor speed was controlled through the parallel port of a laptop computer by means of custom designed electronics, and was played out at 2 ms intervals. The data presented here are from an experiment in which the setup was placed outside on a sunny day, in a wooded environment not far from where the fly was caught. A simple, but crucial, control is necessary: H1 does not respond if the fly is rotated in the dark, or if the visual scene surrounding it rotates together with the fly. We can thus be confident that H1 is stimulated by visual input alone, and not by other sensory modalities, and also that electronic crosstalk between the motor and the neural recording is negligible.

The motion trace $v(t)$ was derived from a concatenation of body angle readings over the course of the flight paths of a leading and a chasing fly as depicted in figure 4 of Land and Collett (1974). For technical reasons we had to limit the velocity values to half those derived from that figure, but we have no reason to believe that this will affect the main result very much. Translational motion components were not present, representing a situation with objects only at infinity. Padded with a few zero velocity samples, this trace was 2.5 s long. That sequence was repeated with the sign of all velocity values changed, to get a full 5 s long sequence. This full sequence was played 200 times in succession while spikes from the axon terminals of H1 were recorded as an analog waveform at 10 kHz sampling rate. In off line analysis spike occurrence times were derived by matched filtering and thresholding.

Before looking at the responses of H1, we emphasize several aspects of the stimulus conditions:

1. The motion stimulus is obtained from direct measurement of flies in free flight, not from a torque measurement of a tethered fly watching a CRT monitor. As argued above, the electromechanical properties of the setup used by Warzecha and Egelhaaf are likely to have drastic effects on the frequency and amplitude characteristics of the motion.

2. The field of view experienced by the fly in our setup is almost as large as that for a free flying fly. Most of the visual field is exposed to movement, with the exception of a few elements (e.g. the preamplifier) that rotate with the fly, and occupy just a small portion of the visual field.

3. The experiment is done outside, in an environment close to where our experimental flies are caught, so that almost by definition we stimulate the fly with natural scenes.

4. The experiment is performed in the afternoon on a bright day. From dim to bright patches of the visual scene the effective estimated photon flux for fly photoreceptors under these conditions varies from $5 \cdot 10^5$ to $5 \cdot 10^6$ photons per second per receptor. Warzecha and Egelhaaf's experiments (as many experiments of ours) were done with a fly watching a Tektronix 608 cathode ray tube, which has an estimated maximum photon flux of about 10^5 photons per second per receptor.

Figure 2 shows the spike trains generated by H1 in the "outdoor" experiment, focusing on a short segment of the experiment just to illustrate some qualitative points. The top trace shows the velocity waveform $v(t)$, and subsequent panels show the spikes generated by H1 in response to this trajectory (H1+) or its sign reverse (H1−). Visual inspection reveals that some aspects of the response are very reproducible, and further that particular events in the stimulus can be associated reliably with small numbers of spikes. The first stimulus zero crossing at about 1730 ms is marked by a rather sharp drop in the activity of H1+, with a sharp rise for H1−. This sharp switching of spike activity is not just a feature of this particular zero crossing, but occurs in other instances as well. Further, the small hump in velocity at about 2080 ms lasts only about 10 ms, but induces a reliable spike pair in H1+ together with a short pause in the activity of H1−. The first spike in H1− after this pause (Fig. 2c) is timed quite well; its probability distribution (Fig. 2e) has a standard deviation of 0.73 ms. Thus, under natural stimulus conditions individual spikes can be locked to the stimulus with millisecond precision.

In fact the first few spikes after the pause in H1− have even greater internal or relative temporal precision. The raster in figure 2c shows that the first spike meanders, in the sense that the fluctuation in timing from trial to trial seems to be slow. This suggests that much of the uncertainty in the timing of this spike is due to a rather slow process, perhaps metabolic drift. To outside observers, like us, these fluctuations just add to the spike timing uncertainty, which even then is still submillisecond. Note, however, that to some extent the fly may be able to compensate for that drift. If the effect is metabolic, then different neurones might drift more or less together, and the time interval between spikes from different cells could be preserved quite well in spite of temporal drift of individual spikes. Similarly, within one cell, spikes could drift together (Brenner et al. 2000), and this indeed is the case here. As a result the interval between the first spike and the next is much more precise, with a 0.18 ms standard deviaton, and it does not seem to suffer from these slow fluctuations (Fig. 2d). The timing accuracy of ensuing intervals from the first spike to the third and fourth, although becoming gradually less well defined, is still submillisecond (Fig. 2f). So it is clear that some identifiable patterns of spikes are generated with a timing precision of the order of a millisecond or even quite a bit better.

Although we have emphasized the reproducibility of the responses to natural stimuli, there also is a more qualitative point to be made. All attempts to characterize the input/output relation of H1 under laboratory conditions have indicated that the maximum spike rate should occur in response to velocities below about 100°/s, far below the typical velocities used in our experiments. Indeed, many such experiments suggest that H1 should shut down and not spike at all in response to these extremely high velocities. In particular, Warzecha and Egelhaaf (1996) claim that steady state spike rates in H1 are decreased dramatically at high velocities, that this lack of sensitivity to high speeds is an essential result of the computational strategy used by the fly in computing motion, and further that this behaviour

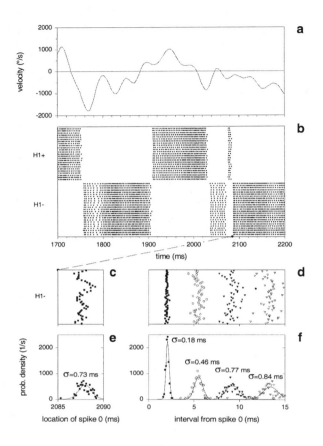

Fig. 2 Direct observations of H1 spike timing statistics in response to rotational motion derived from Land and Collett's (1974) free flight data (see Fig. 1a). **a** A 500 ms segment of the motion trace. **b** Top: raster plot with 25 traces representing spike occurrences measured from H1. Bottom: raster plot of 25 traces of spike occurrences from the same cell, but in response to a velocity trace that was the negative of the one shown in (a). For ease of reference we call these traces H1+ and H1− respectively. **c** Raster plot of 25 samples of the occurrence time of the first spike fired by H1− after time t = 2080 ms in the stimulus sequence (indicated by the dashed line connecting the axis of b to panel c). **d** Raster plots of 25 samples of the interval from the spike shown in (c) to the first (filled circles), second (open circles), third (filled triangles), and fourth (open triangles) consecutive spike. Note the time axes: The rasters in (c) and (d) are plotted at much higher time resolution than those in (b). **e** Probability density for the timing of the spike shown in (c). The spread is characterized by σ = 0.73 ms, where σ is defined as half the width of the peak containing the central 68.3% of the total probability. If the distribution were Gaussian, then this would be equivalent to the standard deviation. Here we prefer this definition instead of one based on computing second moments. The motivation is that there can be an occasional extra spike, or a skipped spike, giving a large outlier which has a disproportionate effect on the width if it is calculated from the second moment. Filled squares represent the experimental histogram, based on 200 observations, while the solid line is a Gaussian fit. **f** Probability densities for the same interspike interval shown in (d). The definition of σ is the same as the one in (e).

can be used to advantage in optomotor course control. In the data they present, the response of H1 peaks at about 60°/s, and its response is essentially zero above 250°/s. The outdoor experiment demonstrates that none of these conclusions are relevant to more natural conditions, where H1's steady state response peaks at about 1000°/s and is robust and reliable up to angular velocities of over 2000°/s.

The arguments presented here rested chiefly on visual inspection of the spike trains, and this has obvious limitations. Our eyes are drawn to reliable features in the response, and one may object that these cases could be accurate but rare, so that the bulk or average behaviour of the spike train is much sloppier. To proceed we must turn to a more quantitative approach.

3. How do we analyse the responses to natural stimuli?

When we deliver simple sensory stimuli it is relatively easy to analyse some measures of neural response as a function of the parameters that describe the stimulus. Faced with the responses of a neurone to the complex, dynamic signals that occur in nature – as in figure 1 – what should we measure? How do we quantify the response and its relation to the different features of the stimulus? The sequence of spikes from a motion sensitive neurone constitutes an encoding of the trajectory $v(t)$. Of course, this encoding is not perfect: there is noise in the spatio-temporal pattern of the photon flux from which motion is computed, the visual system has limited spatial and temporal resolution, and inevitably there is internal noise in any physical or physiological system. This may cause identical stimuli to generate different responses. The code also may be ambiguous in the sense that, even if noise were absent, the same response can be induced by very different stimuli. Conceptually, there are two very different questions we can ask about the structure of this code. First, we can ask about the features of the spike train that are relevant for the code: Is the timing of individual spikes important, or does it suffice to count spikes in relatively large windows of time? Are particular temporal patterns of spikes especially significant? Second, if we can identify the relevant features of the spike train then we can ask about the mapping between these features of the response and the structure of the stimulus: What aspects of the stimulus influence the probability of a spike? How can we (or the fly) decode the spike train to estimate the stimulus trajectory, and how precisely can this be done?

There are two general approaches to these problems. One is to compute correlation functions. A classic example is the method of "reverse correlation" in which we correlate the spike train with the time varying input signal (see Section 2.1 in Rieke et al. 1997). This is equivalent to computing the average stimulus trajectory in the neighbourhood of a spike. Other possibilities include correlating spike trains with themselves or with the spike trains of other neurones. A more subtle possibility is to correlate spike trains that occur on different presentations of the same time dependent signal, or the related idea of computing the coherence among responses on different presentations (Haag and Borst 1997). All of these methods have the advantage that simple correlation functions can be estimated reliably even from relatively small data sets. On the other hand, there are an infinite number of possible correlation functions that one could compute, and by looking only at the simpler ones we may miss important structures in the data.

An alternative to computing correlation functions is to take an explicitly probabilistic point of view. As an example, rather than computing the average stimulus trajectory in the neighbourhood of a spike, as in reverse correlation, we can try to characterize the whole distribution of stimuli in the neighbourhood of a spike (de Ruyter van Steveninck and Bialek 1988). Similarly, rather than computing correlations among spike trains in different presentations of the same stimulus, we can try to characterize the whole *distribution* of spike sequences that occur across multiple presentations (de Ruyter van Steveninck et al. 1997; Strong et al. 1998). The probability distributions themselves can be difficult to visualize, and we often want to reduce these rather complex objects to a few sensible numbers, but we must be sure to do this in a way that does not introduce unwarranted assumptions about what is or is not important in the stimulus and response. Shannon (1948) showed that there is a unique way of doing this, and this is to use the entropy or information associated with the probability distributions. Even if we compute correlation functions, it is useful to translate these correlation functions into bounds on the entropy or information, as is done in the stimulus reconstruction method (Bialek et al. 1991; Haag and Borst 1997; Rieke et al. 1997; Borst and Theunissen 1999). Although the idea of using information theory to discuss the neural code dates back nearly to the inception of the theory (MacKay and McCulloch 1952), it is only in the last ten years that we have seen these mathematical tools used widely for the characterization of real neurones, as opposed to models.

3.1 Correlation functions

Although we believe that the best approach to analysing the neural response to natural stimuli is grounded in information theory, we follow Warzecha and Egelhaaf and begin by using correlation functions. From an experiment analogous to the one in our figure 2, Warzecha et al. (1998) compute the correlation function of the spike trains of simultaneously recorded H1 and H2 cells, $\Phi_{\text{spikeH1-spike H2}}(\tau)$, and also the average crosscorrelation function among spike trains from different presentations (trials) of the same stimulus trajectory, $\Phi_{\text{crosstrial H1-H2}}(\tau)$. If the spike trains were reproduced perfectly from trial to trial, these two correlation functions would be identical; of course this is not the case. Warzecha and Egelhaaf conclude from the difference between the two correlation functions that the spikes are not "precisely time coupled" to the stimulus, and they argue further that the scale which characterizes the precision (or imprecision) of spike timing can be determined from the width of the crosstrial correlation function $\Phi_{\text{crosstrial H1-H2}}(\tau)$. This is one of their arguments in support of the notion that the time resolution of the spike train under behaviourally generated conditions is in the range of 40-100 ms, one or two orders of magnitude less precision than was found in previous work.

The crosstrial correlation function obviously contains information about the precision of the neural response, but there is no necessary mathematical relation between the temporal precision and the width of the correlation function. To make the discussion concrete, we show in figure 3a the autocorrelation $\Phi_{\text{spike-spike}}(\tau)$ and in figure 3b the crosstrial correlation function $\Phi_{\text{crosstrial}}(\tau)$ computed for the outdoor experiment. We see that $\Phi_{\text{crosstrial}}(\tau)$ is very broad, while $\Phi_{\text{spike-spike}}(\tau)$ has structure on much shorter time scales, as found also by Warzecha and Egelhaaf. But the characterization of the crosstrial correlation function as broad does not capture all of its structure: rather than having a smooth peak at $\tau = 0$, there seems to be a rather sharp change of slope or cusp, and again this is seen in the data presented by Warzecha and Egelhaaf, even though the stimulus conditions are very different. This cusp is a hint that the width of the correlation function is hiding structure on much finer time scales.

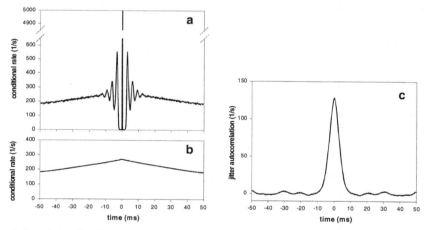

Fig. 3 Correlation functions for H1 during stimulation with natural motion, all computed at 0.2 ms resolution. **a** The spike-spike autocorrelation $\Phi_{\text{spike-spike}}(t)$, normalized as a conditional rate. There are strong oscillations in the conditional rate, due to neural refractoriness. **b** The cross-trial correlation function $\Phi_{\text{crosstrial}}(t)$, computed as the correlation function of the estimated time dependent rate minus a contribution from $\Phi_{\text{spike-spike}}(t)$ scaled by 1/N (N is the number of trials) to correct for intratrial correlations. **c** Autocorrelation of the assumed underlying distribution of spike jitter times, computed by deconvolving the data in (b) by those in (a). See text for further explanation.

Before analysing the correlation functions further, we note some connections to earlier work. Intuitively it might seem that by correlating the responses from different trials we are probing the reproducibility of spike timing in some detail. But because $\Phi_{\text{crosstrial}}(\tau)$ is an average over pairs of spikes (one from each trial), this function is *not* sensitive to reproducible patterns of spikes such as those we have seen in figure 2. In fact, the crosstrial correlation function is equal (with suitable normalization) to the autocorrelation function of the time dependent rate r(t) that we obtain by averaging the spike train across trials. Thus the crosstrial correlation does not contain information beyond the usual poststimulus time histogram or PSTH, and the time scales in the correlation function just measure how rapidly the firing rate can be modulated; again, there is no sensitivity to spike timing beyond the rate, and hence no sensitivity to spike patterns. Since the crosstrial correlation function is equal to the autocorrelation of the rate, the Fourier transform $\Phi_{crosstrial}$ (τ) is equal to the power spectrum of the rate, which has been used by Bair and Koch (1996) to discuss the reproducibility of responses in the motion sensitive neurones of monkey visual cortex. If we Fourier transform both the crosstrial correlation function and the spike-spike correlation, their ratio is proportional to the crosstrial coherence considered by Haag and Borst (1997) in their analysis of H1.

Even granting the limitations of the correlation function as a probe of spike timing, we would like to reveal the finer time scale structure that seems to be hiding near $\tau = 0$. To do this we consider a simple model that can be generalized without changing the basic conclusions. Imagine that each spike has an "ideal" time $\langle t_i \rangle$ relative to the stimulus, and that from trial to trial the actual arrival time of the ith spike fluctuates as $t_i = \langle t_i \rangle + \delta t_i$. The meandering of spikes from trial to trial in figure 2c suggests that the δt_i and δt_j of nearby spikes i and j are correlated, and if these correlations extend over a sufficiently long time (roughly 10 ms is sufficient) then there is a simple approximate equation relating the crosscorrelation among trials to the autocorrelation and the distribution of time jitter, $P(\delta t_i)$: $\Phi_{crosstrial}(\tau) = \Phi_{PP}(\tau) \otimes \Phi_{spike-spike}(\tau)$, where $\otimes$ denotes convolution and $\Phi_{PP}(\tau)$ is the autocorrelation of the distribution $P(\delta t_i)$. Thus $\Phi_{PP}(\tau)$ can be computed from the measured correlation functions by deconvolution. For our outdoor experiment we find that $\Phi_{PP}(\tau)$ has a width of 3.1 ms (Fig. 3c), so that a reasonable estimate for the width of the underlying jitter distribution is $\delta t_{rms} = 3.1/\sqrt{2} \approx 2.2$ ms. This analysis shows that the difference between the crosstrial and the spike-spike correlation functions is consistent with jitter in the range of a few milliseconds, not the many tens of milliseconds claimed by Warzecha and Egelhaaf[2].

---

[2]There are further difficulties in the interpretation of correlation functions offered by Warzecha and Egelhaaf. One of their arguments for the irrelevance of high frequency stimuli is based on a comparison of the velocity spectrum with the spectrum of fluctuations in the time dependent rate (Fig. 2 of Warzecha and Egelhaaf 1997); the spectrum of the time dependent rate should be the Fourier transform of the crosstrial correlation function, as noted above. On the down going slope, across a decade of frequency the decline in the response spectrum is *slower* than the decline in the stimulus spectrum. If we define a transfer function by taking the ratio of the response and stimulus spectra, then the cell is amplifying the higher frequency components, not attentuating them as Warzecha and Egelhaaf claim. This is consistent with the experiments of Haag and Borst (1998) demonstrating that the motion sensitive neurones have active membrane mechanisms to achieve such amplification.

Because the interpretation of correlation functions is a crucial issue, let us give an example from spatial vision, where it is clear that the width of the correlation function (correlation length) is not a good indicator of the precision required to read out a signal. It is well documented that natural scenes typically have broad spatial correlations, often associated with $1/f$-like power density spectra (Srinivasan et al. 1982; Field 1987; Ruderman and Bialek 1994). Using the same reasoning that Warzecha and Egelhaaf apply to spike trains, one would conclude that the visual system should not bother to use high spatial resolution. This would be true for environments with Gaussian statistics, where second order descriptions – the simplest correlation functions – are sufficient. But the world we live in definitely is not Gaussian. It is made out of objects that typically have well defined edges, and these edges are important to us, not least because they are often associated with rigid objects. The width of the spatial correlation function is defined, very roughly, by the apparent size of the objects in our visual field. But this width has nothing to do with the precision with which we can estimate the position of edges and hence the location of object boundaries. Just as for spatial edges, the location of temporal edges may also be important, and we can look at horse racing for an example: In Warzecha and Egelhaaf's interpretation we would not need to time horses any more precisely than the width of the "horse density" correlation function, which corresponds roughly to the time required for the entire horse to cross the finish line. Yet fortunes are won and lost over differences corresponding to a fraction of a horse's nose. What matters here is that we attach importance to features that are defined very sharply in time, and this temporal precision cannot be measured from the width of one simple correlation function. For precisely the same reason one cannot equate the relevant time scale of retinal image motion to spike timing precision, as Warzecha and Egelhaaf argue in this volume (Section 7). Let us then turn to an information theoretic approach.

3.2 Information

Looking at the responses to repeated presentations of a natural complex dynamic stimulus, as in figure 2, we see many different features, some of which have been noted above: there are individual spikes which are reproduced from trial to trial with considerable accuracy; there are patterns of spikes in which the intervals between spikes are reproduced more accurately than the absolute spike times, so that the patterns appear to "meander" from trial to trial; there are trials in which spikes are deleted, apparently at random, and trials in which extra spikes appear. How are we to make sense out of this variety of phenomena? Specifically, we want to know whether the detailed timing of spikes is important for the encoding of naturalistic stimuli. How can we analyse data of this sort to give us a direct answer to this question about the structure of the neural code?

Intuitively, the sequence of action potentials generated by H1 "provides information" about the motion trajectory. If the response of H1 were always the same, independent of the trajectory, of course no information would be provided. Generally, then, the greater the range of possible responses the greater is the capacity of the cell to provide information: if we think of segments of the neural response as being like words in a language, then the ability of the neurone to "describe" the input is enhanced if it has a larger vocabulary. On the other hand, it clearly is not useful to generate words at random, no matter how large our vocabulary, and so there must be a reproducible relationship between the choice of words and the form of the motion trajectory. These intuitive ideas have a precise formulation in Shannon's information theory (Shannon 1948): the size of the neurone's "vocabulary" is measured by the entropy of the distribution of responses, the (ir)reproducibility of the relation between stimulus and response is related to the conditional or noise entropy computed from the distribution of responses seen in multiple trials, and the information that the response conveys about the stimulus is the difference between the entropy and the noise entropy (see also de Ruyter van Steveninck et al. 1997). These measures from information theory are not just one of many possible ways of quantifying the neural response; Shannon proved that these are the only measures of variability, reproducibility and information that are consistent with certain simple and intuitively plausible constraints.

If we believe that the neural code makes use of a time resolution Δt, then we can describe the neural response in discrete time bins of this size. If Δt is very large this amounts to counting the number of spikes in each bin, while as Δt becomes small this description becomes a binary string in which we record the presence or absence of individual spikes in each bin. As our time resolution improves (smaller Δt) the size of the response "vocabulary" increases because we are distinguishing as different responses that were, at larger Δt, lumped together as being the same. Quantitatively, the entropy of the responses is a function of time resolution, so that the capacity of the neurone to convey information is greater at smaller Δt, as first emphasized by MacKay and McCulloch (1952). The question of whether spike timing is important to the neural code is then whether neurones make efficient use of this extra capacity (Rieke et al. 1993, 1997). In the next section we address precisely this question in the context of the "outdoor" experiment on H1, reaching conclusions that parallel closely those from our earlier work (de Ruyter van Steveninck et al. 1997; Strong et al. 1998). First we consider the results of Egelhaaf and collegues, who have drawn nearly opposite conclusions.

Warzecha and Egelhaaf (1997) and Egelhaaf and Warzecha (1999) set out to study the dependence of information transmission on time resolution, along the lines indicated above. Specifically, they count spikes in bins of size Δt and then ask how much information this spike count on a single trial provides about the local firing rate, or "Stimulus Induced Response Component" (SIRC) computed as an average over many trials. Their information measure shows a peak for a window width of $\Delta t = 80$ ms (Warzecha and Egelhaaf 1997, Fig. 3), from which they conclude that this is the time resolution at which signals are best represented by H1. It is not clear what measure of information Warzecha and Egelhaaf (1997) are using to find the optimum: the rate at which the spike train provides information about the stimulus must be a monotonic function of the time resolution. By marking spike arrival times more accurately we can only gain, and never lose, information. Thus a proper measure of information rate vs. time resolution cannot show the behaviour reported by Warzecha and Egelhaaf.

In the present volume they substitute the information theoretical analysis by one in which they quantify the same difference (that is, between the SIRC and a running window average count of the single trial spike train) by a standard deviation. This standard deviation reaches a minimum for a window width Δt of about 50 ms. This analysis, as their correlation function analysis, is based on a consideration of second order statistics, and is therefore subject to the same shortcomings discussed before.

Both these approaches suffer from the same fundamental problem: Warzecha and Egelhaaf do not quantify the relation between the neural response and the stimulus, but instead between spike counts and the SIRC. Implicitly, then, they postulate that the stimulus is encoded exclusively in the time dependent firing rate, or the SIRC as they prefer to call it, and further that all information about the local rate can be "read out" by counting spikes[3]. As in the analysis of crosstrial correlation functions, this ignores by construction the possibility that temporal patterns of spikes may play a special role in the code, and their reasoning is therefore circular. For many investigators this issue of whether patterns are important is *the* question about the structure of the neural code, and in the case of H1 it is now more than a decade since de Ruyter van Steveninck and Bialek (1988) reported that patterns with short interspike intervals carry a considerable excess of information about the stimulus (see also Rieke et al. 1997 and Brenner et al. 2000a). The approach taken by de Ruyter van Steveninck et al. (1997) and by Strong et al. (1998) describes the neural response at fine time resolution as a binary string, marking the presence or absence of spikes in each small time bin, and hence all patterns of spikes are included automatically. This is the approach that we will use below for the analysis of the outdoor experiment.

In principle, the methods used by de Ruyter van Steveninck et al. (1997) and by Strong et al. (1998) are independent of any model for the structure of the neural code: we do not need to assume that we know which features of the neural response are relevant, nor do we need to assume which features of the stimulus are most important for the neurone. A number of results on information transmission by H1 have been obtained with a less direct method, in which we use the spike train to reconstruct the stimulus and then measure the mutual information between the stimulus and the reconstruction (Bialek et al. 1991; Haag and Borst 1997,

---

[3]Even if the changing stimulus serves only to modulate the spike rate, it might be that different rates can be distinguished more easily because, for example, the shape of the interspike interval distributon changes as function of rate. This is known to occur in many cells. Mathematically, counting spikes is the optimal way of recovering rate information only if the spike train is a modulated Poisson process.

1998). Warzecha and Egelhaaf emphasize that errors in the reconstruction result only in part from noise, and they claim that one therefore cannot conclude anything about the reliability of neurones from the quality of reconstructions (Warzecha and Egelhaaf 1997; see also their contribution to this volume). The thrust of their argument is that there need be no conflict between their claim of imprecision in the coding of behaviourally relevant stimuli and previous work demonstrating precise reconstruction of the velocity waveforms, because the reconstruction doesn't really measure the precision of the neural system. But this discussion ignores the fact that the reconstruction method provides a lower bound on the performance of the neurone (Rieke et al. 1997; Borst and Theunissen 1999). Thus it is possible that reconstruction experiments underestimate the precision of neural coding and computation, but properly done the reconstruction method cannot overestimate neural performance.

Since the reconstruction procedure is a bound on performance and not a direct measurement, it is reasonable to ask how tight this bound will be. Warzecha and Egelhaaf state that the reconstruction of velocity signals would underestimate the performance of the neurone if the cell is sensitive to derivatives of the velocity; specifically they claim that the coherence between the stimulus and reconstruction would be reduced if the neurone were sensitive to derivatives. In fact, the particular reconstruction procedure of Bialek et al. (1991) is invariant to linear transformations of the signal such as differentiation and integration, and the computation of coherence *always* is invariant to these transformations (Lighthill 1958). Is there any independent way to assess the efficacy of the reconstruction method? One approach is to try different reconstruction algorithms (Warland et al. 1997). Another is to check for consistency among different measures of coherence (Haag and Borst 1997).

Finally, we can compare the noise levels in the reconstructions with the noise levels that would be generated by an ideal observer who is limited only by noise in the photoreceptors. In the high frequency limit (of order 30 Hz), where the ideal observer's performance can be calculated from photoreceptor measurements, the ideal observer does not perform substantially better than the reconstruction (Bialek et al. 1991), which demonstrates that H1's response approaches ideal observer performance. This can only be true if the fly's visual brain makes efficient use of the information present in the array of photoreceptors, and does not add a substantial amount of noise to the computation of motion. This finding is confirmed by measurements of neural performance that do not depend on reconstructions (de Ruyter van Steveninck and Bialek 1995). Of course, the accuracy and efficiency of the reconstructions also imply the functional correctness of the reconstruction algorithm. Criticism of the reconstruction algorithm itself cannot invalidate the demonstration of accurate reconstructions.

4. Information transmission with natural stimuli

In the following we use methods described in detail by Strong et al. (1998) to quantify information transmission in our natural motion experiment. Briefly, we analyse the statistics of firing patterns that H1 produces in response to the stimulus used in our experiment, and consider segments of the spike train with length T divided in a number of bins of width Δt, where Δt will range from very small (order of a millisecond) up to $\Delta t = T$. Each such bin may hold a number of spikes, and within a bin no distinction is made on where the spikes appear. However, two windows of length T that have different combinations of filled bins are considered to be different firing patterns, and are therefore distinct. From an experiment in which we repeat a reasonably long natural stimulus a number of times (here 200 repetitions of a 5 s long sequence) we get a large number of these firing patterns, and from that set we compute two entropies:

1. The total entropy, which characterizes the probability distribution of all spike firing patterns of length T that consist of n adjacent bins each Δt wide (that is, $T = n \cdot \Delta t$). This entropy measures the richness of the "vocabulary" used by H1 under these experimental conditions, hence the time of occurrence of the pattern within the experiment is irrelevant.

2. The noise entropy, which gives us an estimate of how variable the response to identical stimuli can be. We first accumulate, for each point in time in the stimulus sequence, the distribution across all trials of firing patterns that begin at that point. The entropy of this distribution measures the (ir)reproducibility of the response at each instant. Calculating this for each point in time and averaging all these values we obtain the average noise entropy.

The information contained in firing patterns of length T and resolution Δt is the total entropy minus the average noise entropy (Shannon 1948). One interesting measure is to estimate this information as we let T become very long, and Δt very short. This limit is the average rate of information transmission, as discussed by Strong et al. (1998). Here, instead, we will just calculate the information transmitted in constant time windows, $T = 30$ms, as a function of Δt. We choose $T = 30$ms because that amounts to the delay time with which a chasing fly follows turns of a leading fly during a chase (Land and Collett 1974); the end result, namely the dependence of information transmission on Δt, does not depend critically on the choice of T. The data in figure 4a show that the information contained in a 30 ms window depends strongly on Δt, increasing from about 2 bits to about 5 bits when the resolution increases from $\Delta t = 30$ms to $\Delta t = 1$ms. Although in the limit of arbitrarily fine time resolution ($\Delta t \rightarrow 0$), the information must reach a finite limit, we see no evidence for a plateau at $\Delta t = 1$ ms. For shorter time windows ($T=12$ms) we find that the information keeps on increasing up to $\Delta t = 0.25$ms. This lack of a clear plateau makes sense: the motion stimuli themselves have a distribution of temporal features so it is not surprising that there is not a sharply defined single timescale in the response. We also note that, as in earlier work with less natural stimuli (Strong et al. 1998), the information rate is a bit more than half the total entropy, even at millisecond resolution (see Fig. 4b), so the neurone utilizes a significant fraction of its coding capacity even on this fine time scale.

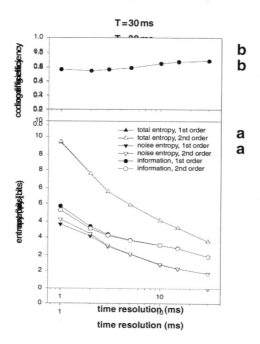

Fig. 4 Information and coding efficiency in firing patterns for naturalistic motion stimuli (see legend for Fig. 2). **a** Total entropy, noise entropy and information in an observation window of T = 30ms, as a function of time resolution, Δt. From the trends observed in partitioning the finite dataset we estimate first and second order extrapolations to the entropies for an infinite dataset. Filled symbols are first order-, open symbols are second order extrapolations. The deviation between first and second order extrapolations is small, indicating that systematic errors in the entropy estimates are small (for details see Strong et al. 1998). Statistical errors were estimated from the spread in the different partitions of the original dataset. These errors are smaller than the size of the symbols. **b** Coding efficieny (information divided by total entropy) as a function of Δt.

The question of whether spike timing is important in the neural code has been debated for decades, and our present experiment addresses the importance of millisecond resolution in information transmission by a single cell. Ultimately one would like to connect the responses of neurones to animal behaviour. Thus, one way to demonstrate the importance of spike timing would be to search for experimental conditions in which the timing of just a few spikes would be correlated with a behavioural decision, in the spirit of the work by Newsome and colleagues (Newsome et al. 1995). Another approach is to look for other neurones that can "read" the temporal structure, for example along the lines of recent work from Usrey et al. (1998). Here we focus on the response of a single neurone, and ask if the precise timing of spikes carries information under natural stimulus conditions. The answer is yes.

5. Responses to static and dynamic stimuli

The measured precision of responses in H1 to dynamic stimuli seems to suggest that the behaviour of the fly visual system might be very different from other systems, especially the mammalian cortex. Neurones in visual cortex, for example, commonly show a large variance in the responses across repeated presentations of the same visual stimulus (Tolhurst et al. 1983). To quantify this observation several groups have studied the variance in the number of spikes that are counted in a window of fixed size, and then manipulated the stimulus conditions to find the relation between the variance of the response and its mean. Typically, the variance in spike count is found to be close to or somewhat larger than the mean over a wide range of conditions; there is a tendency for the ratio variance/mean (the Fano factor) to be larger in larger time windows.

More recently several groups are investigating to what extent accurate spike timing, such as observed in H1, can be consistent with the variability of neural responses observed in cortex.

Almost all experiments on the variability of responses in visual cortex had been done with static or slowly varying stimuli, while all the work indicating precise responses and the importance of spike timing in H1 had been done using complex, dynamic inputs. Newsome and collaborators studied the responses of motion sensitive neurones in the monkey visual cortical area MT using dynamic random dot stimuli, but their work focused on the connection of neural responses to the monkey's perception of coherent motion in the entire display (Newsome et al. 1995). Bair and Koch (1996) reanalysed some of these data to show that when the monkey saw exactly the same dynamic dot movies the neural response showed significant modulations on a time scale of 30 ms or less. Strong analysed the same data to show that the spike train of a cortical neurone could provide information about the movie at a rate of ~2 bits/spike, comparable to the results in H1 (see note 19 in Strong et al. 1998), and in unpublished analyses he found that the variance of the spike count in windows of 30 ms or less could be significantly less than the mean.

Mainen and Sejnowski (1995) found that they could produce irregular spike trains in a slice of cortex if they injected constant current into a neurone: after some time the cell "forgets" the time at which the current was turned on and the spikes drift relative to the stimulus. With dynamic currents, however, there can be precise temporal locking of spikes to particular events in the input signal. Berry et al. (1997) found that ganglion cells in the vertebrate retina – which are known to generate irregular and highly variable spike trains in response to static or slowly varying images – generate highly reproducible spike trains in response to more dynamic movies. A hint in the same direction had been found earlier by Miller and Mark (1992), who showed that primary auditory neurones in the cat give less variable responses to complex speech stimuli than to pure tones. Finally, de Ruyter van Steveninck et al. (1997) showed explicitly that the low variance, reproducible response of H1 to dynamic stimuli coexists with a much more variable response to constant velocity inputs: studying a range of constant velocities that drove H1 to average firing rates up to about 70 spikes per second (which corresponds to the time average rate elicited by dynamic stimuli in comparable stimulus conditions), mean counts and variances in 100 ms windows straddled the line at which variance is equal to mean, and fell well within a cloud of points obtained from experiments in visual cortex.

Taken together, all of these different results point to the conclusion that the statistical structure of the neural response to static stimuli may be very different from that in response to dynamic or naturalistic stimuli. The crucial conclusion is that we cannot extrapolate from the observation of highly variable responses under one set of conditions to reach conclusions about the structure of the neural code under more natural conditions. This fits very well with the ethological perspective that we introduced at the beginning of this contribution, and indeed many of the analysis methods that we have discussed here were developed to meet the challenges of quantifying the neural response to more naturalistic stimuli. From a more mechanistic point of view there is now considerable interest in understanding why neurones seem to respond so differently to static and dynamic inputs (Jensen 1998; Schneidman et al. 1999).

Against this background it came as a surprise when Warzecha and Egelhaaf (1999) claimed that the variance of H1's response to constant velocity is no different from that in response to dynamic stimuli when the responses are compared at equal mean firing rates. It would appear that they have done an experiment very similar to that described by de Ruyter van Steveninck et al. (1997) but reached the opposite conclusion: while Warzecha and Egelhaaf confirm the highly reproducible, low variance response to dynamic stimuli, they find similarly reproducible responses to constant velocities. There are many issues here, but we focus first on the explicit disagreement regarding the variability of responses to constant velocity. Warzecha and Egelhaaf themselves offer several possible explanations for the discrepancy, but they do not draw attention to the fact that the stimuli used in the two sets of experiments differ substantially; these differences exist along every stimulus dimension known to affect the response of H1 – velocity, image contrast, spatial pattern, and size of the visual field. Further, in the crucial comparison of static to dynamic stimuli, it is not clear what is being held constant in the Warzecha and Egelhaaf experiments. In the dynamic experiments changes in spike rate are of course driven by variations in angular velocity, but in their static experiments they hold the velocity fixed and vary the image size. At best these experiments show that H1 responds with different statistics under different conditions, but we still find the discrepancies disturbing.

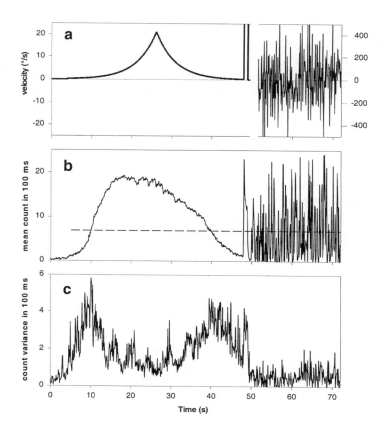

Fig. 5 Mean count and variance compared for quasistatic and dynamic velocity stimuli. **a** Stimulus. For the first 48 s the velocity is slowly ramped up and down. From 50 to 72 s the stimulus is dynamic with a standard deviation of 100°/s, and a cutoff frequency of 250 Hz. Note that the vertical scales (left for the quasistationary- and right for the dynamic stimulus) are different. The peak at 50 s is a reset phase in which the pattern is moved at maximum speed so as to bring the stimulus pattern into exact register on every trial. **b** Trial average spike count in 100 ms windows, as a function of time. The dashed line represents the time averaged count in response to the dynamic stimulus. **c** Spike count variance in 100 ms windows, as a function of time.

In an attempt to resolve the issue, we have gone back over several years of experiments to collect all the data which may be relevant to relation between variance and mean in static experiments, we have done new experiments that come close to the conditions of the Warzecha and Egelhaaf work, and we have designed new stimuli that highlight the differences between static and dynamic responses in a single experiment. In brief, studying 20 flies under a wide variety of static stimulus conditions, we find a broad distribution of variances at each value of the mean, but up to rates of about 100 spikes/s there is no overlap with the results of Warzecha and Egelhaaf (1999). Further, when we match the conditions of their experiments we cannot reproduce even the mean spike counts, let alone the variances. For example, their figure 4a in this volume shows a mean spike count of about 6.5 in 100 ms windows for a high contrast large field ($91° \times 7.5°$) pattern moving at about $36°/s$, and for the same experimental conditions Warzecha and Egelhaaf (1999) report a mean count of about 4. These values correspond to mean firing rates of 65 and 40 spikes/s respectively. In 8 flies tested under comparable conditions (contrast, velocity and stimulated area) we never get rates below 120 spikes/s, consistent with the findings of Lenting et al. (1984).

In figure 5 we show the response of H1 to a slowly varying velocity ramp, and contrast this response to that obtained with dynamic velocities. Computing the mean spike counts in 100 ms windows across 50 trials, we see that the static and dynamic stimuli give the same range of mean responses, yet when we compute the variances there are huge differences that are obvious to the eye. The count variance during quasistatic stimulation peaks for mean counts that are about equal to the average count during dynamic stimulation (that is, in the two places where the dashed line in figure 5b intersects the smooth curve). This is not just a coincidence of our choice of standard deviation of the dynamic stimulus; it turns out that the fly's visual system adapts such that the mean firing rate during dynamic stimulation is rather insensitive to the standard deviation of the dynamic stimulus (Brenner et al. 2000b). For higher values of the mean count the variance decreases strongly, due to the effects of refractoriness (Hagiwara 1954). So H1 has relatively low count variance both for low and high rates, but its count variance is high for intermediate rates. Loosely, one may think of the dynamic stimulus as switching the cell rapidly back and forth from a state of low rate and low variance to a state of high rate and low variance. By switching fast, the cell effectively bypasses the intermediate condition of high variance, so that its count variance for dynamic stimuli always remains low, as can be seen directly from figure 5c. Thus, if we match windows with the same mean count, up to a count of about 10 for 100 ms windows, we find that H1's count variance is lower in response to dynamic than to static stimuli, which was precisely the point of the original work by de Ruyter van Steveninck et al. (1997).

6. Conclusion

Most of what we know about the nervous system has been learned in experiments that do not even approach the natural conditions under which animals normally operate. Much of our recent work, and the core of the debate between our groups and Warzecha and Egelhaaf, concerns the structure of the neural code under natural conditions. We emphasize that this is not an easy problem, and by no means are the issues specific to flies or even the visual system; in many different sensory and motor systems we would like to design and analyse experiments on the coding and processing of more natural signals.

Our approach has been to break this large question into (hopefully) manageable pieces, and then to use information theory as a framework to pose these questions in a form such that suitable experiments should yield precise quantitative answers. In particular, we endeavour to make statements that do not depend on multiple prior assumptions, and to develop methods which can be used in analysing many different kinds of experiments. Thus, we have used stimulus reconstruction techniques to give lower bounds on the performance of fly motion sensitive neurones, and we have been able to measure the average information carried by single spikes, patterns of spikes, and continuous segments of the spike train, all without assumptions regarding the "important" features of the stimulus or neural response. Many of the results obtained in this way point clearly toward a picture of the fly's visual system as close to optimal in extracting motion information from the photoreceptor cell array, and then encoding this signal efficiently in the timing of action potentials of motion sensitive neurones.

In contrast to the view developed over the past decade, the recent papers from Egelhaaf and collegues, including their contribution to this volume, make the explicit claim that the system is very noisy and that meaningful information is contained only in averages over time windows containing many spikes. In many cases these claims are introduced with plausible qualitative arguments. As emphasized long ago by Bullock (1970), however, the challenge is to *quantify* the degree of noisiness or precision in the nervous system, and there is a danger that a neurone may appear noisy because we have an incomplete understanding of its function. Thus we have grown sceptical about qualitative or even semiquantitative arguments for the imprecision of neural responses. The interpretation of correlation functions, discussed above, provides a good example: Although there may be an obvious "correlation time" in one correlation function, the hint that other time scales are relevant is hidden in the cusp of the correlation function at short times. More detailed analysis shows that the relations among correlation functions are consistent with temporal precision on scales a factor of 30 smaller than the nominal correlation time. Further, this measure of temporal precision is consistent with the results of a rigorous information theoretic analysis.

Because this Chapter is intended (by the editors) as a response to the contribution of Warzecha and Egelhaaf, we have tried to understand how they have reached conclusions so nearly opposite from our own. As emphasized at the outset, there are two different questions. First there is the problem of constructing an approximately natural stimulus, and then there is the problem of analysing the response to such a complex signal. Although there is a whole generation of quantitative observations on insect flight trajectories, Warzecha and Egelhaaf present as the stimulus they analyse a signal that is substantially impoverished both in amplitude and in frequency content, as is clear from figure 1. Further, their visual stimulus is very dim compared to daytime natural conditions, and has a visible area much smaller than what is experienced by a free flying animal. They repeatedly stress that motion induced responses depend on many stimulus variables in addition to velocity, but never present a critical quantitative discussion on the extrapolation from their experiments to natural behaviour, in spite of the large differences in many of the crucial variables. They repeatedly stress that motion induced responses depend on many stimulus variables in addition to velocity, but never discuss critically the extrapolation from their experiments to natural behaviour, in spite of the large differences in many of the crucial variables. Similarly, although there is now a decade of papers concerning the quantitative information theoretic analyses of neural spike trains, and of the cell H1 in particular, Warzecha and Egelhaaf do not present their results on information transmission in absolute units (bits); closer examination suggests that there are more basic mathematical problems in their approach, as outlined above.

In this contribution we have presented the results from a new experiment which brings us much closer to the natural conditions of fly vision. Visual inspection of the responses of H1 under these conditions indicates that individual spikes are reproducible on a millisecond time scale, and aspects of temporal pattern in the spike train can be reproducible on a substantially *sub*millisecond time scale. This impression is borne out by the quantitative demonstration that the spike train conveys information with nearly constant efficiency down to millisecond time resolution; indeed, the information provided by the spike train shows no sign of saturation as we approach millisecond resolution. These responses to natural stimuli thus are even more precise than suggested by our earlier work.

In the early 1960's Reichardt and his collegues started working on flies, with a special emphasis on motion detection (Reichardt 1961). One of their motivations was that motion detection in flies represents a good compromise between a reasonable complexity of information processing properties and an amenability to quantitative analysis (Reichardt and Poggio 1976). Over the years this intuition has proved to be very fruitful, and the fly has turned out to be a system in which many issues could be studied, often with unprecedented quantitative detail. In particular, the fly's motion sensitive neurones have been an important testing ground for ideas about the neural code, especially in the ongoing effort to characterize the coding of more natural stimuli. Approached with proper mathematical tools, the fly visual system can continue to provide answers to many fundamental and quantitative questions in real time neural information processing.

References

Bialek W, Rieke F, de Ruyter van Steveninck RR, Warland, D (1991) Reading a neural code. Science 252: 1854-1857

Bair W, Koch C (1996) Temporal precision of spike trains in extrastriate cortex of the behaving Macaque monkey. Neural Comp 8: 1185-1202

Berry MJ, Warland DK, Meister M (1997) The structure and precision of retinal spike trains. Proc Natl Acad Sci USA 94: 5411-5416

Borst A, Egelhaaf M (1987) Temporal modulation of luminance adapts time constant of fly movement detectors. Biol Cybern 56: 209-215

Borst A, Theunissen, FE (1999) Information theory and neural coding. Nature Neurosci 2: 947-957

Brenner N, Strong SP, Koberle R, Bialek W, de Ruyter van Steveninck R (2000a) Synergy in a neural code. Neural Comp: in press

Brenner N, Bialek W, de Ruyter van Steveninck R (2000b) Adaptive rescaling maximizes information transmission. Neuron: in press

Bullock TH (1970) The reliability of neurons. J Gen Physiol 55: 565-584

Dethier VG (1976) The hungry fly. A physiological study of the behavior associated with feeding. Harvard University Press, Cambridge, MA

Egelhaaf M, Warzecha A-K (1999) Encoding of motion in real time by the fly visual system. Curr Opinion Neurobiol 9: 454-460

Field D (1987) Relations between the statistics of natural images and the response properties of cortical cells. J Opt Soc Am A 4: 2379-2394

Gauck V, Borst A (1999) Spatial response properties of contralateral inhibited lobula plate tangential cells in the fly visual system. J Comp Neurol 406: 51-71

Haag J, Borst A (1997) Encoding of visual motion information and reliability in spiking and graded potential neurons. J Neurosci 17: 4809-4819

Haag J, Borst A (1998) Active membrane characteristics and signal encoding in graded potential neurons. J Neurosci 18: 7972-7986

Hagiwara S (1954): Analysis of interval fluctuations of the sensory nerve impulse. Jpn J Physiol 4: 234-240

van Hateren JH, Schilstra C (1999) Blowfly flight and optic flow II. Head movements during flight. J Exp Biol 202: 1491-1500

Hecht S, Shlaer S, Pirenne MH(1942) Energy, quanta and vision. J Gen Physiol 25: 819-840

Jensen, RV (1998) Synchronization of randomly driven nonlinear oscillators. Phys Rev E 58: 6907-6910

Land MF (1981) Optics and vision in invertebrates. In: Autrum H (ed) Handbook of Sensory Physiology VIII/6b. Springer, Berlin, Heidelberg, New York, pp 472-592

Land MF, Collett TS (1974) Chasing behavior of houseflies (*Fannia canicularis*). A description and analysis. J Comp Physiol 89:331-357

Lenting, BPM, Mastebroek HAK, and Zaagman WH (1984) Saturation in a wide-field, directionally selective movement detection system in fly vision. Vision Res 24:1341-1347

Lewen GD, Bialek W, de Ruyter van Steveninck RR Neural coding of natural stimulus ensembles. submitted

Lighthill, MJ (1958) An introduction to Fourier analysis and generalised functions. Cambridge University Press, Cambridge, UK

MacKay D, McCulloch WS (1952) The limiting information capacity of a neuronal link. Bull Math Biophys 14: 127-135

Maddess T, Laughlin SB (1985) Adaptation of the motion-sensitive neuron H1 is generated locally and governed by contrast frequency. Proc R Soc Lond B 225: 251-275

Mainen ZF, Sejnowski TJ (1995) Reliability of spike timing in neocortical neurons. Science 268: 1503-1506

Miller MI, Mark KE (1992). A statistical study of cochlear nerve discharge patterns in response to complex speech stimuli, J Acoust Soc Am 92: 202-209

Newsome WT, Shadlen MN, Zohary E, Britten KH, Movshon JA (1995) Visual motion: linking neuronal activity to psychophysical performance. In: Gazzaniga M (ed) The cognitive neurosciences. MIT Press, Cambridge, MA, pp 401-414

Reichardt, W (1961) Autocorrelation, a principle for the evaluation of sensory information by the central nervous system. In: Rosenblith WA (ed) Principles of sensory communication. Wiley, New York, NY, pp 303-317

Reichardt W, Poggio T (1976) Visual control of orientation behavior in the fly. Part I: A quantitative analysis. Q Rev Biophys 9: 311-375

Rieke F, Warland D, Bialek W (1993) Coding efficiency and information rates in sensory neurons. Europhys Lett 22: 151-156

Rieke F, Warland D, de Ruyter van Steveninck RR, Bialek W (1997) Spikes: exploring the neural code. MIT Press, Cambridge, MA

Ruderman DL, Bialek W (1994) Statistics of natural images: Scaling in the woods. Phys Rev Lett 73: 14-817

de Ruyter van Steveninck RR, Zaagman WH, Mastebroek HAK (1986) Adaptation of transient responses of a movement sensitive neuron in the visual system of the blowfly *Calliphora erythrocephala*. Biol Cybern 54: 223-236

de Ruyter van Steveninck R, Bialek W (1988) Real-time performance of a movement sensitive neuron in the blowfly visual system. Proc Roy Soc Lond B 234: 379-414

de Ruyter van Steveninck R, Bialek W (1995) Reliability and statistical efficiency of a blowfly movement-sensitive neuron. Phil Trans Roy Soc Lond B 348: 321-340

de Ruyter van Steveninck RR, Bialek W, Potters M, Carlson RH, Lewen GD (1996) Adaptive movement computation by the blowfly visual system. In: Waltz DL (ed) Natural and artificial parallel computation: Proc Fifth NEC Res Symp, SIAM, Philadelphia, pp 21-41

de Ruyter van Steveninck RR, Lewen GD, Strong SP, Koberle R, Bialek W (1997) Reproducibility and variability in neural spike trains. Science 275: 1805-1808

Schilstra C, van Hateren JH (1998) Stabilizing gaze in flying blowflies. Nature: 395:654

Schneidman E, Freedman B, Segev I (1998) Ion channel stochasticity may be critical in determining the reliability and precision of spike timing. Neural Comp 10: 1679-1703

Shannon CE (1948) A mathematical theory of communication. Bell Syst Techn J 27: 379-423 and 623-656

Srinivasan MV, Laughlin SB, Dubs A (1982) Predictive coding: a fresh view of inhibition in the retina. Proc Roy Soc Lond B 216: 427-459

Strong SP, Koberle R, de Ruyter van Steveninck RR, Bialek W (1998) Entropy and information in neural spike trains. Phys Rev Lett 80: 197-200

Tolhurst DJ, Movshon JA, Dean AF (1983) The statistical reliability of signals in single neurons in cat and monkey visual cortex. Vision Res 23: 775-785

Usrey WM, Reppas JB, Reid RC (1998) Paired-spike interactions and synaptic efficacy of retinal inputs to the thalamus. Nature 395: 384-387

Wagner H (1986a) Flight performance and visual control of flight of the free-flying housefly (*Musca domestica* L.). I. Organization of the flight motor. Phil Trans Roy Soc Lond B 312: 527-551

Wagner H (1986b) Flight performance and visual control of flight of the free-flying housefly (*Musca domestica* L.). II Pursuit of targets. Phil Trans Roy Soc Lond B 312: 553-579

Wagner H (1986c) Flight performance and visual control of flight of the free-flying housefly (*Musca domestica* L.). III. Interactions between angular movement induced by wide- and smallfield stimuli. Phil Trans Roy Soc Lond B 312: 581-595

Warland DK, Reinagel P, Meister M (1997) Decoding visual information from a population of retinal ganglion cells. J Neurophysiol 78: 2336-2350

Warzecha A-K, Egelhaaf M (1997) How reliably does a neuron in the visual motion pathway of the fly encode behaviourally relevant information? Europ J Neurosci 9: 1365-1374

Warzecha A-K, Egelhaaf M (1998) On the performance of biological movement detectors and ideal velocity sensors in the context of optomotor course stabilization. Visual Neurosci 15: 113-122

Warzecha A-K, Egelhaaf M (1999) Variability in spike trains during constant and dynamic stimulation. Science 283: 1927-1930

Warzecha A-K, Kretzberg J, Egelhaaf M (1998) Temporal precision of the encoding of motion information by visual interneurons. Curr Biol 8: 359-368

Wehrhahn C (1979) Sex-specific differences in the chasing behavior of houseflies (*Musca*). Biol Cybern 32: 239-241

Wehrhahn C, Poggio T, Bülthoff H (1982) Tracking and chasing in houseflies (*Musca*). An analysis of 3-D flight trajectories. Biol Cybern 45: 123-130

A Comparison of Spiking Statistics in Motion Sensing Neurones of Flies and Monkeys

Crista L. Barberini, Gregory D. Horwitz and William T. Newsome

Howard Hughes Medical Institute and Department of Neurobiology, Stanford University School of Medicine, Stanford, USA

1. Introduction

The information processing strategies employed by the brain are constrained by the reliability with which individual neurones encode external stimuli. If repeated presentations of a particular stimulus elicit extremely reliable responses, relatively few neurones are needed to provide an adequate representation of the stimulus. If, on the other hand, responses are highly variable, more neurones and additional processing stages (e.g. averaging) may be necessary to encode the stimulus with equal precision. In their article in this volume, Warzecha and Egelhaaf conclude provocatively that the motion sensing neurone, H1, of the fly responds to stimuli with considerable reliability, a finding that stands in contrast to the highly variable responses of motion sensing neurones in the middle temporal visual area (MT) of the monkey cerebral cortex. This difference may make a great deal of sense. Invertebrate nervous systems contain far fewer neurones than those of vertebrates; accurate representation of the sensory world with a relatively small number of neurones may therefore be of great importance to invertebrates. Vertebrate nervous systems, on the other hand, may sacrifice fidelity at the single neurone level to obtain the increased computational power afforded by a more densely interconnected nervous system. Because of the explosion of neurone number in the vertebrate central nervous system, fidelity sacrificed at the single neurone level may be preserved by redundant representation and signal averaging (Buracas et al. 1998; Shadlen and Newsome 1998).

   If sensory neurones of invertebrates indeed respond to stimuli with greater fidelity than do those of vertebrates, comparison of the two systems may provide important insights into the cellular mechanisms and network properties governing response variability. Unfortunately, direct comparison of response variability between fly H1 and MT neurones has been hampered by differences in experimental methodology and analytic approaches. In the next section we present a theoretical framework for analysing the response statistics of sensory neurones; this framework reveals pitfalls in such analyses that have not been commonly recognized in prior studies. We then make a direct comparison of H1 and MT response statistics, concluding that they are indeed remarkably different.

2. Theory

To assess the reliability of neuronal firing, investigators measure responses to repeated presentations of a particular stimulus so as to eliminate variability induced by changes in the stimulus itself. The response variability that remains represents noise associated with the neuronal representation of the stimulus. A common measure of response variability, employed by Warzecha and Egelhaaf and by many others, is the relationship between the variance and the mean of the spike count. For many sensory neurones, the variance changes systematically as a function of the mean; thus a relationship, rather than a single value, must be determined.

To compute this metric, the spike count (the number of spikes that occurs within an epoch of a specified duration) is tallied for each trial, and the mean and variance of this set of counts is computed. The trials are aligned on stimulus onset, and the measurement epoch (hereafter time window) is placed at a specific time relative to the start of the trial. Because the time window usually has a much shorter duration than the whole trial, this process can be repeated for many time-points relative to stimulus onset, a variance and mean calculation of the spike count across trials being made for each position of the time window. The resulting data form a cloud of points on a scatterplot of variance versus mean count, which can be profitably compared to various theoretical predictions.

We will consider two theoretical models that lie along a continuum of possible spiking regimes. At one end is a neurone that responds with extreme regularity, firing spikes at nearly constant intervals, akin to the ticking of a clock. For neurones operating in this firing regime (e.g., neurones of the pontine oculomotor structures: Keller and Robinson 1972; Keller 1977; King et al. 1986; Everling et al. 1998), the relationship between the variance and mean of the spike count will take the form of a theoretical minimum curve (see below). The second model we consider, a modification of the Poisson process, represents a much more irregular mode of firing. While other models could produce even more variable spike patterns, we employ the Poisson process because it is a prototype of randomness, and has been used extensively in prior models of neuronal firing. In the Poisson model, neurones discharge action potentials at a particular rate, but the exact timing of individual action potentials is random given the underlying rate. The probability of firing at any instant in time is independent of the recent history of spiking. The spike trains of real neurones, however, do not obey Poisson statistics in one particularly salient respect: each spike is followed by a refractory period during which no other spikes can occur. In contrast to a true Poisson process, then, the probability of firing is not constant across time, but depends upon the recent history of spike firing. We therefore modified a Poisson model by incorporating a refractory period to determine its effect on the expected variance-to-mean relationship. For this "quasi-Poisson" neurone, the probability of firing remains constant throughout the trial, except for a brief refractory period following each spike, for which the probability of firing is zero.

Figure 1a illustrates simulated spike trains from a quasi-Poisson neurone with a 1 ms refractory period for three average firing rates. The thick line in figure 1b depicts the expected variance-to-mean curve[1] for spike counts calculated within 10 ms time windows; the arrows indicate the points on the curve (emphasized with bullets) that correspond to the three rasters in figure 1a. For comparison, the identity line shows the relationship expected of a pure Poisson process (variance equal to the mean). The shaded boxes and the bracket in figure 1a indicate the time window used in these simulations. At low firing rates (Fig. 1a, top panel), spike times are irregular despite the presence of a refractory period, and the quasi-Poisson neurone behaves much like a Poisson process – the data points in figure 1b lie near the identity line. As the average firing rate increases (middle panel), the refractory period causes spike times to become more regular. The variance-to-mean curve diverges from the Poisson diagonal because regularity in spike timing limits the variance progressively more as the mean increases. At high firing rates (bottom panel), the spike train approaches perfect regularity because the neurone fires a spike as soon as the refractory period is over. The number of spikes occurring in an epoch approaches the maximum possible value, and the variance, accordingly, approaches zero. (The small bump at the right end of the curve will be explained shortly.)

Importantly, the quantitative relationship between variance and mean count for a quasi-Poisson neurone depends on the length of the refractory period. In a given time window, a neurone with a longer refractory period will have a lower maximum firing rate and a lower expected spike count variance. The variance-to-mean curve will depart from the Poisson diagonal more quickly and return to a variance of zero at a lower spike count – the maximum imposed by the longer

---

[1] The variance-to-mean relationship for the quasi-Poisson neurone was calculated by a brute force method using results from the theory of renewal processes (Cox 1962). Complete spike count distributions were obtained by the sequential convolution of interspike interval densities. The distribution of the time-to-first-spike was exponential. The distribution of all subsequent interspike intervals were identical exponential functions shifted on the time axis by an amount corresponding to the refractory period. The distribution of the time-to-nth-spike was thus a gamma density shifted on the time axis by an amount corresponding to the cumulative refractory periods of the previous spikes. The parameter of the exponential distribution was varied to simulate different firing rates. Note that an equilibrium renewal process might be more closely analogous to our data analysis method, but is less convenient analytically.

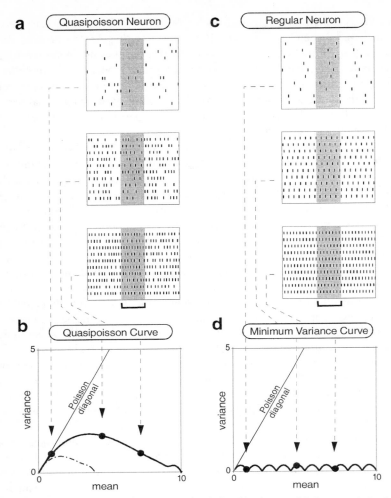

Fig. 1 Predictions for the variance-to-mean relationship for quasi-Poisson and regularly-firing model neurones. **a** Sample spike trains for a quasi-Poisson model neurone, generated by assigning spikes at random and with a constant probability, except for the first ms after each spike during which no additional spikes could occur. To achieve this, interspike intervals were drawn from a continuous exponential distribution with a zero probability region in the 1 ms time bin; different exponential distributions produced different spike rates. The three panels represent three average firing rates (three different constant probabilities): 92, 440, and 714 spikes/second. Grey regions and the bracket below the third panel delineate a time window (here 10 ms wide) in which spikes are counted. **b** The variance-to-mean relationship for the neurone in (a). Sample data points are computed from 200 trials. Dashed line indicates the variance-to-mean relationship for a similar quasi-Poisson neurone but with a 2.5 ms refractory period. **c** Spike trains for a highly regular model neurone, generated by assigning spikes at perfectly regular time intervals, but randomly varying the time of the first spike. Firing rates as in (a). **d** The variance-to-mean relationship for the neurone in (c). b, d: Data points on the curve represent the mean and variance of the count for the corresponding spike trains in (a) and (c); the corresponding panel is indicated with an arrow.

refractory period. The dashed curve in figure 1b, for example, illustrates the expected relationship between variance and mean for a quasi-Poisson neurone with a refractory period of 2.5 ms. For neurones with different refractory periods, plots of the variance-to-mean relationship are not sufficient to distinguish regular and irregular firing regimes. Rather, the variance-to-mean data must be examined relative to the theoretical curves appropriate for the two refractory periods. This is an important concern when comparing spike train statistics of fly H1 with those of MT neurones, because H1 appears to have a significantly longer refractory period (about 2.5 ms; Warzecha, personal communication) than MT neurones do (1 ms or less; Barberini, Horwitz and Newsome, unpublished observations). In the data analysis below, we plot both H1 and MT data relative to the theoretical predictions for a quasi-Poisson neurone using these absolute refractory periods.

Figures 1c and d illustrate spike trains from a simulated neurone with perfectly regular discharge. The variance-to-mean relation of this neurone differs substantially from that of the quasi-Poisson neurone, following what has been called the minimum variance curve (Berry et al. 1997; de Ruyter van Steveninck et al. 1997; Warzecha and Egelhaaf 1999). Figure 1c shows spike trains of this neurone for three firing rates, and the thick curve in figure 1d depicts its variance-to-mean relationship. The minimum variance is not always zero: because spike counts are necessarily integers, non-integer mean counts cannot have zero variance. Thus the curve is scalloped, rising a small distance above zero for non-integer mean counts. Note that the minimum variance curve is a lower bound derived directly from the formula for variance, in contrast to the quasi-Poisson curve, which represents a prediction of a statistical model of neural spiking behaviour. Indeed, the minimum variance curve accounts for the small bump at the right end of the quasi-Poisson curve in figure 1b. Both curves will serve as useful points of comparison for evaluating the neural data from H1 and MT in the next section.

3. A word about stimuli

What stimuli should be employed in studies of spiking variability in sensory neurones has been a subject of much debate (de Ruyter van Steveninck et al. 1997; Warzecha and Egelhaaf 1999). Traditionally, a "constant" stimulus is presented for a few seconds, and this process is repeated over many trials to obtain a database adequate for statistical analysis. The adjective "constant" does not mean that the stimulus is static. A constant motion stimulus, for example, would move in a constant direction with constant speed for the duration of the trial. In contrast, a dynamic (or "naturalistic") motion stimulus would change directions and speeds during the course of a trial, presumably mimicking more closely stimuli that appear in nature. Unfortunately, the use of dynamic stimuli can obscure the difference between the predictions of our two models – regularly spiking versus quasi-Poisson. As figures 1b and d show, the predictions of the two models converge at a variance of zero at either end of the neurone's dynamic range. A highly dynamic stimulus that drives a neurone rapidly from a near-maximum to a near-minimum firing rate will elicit responses that are equally well accounted for by both models. For this reason, our analysis focuses on relatively constant stimuli because our goal is to determine whether the spiking statistics of H1 and MT follow fundamentally different rules.

The constant stimulus used by Warzecha and Egelhaaf was a square wave grating moving in the preferred direction at specified speed. By varying the size of the grating stimulus in different blocks of trials, they acquired data over a considerable portion of the dynamic range of the neurone. Unfortunately, we do not possess MT data collected with a truly constant motion stimulus. Instead, we analyse data collected using dynamic random dot stimuli of 0% coherence (the random number seed used on each trial to generate the motion stimulus was held constant such that the stimulus was identical across trials; see Britten et al. 1992). While this stimulus is dynamic, the variability in direction and speed of the motion signal over time is modest, and the firing rates of MT neurones to the stimulus rarely approach either end of their dynamic range. However, under the quasi-Poisson model, we would expect that a dynamic stimulus would tend to reduce the variance-to-mean ratio[2].

4. H1 and MT: spiking statistics compared

As reported by Warzecha and Egelhaaf (this volume), H1 spike statistics are poorly described by a Poisson model. Figure 2b replots the data from their figure 4a, which is the variance-to-mean relationship for a single H1 neurone obtained using a 100-ms counting window. Figure 2a displays the corresponding data for a 10-ms counting window; this fuller version of the data set can be found in Warzecha and Egelhaaf (1999). The Poisson diagonal, the quasi-Poisson curve, and the minimum variance curve are included in both plots. Note that each plot is

---

[2] At first glance, it may seem counter-intuitive that a more dynamic stimulus will reduce the variance-to-mean ratio. How can a dynamic stimulus, which modulates the neural firing rate more vigorously than a constant stimulus, cause the neural response to be less variable? The key insight is that the variance and mean of the spike count are assessed across trials, at the same time within each trial. Because the dynamic stimulus is identical on each trial, the firing rate within any given time window will be stereotyped. Thus the dynamic stimulus will not lead to increased response variance across trials. In fact, just the opposite is likely to occur. Unlike a constant stimulus, a dynamic stimulus can drive a neurone rapidly between very high and very low firing rates within a single time window. The mean spike count across the window may fall in the middle of the neurone's dynamic range, while the count variance is considerably reduced because of the stimulus-driven excursions toward maximum and minimum firing rates where response variance approaches zero (see Fig. 1b).

scaled so that the quasi-Poisson curve spans the abscissa; this change in scale reflects the larger spike counts obtained in larger time windows. The right end-point of the quasi-Poisson curve represents the (estimated) maximum spike count for that time window given the refractory period of the neurone. Thus the quasi-Poisson curve reflects the dynamic range of the neurone, and the position of any data point along the abscissa indicates where a given firing rate falls within this dynamic range. The minimum variance curve, which does not depend on refractory period or window size, is identical in the two panels.

The H1 data fall far below the Poisson diagonal as well as the quasi-Poisson curve for both time windows. Thus the departure of H1 from Poisson statistics is not due simply to regularity imposed by its rather long refractory period. For the 10-ms counting interval, in fact, H1 statistics actually follow the minimum variance curve closely over a wide range of firing rates. The spike train is highly regular; the variance does not covary with the mean, as it does for both Poisson and quasi-Poisson model neurones. The variance rises above the minimum variance curve for the longer counting window (Fig. 2b), but the data still lie far below the quasi-Poisson curve.

In contrast, analogous data from MT neurones exhibit much greater variability. Figures 2c and d depict data from one MT neurone, obtained in response to repeated presentations of a random dot motion stimulus, analysed in the same manner as the H1 data in panels a and b. For the 10-ms time window (Fig. 2c), the MT neurone follows the quasi-Poisson prediction rather closely, suggesting that most of the departure from Poisson statistics can be accounted for by regularity imposed by the refractory period. The variance-to-mean ratio increases for the 100-ms time window (Fig. 2d). For this window, in fact, the data correspond more closely to the Poisson prediction. We will discuss below possible reasons for an effect of time window size on the spike statistics.

Figures 2e and f depict variance as a function of mean spike count averaged across 8 H1 neurones and 8 MT neurones. The differences in spiking statistics evident in the single neurone examples are also obvious in the averaged data. Thus the pronounced difference between H1 and MT is robust across neurones.

5. Discussion

We conclude, in agreement with Warzecha and Egelhaaf, that the spike train statistics of H1 neurones differ strikingly from those of MT neurones (see also Buracas et al. 1998). The discharge of H1 neurones appears to be substantially more regular than that of MT neurones, and this difference is not accounted for by the longer refractory period of H1. Small eye movements could potentially contribute to variance measurements made in awake monkeys (Gur et al. 1997), but the available data suggest that this contribution is small for MT neurones (Bair and O'Keefe 1998). Our use of modestly dynamic stimuli for the MT measurements

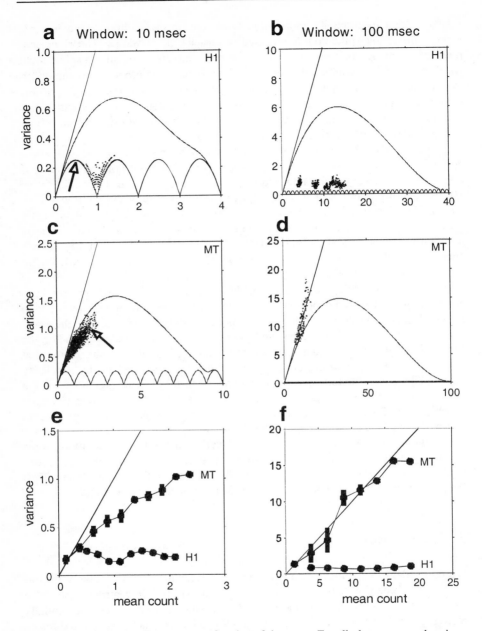

Fig. 2 Variance of the spike count as a function of the mean. For all plots, mean and variance across trials in the number of spikes within an epoch of time is plotted for time windows that are placed at different points in time relative to stimulus onset; the length of the time window is a fixed duration for each plot. **a** Data for H1, 10-ms time window. **b** Data for H1, 100-ms time window **c** Data for MT, 10-ms time window. **d** Data for MT, 100-ms time window (Note that the compressed vertical scale renders the minimum variance curve nearly invisible). **e** Data for H1

was not ideal for the comparison of H1 and MT, but any artefact introduced by our stimuli is likely to be conservative with respect to our conclusions, making MT responses more reliable and thus more similar to H1. Thus the striking difference in spiking statistics between H1 and MT probably reflects fundamental differences in the neural circuits in which they are embedded.

5.1 Analysis of spiking statistics

Importantly, our analysis shows that the variance-to-mean ratio, in and of itself, must be interpreted very cautiously. When spike train statistics from different brain structures or different organisms are compared, the data must be located relative to the theoretical predictions of pertinent models. Because these predictions vary substantially with the refractory period, and thus the dynamic range, of the neurones in question, it is essential to take this factor into account. For example, a variance-to-mean ratio of 0.5 can be generated by neurones operating under both highly regular and highly irregular firing regimes. The data point indicated by the arrow in figure 2a corresponds to a variance-to-mean ratio of 0.5, as does the data point indicated by the arrow in figure 2c. Yet the former point lies on the minimum variance curve, reflecting exceedingly regular firing, while the latter point falls very near the quasi-Poisson curve, reflecting an irregular firing pattern. In addition, care must be taken in interpreting variance-to-mean data from the same neurone operating at different points within its dynamic range. Low variance-to-mean ratios reflect regular, substantially non-Poisson statistics only in the middle of the neurone's dynamic range; at the low and high end of the dynamic range, low variance-to-mean ratios do not distinguish between regular and irregular firing regimes.

5.2 Effect of the time window

As figure 2 shows, the variance-to-mean relations for both H1 and MT depend on the length of the time window over which spikes are counted: the variance increases disproportionately with respect to the mean in the long (100 ms) counting window. Similar effects of counting window length have been previously documented in MT of awake monkey (Buracas et al. 1998) and in eighth nerve afferents of anesthetized cat (Teich et al. 1990). The MT neurone shown in figures 2c and d exhibits this effect particularly strongly. The variance-to-mean relationship of this neurone agrees reasonably well with the quasi-Poisson prediction for

and MT, 100-ms window **f** Data for H1 and MT, 10-ms window. a, b, c, d: Data from a single H1 and a single MT cell. e, f: Data from 8 H1 and 8 MT cells, averaged together as described by Warzecha and Egelhaaf (this volume). The 8 MT cells with the greatest number of trials were chosen from a larger data set. Error bars are S.E.M. For all data presented here (both H1 and MT), neighbouring time windows overlap by 90%; for a 100-ms window, for example, they have 90 ms in common.

the 10-ms window, but changes to match the Poisson prediction for the 100-ms window. Obviously, the choice of window length does not alter the spiking statistics of a neurone – the same set of spike trains is being analysed in all cases. Rather, this time window dependence hints at aspects of the neuronal discharge that are captured neither by the Poisson model nor the quasi-Poisson model.

The observed dependence of the variance-to-mean relationship on window length would be expected if spike counts in successive time epochs were correlated on a time-scale longer than the 10-ms window. Expressed quantitatively, if "A" is the number of spikes fired in a 10 ms epoch and "B" is the number of spikes fired in the next 90 ms, then "A+B", the number of spikes fired over the 100 ms interval has a mean:

$$mean(A+B) = mean(A) + mean(B)$$

and a variance:

$$var(A+B) = var(A) + var(B) + 2*cov(A,B).$$

Thus the mean over the full 100 ms is simply the sum of the component means, whereas the variance is the sum of the two component variances, plus twice the covariance between them. For a Poisson process, the covariance is zero. For our quasi-Poisson model neurone, the covariance is actually negative, owing to the refractory period, and the variance is reduced relative to the mean as we have shown above. For neurones that exhibit positively correlated spike counts over time, however, the covariance is positive and the variance will therefore increase disproportionately with respect to the mean. We have detected such spike count correlations in a substantial fraction of MT neurones, and we suspect strongly that this accounts for the effect of time window length in figures 2c and d (Horwitz and Newsome, unpublished observations). A number of physiological processes, such as attention, could modulate the average firing rate of the neurone to produce such nonstationarity across trials. Modest modulations of the average firing rate can easily create the difference between figures 2c and d[3].

An important corollary of our analysis is that the apparent adherence of the MT neurone in figure 2d to Poisson statistics is misleading; it does not mean that the MT neurone is emitting spikes in the random, independent fashion expected of

---

[3] Simulations of quasi-Poisson spike trains (with a 1 ms refractory period) were made in which the average firing rate was alternately increased or decreased by a given percent from trial to trial. The interspike interval distributions were exponential, with a zero probability region for the first 1 ms. Many such rasters were generated to create a cloud of points that spanned a range of firing rates. Modulations of the firing rate of as little as 20% had barely visible impact on this cloud in the 10-ms time window, but drove the cloud to the Poisson diagonal in the 100-ms time window. Simulation data are not shown, but can easily be generated using the information here. Troy and Robson (1992, p 549) expressed a similar concern about the effects of nonstationarity.

a Poisson process. Incorporation of a refractory period into a Poisson model drives the expected variance-to-mean relationship to the quasi-Poisson curve, well below the Poisson diagonal (Fig. 2c), but temporal correlation in spike counts over a longer time scale increases the variance-to-mean ratio above the quasi-Poisson prediction. Thus the observed variance-to-mean relationship represents an interplay between two opposing factors: negative correlation on a short time scale (refractoriness) and positive correlation on a long time scale. Whether the resulting data points fall above, on, or below the Poisson diagonal depends upon the relative strength of these effects. Most prior studies of the spiking statistics of cortical neurones have employed long time windows and are therefore subject to these concerns.

5.3 H1 spiking statistics: an alternative view

We should note that some disagreement appears in the literature about the spiking statistics of H1. De Ruyter van Steveninck et al (1997), for example, agree that H1 spike discharge is highly regular in response to strongly dynamic ("naturalistic") stimuli. In striking contrast to Warzecha and Egelhaaf, however, de Ruyter van Steveninck and colleagues suggest that the spiking responses of H1 to constant stimuli conform closely to Poisson statistics. Warzecha and Egelhaaf have discussed elsewhere several possible reasons for this discrepancy (Warzecha and Egelhaaf 1999), including the fact that the data from de Ruyter van Steveninck and colleagues appear to have been obtained from only a single H1 neurone. Our conclusions in this paper are based on the more robust data set of Warzecha and Egelhaaf, but could obviously change if future findings lead to a different view of H1 spiking statistics.

5.4 Why are H1 and MT different?

Despite the similarities in the stimulus selectivity of H1 and MT neurones, their spiking statistics differ substantially. What accounts for the difference? Having ruled out the length of the refractory period, we are compelled to search for more interesting explanations. As many investigators have pointed out, the sources of response variance are likely to lie outside the neurone because the spike generating mechanism within the neurone appears to be extremely reliable (Calvin and Stevens 1968; Mainen and Sejnowski 1995), and because response variance appears to be correlated within groups of neighbouring neurones (Scobey and Gabor 1989; Zohary et al. 1994; Arieli et al. 1996; Gawne and Richmond 1993; Buracas et al. 1998). Thus response variance is probably inherited from the inputs to each neurone, and common input likely accounts for correlated response variance between neurones.

Shadlen and Newsome (1994, 1998) have developed a quantitative model of response variance that may inform comparisons of H1 and MT. In cortical sensory neurones of vertebrates, one major source of variance is likely to be the interplay of excitatory and inhibitory potentials arriving at the postsynaptic neurone. If EPSPs and IPSPs arrive randomly in time and are roughly balanced in their net effect on the neurone, their interplay results in an irregular interspike interval and substantial variance in the spike count. A second major source of variance is correlated activity among the pool of afferent neurones that provides input to a given neurone. If the input neurones fire in tandem slightly above or below their mean firing rates, the responses of the postsynaptic neurone will be even more variable than expected from the interplay of EPSPs and IPSPs. Finally, response variance is likely to be increased by additional modulating inputs to the neurone, perhaps in the form of attentional or other "cognitive" systems, as indicated above in the discussion of time window size.

From the point of view of this model, the difference in spiking statistics of H1 and MT neurones could arise from any one, or all three, of these sources. Our knowledge of the interplay between excitatory and inhibitory inputs, and our measurements of correlated activity among afferent neurones, are woefully incomplete for both cortical neurones and for H1 (but see Single et al. 1997). These gaps in our knowledge should be remedied.

In contrast to H1 and MT, the anatomy and functional connectivity of the retina is well understood. Applying our analytic framework to the study of retinal spiking statistics might yield useful insights into the sources of response variability. Retinal circuitry is stereotyped and simpler than cortical circuitry, and ganglion cells of most species receive no descending feedback from the brain. If spiking statistics simply reflect circuit complexity, responses should be more reliable in the retina than in the cortex. Berry et al. (1997) and Croner et al. (1993) have reported that the responses of retinal ganglion cells, like H1, are highly regular, but the stimuli used in these studies were probably too dynamic to permit a clean test of the response regimes outlined in this paper. Troy and Robson (1992) and Troy and Lee (1994) have shown that under constant and uniform illumination, the variability of retinal ganglion cells appear to undergo only modest changes with rate, though in both studies the analysis methods are sufficiently different to preclude quantitative comparison with the H1 and MT data presented here. Additional experiments will probably be necessary to determine whether retinal ganglion cells conform to "regular" or "quasi-Poisson" spiking regimes.

Resolution of this issue will contribute to our understanding of the sources of response variability, how response variability limits psychophysical sensitivity, and whether response variability reflects multiplexing of signals within the spike trains of single neurones. The answer will also have strong implications concerning the degree of diversity or conservation of neural coding and processing strategies across species.

Acknowledgments

We are grateful to A. Warzecha and M. Egelhaaf for providing H1 data for our analysis. We also thank M.N. Shadlen and E.J. Chichilnisky for comments on the manuscript. C.L. Barberini is supported by a Predoctoral Fellowship from the Howard Medical Institute (HHMI). W.T. Newsome is an HHMI Investigator.

References

Arieli A, Sterkin A, Grinvald A, Aertsen A (1996) Dynamics of ongoing activity: explanation of the large variability in evoked cortical responses. Science 273: 1868-1871

Bair W, O'Keefe LP (1998) The influence of fixational eye movements on the response of neurons in area MT of the macaque. Visual Neurosci 15:779-786

Berry MJ, Warland DK, Meister, M (1997) The structure and precision of retinal spike trains. Proc Natl Acad Sci USA 94: 5411-5416

Britten KH, Shadlen MN, Newsome WT, Movshon JA (1992) A comparison of neuronal and psychophysical performance. J Neurosci 12: 4745-4765

Buracas GT, Zador AM, DeWeese MR, Albright TD (1998) Efficient discrimination of temporal patterns by motion-sensitive neurons in primate visual cortex. Neuron 20: 959-969

Calvin WH, Stevens CF (1968) Synaptic noise and other sources of randomness in motoneuron interspike intervals. J Neurophysiol 31: 574-587

Cox DR (1962) Renewal Theory. Methuen, London; Wiley, New York

Croner LJ, Purpura K, Kaplan E (1993) Response variability in retinal ganglion cells of primates. Proc Natl Acad Sci USA 90: 8128-8130

Everling S, Pare M, Dorris MC, Munoz DP (1998) Comparison of the discharge characteristics of brain stem omnipause neurons and superior colliculus fixation neurons in monkey: implications for control of fixation and saccade behavior. J Neurophysiol 79: 511-528

Gawne TJ, Richmond BJ (1993) How independent are the messages carried by adjacent inferior temporal cortical neurons? J Neurosci 13: 2758-2771

Gur M, Beylin A, Snodderly DM (1997) Response variability of neurons in primary visual cortex (V1) of alert monkeys. J Neurosci 17: 2914-2920

Keller EL (1977) The role of the brain stem reticular formation in eye movement control. In: Brooks BA, Bajandas FJ (eds) Eye movement. Plenum, New York, pp 105-126

Keller EL, Robinson DA (1972) Abducens unit behavior in the monkey during vergence movements. Vision Res 12: 369-382

King WM, Lisberger SG, Fuchs AF (1986) Oblique saccadic eye movements of primates. J Neurophysiol 56: 769-784

Mainen ZF, Sejnowski TJ (1995) Reliability of spike timing in neocortical neurons. Science 268: 1503-1506

de Ruyter van Steveninck RR, Lewen GD, Strong SP, Koberle R, Bialek W (1997) Reproducibility and variability in neural spike trains. Science 275: 1805-1808

Scobey RP, Gabor AJ (1989) Orientation discrimination sensitivity of single units in cat primary visual cortex. Exp Brain Res 77: 398-406

Shadlen MN, Newsome WT (1994) Noise, neural codes and cortical organization. Curr Opinion Neurobiol 4: 569-579

Shadlen MN, Newsome WT (1998) The variable discharge of cortical neurons: Implications for connectivity, computation, and information coding. J Neurosci 18: 3870-3896

Single S, Haag J, Borst A (1997) Dendritic computation of direction selectivity and gain control in visual interneurons. J Neurosci 17: 6023-30

Teich MC, Johnson DH, Kumar AR, Turcott RG (1990) Rate fluctuations and fractional power-law noise recorded from cells in the lower auditory pathway of the cat. Hearing Res 46: 41-52

Troy JB, Lee BB (1994) Steady discharges of macaque retinal ganglion cells. Visual Neurosci 11: 111-118

Troy JB, Robson JG (1992) Steady discharges of X and Y retinal ganglion cells of cat under photopic illuminance. Visual Neurosci 9: 535-553

Warzecha A-K, Egelhaaf M (1999) Variability in spike trains during constant and dynamic stimulation. Science 283: 1927-1930

Zohary E, Shadlen MN, Newsome WT (1994) Correlated neuronal discharge rate and its implications for psychophysical performance. Nature 370: 140-143

Dynamic Effects in Real-Time Responses of Motion Sensitive Neurones

Ted Maddess

Visual Sciences Group, Research School of Biological Sciences, Australian National University, Canberra, Australia

1. Adaptation to motion

Bialek et al. (1991) put the theme of the present part on encoding dynamic information very succinctly: "Traditional approaches to neural coding characterize the encoding of known stimuli in average neural responses. Organisms face nearly the opposite task-extracting information about an unknown time-dependent stimulus from short segments of a spike train". Warzecha and Egelhaaf review their elegant experiments dealing with these difficult and topical issues. This article is intended to complement their efforts by reviewing literature and ideas that their work has made important again with a particular focus on adaptive effects.

Neural systems adapt, changing their behaviour according to the recent stimulus history. In the human it is clear that adaptive gain control mechanisms are at work. For example, human observers show a Weber fraction of about 5 to 7% for discriminating image velocity and this is maintained even in the face of quite large random fluctuations of image contrast and temporal frequency (McKee et al. 1986). That is to say in this process gain is regulated to maintain a just noticeable difference of about 5% of the mean velocity. Cats display a similar characteristic of velocity discrimination, albeit with larger Weber fractions (Vandenbussche et al. 1986).

The H1 neurone of flies shows large adaptive changes in the gain, and temporal resolution for the processing of image oscillations. The rate at which new adapted states are obtained is primarily determined by the temporal frequency content of moving images rather than the contrast of the images or their velocity (Maddess and Laughlin 1985), although at low speeds velocity may be more important (de Ruyter van Steveninick et al. 1986). Similar results are found for the human adaptation to image motion (Lorenceau 1987), in cat striate visual cortex (Maddess et al. 1988; Maddess and Vidyasagar 1992; Giaschi et al. 1993), in optomotor neurones of wallabies (Ibbotson et al. 1998) and butterflies (Maddess et al. 1991). Insect optomotor responses also show adaptive gain control effects (Kirschfeld 1989) as do human ocular following responses (Maddess and Ibbotson 1992; Ibbotson and Maddess 1994).

The lack of dependence shown by most of the motion adapting mechanisms mentioned above upon variables such as luminance contrast and response rate is understandable: changes of motion sensitivity should not be based upon information that cannot be reliably attributed to image motion. Low-level biological motion computation is believed to be based on cross-correlation between samples along a baseline (Reichardt 1961; Emerson et al. 1992), a computation that does not yield velocity per se. Thus, the best such a system could do perhaps is to base its adaptation upon signals within an ecologically interesting range of flicker frequencies induced by image motion.

2. Real-time assessment of neural responses

Warzecha and Egelhaaf (this volume) introduce several interesting ways of assessing real-time neuronal responses. Extracting the spike variance as a function of mean spike count within a narrow window (their Fig. 4) indicates that the spike process that is modulated by image motion is non-Poisson. Previous authors have reported H1 responses to less natural stimuli to be not independent of stimulus history (Mastebroek 1974; Gestri et al. 1980). Warzecha and Egelhaaf use an ideal observer model looking at spike rates to assess the neurone's ability to recognize a step in retinal slip and the time needed to detect motion, and to determine the number of stimulus states that can be discriminated by the neuronal response (their Fig. 8). Bruckstein et al. (1983) have shown that decoding the motion signal from the response was greatly improved when prior knowledge about the dynamics of changes in the modulated spike process is included. Evolution could have given the fly optomotor system that is post-synaptic to H1 such knowledge, so that the estimated performance of H1 and H2 as assessed by the ideal observer models of Warzecha and Egelhaaf may represent a lower bound on actual performance.

A third innovation is what Warzecha and Egelhaaf term the Stimulus Induced Response (SIR). Their SIR is the mean neuronal response obtained to many repeated presentations of the same velocity modulation. The authors subtract the SIR from the responses to each trial and describe the resultant residuals as the noise in the system (their Fig. 6). The mean response, however, may not in all cases be a good model of the neuronal response. For example the spike generation process itself may be changing as a function of response rate (see also Mastebroek 1974; Gestri et al. 1980). Even in a stationary system the median response for each time bin might be a better indicator of central tendency than the mean, if the distribution of spike intervals is non-Gaussian.

Figure 1 illustrates another example in which the mean SIR model would not be suitable. Figure 1a shows model responses of an elementary motion detector based on a conventional correlation mechanism (Reichardt 1961). The model has a DC component in one of its two inputs, giving a mixture of response components at the fundamental and second harmonic of the input drift frequency, that have been found experimentally (Ibbotson et al. 1991). Independent noise is added to both inputs. Horizontal slices across figure 1a represent responses to single presentations. A simple exponential adaptive process is included where the time constant of a low pass filter located after the motion computation stage decreases linearly over time. Thus, the response to a velocity impulse declines rapidly at first and then more slowly on subsequent trials. The change in filter characteristics also results in phase shifts shown as a tilt away from the vertical in the stripes of figure 1a. It is worth noting that the only critical feature for the present demonstration is the adaptation: the noise, the DC response component and the phase shifts are only introduced to make the model responses more realistic.

Analysis of Residuals

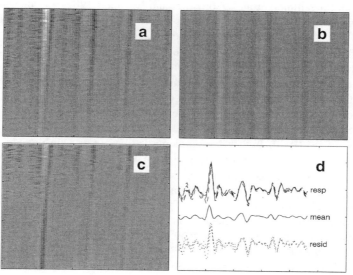

Response Time →

Fig. 1 Analysis of the residuals from a mean SIR model. **a** Model responses of an H1 neurone to a moving periodic grating pattern. The moving grating stimulus is repeated many times (top to bottom) and the responses to each repeated stimulus are shown as horizontal image rows. Brighter regions indicate higher spike (response) rates. Responses slowly decline with stimulus repetition due to adaptation. The abscissa, ordinate and grey scale are the same in **b**, showing the mean response of the SIR model (see text), and **c** showing considerable response components rather than just noise in the residuals from the SIR model. **d** (*resp*) The first 5 response rows from the top of (a); (*mean*) the mean across trials in (a) used to create (b); (*resid*) the first 5 rows of the residuals of (c). The three sets of waveforms are displaced vertically by arbitrary amounts to aid viewing but are otherwise at the same scale.

Figure 1b is the mean of the responses across trials of figure 1a, reproduced repeatedly (vertically) to illustrate what an unchanging SIR would be like. Figure 1c shows the residuals obtained by subtracting figure 1b from figure 1a. Clearly the residuals in this case are not merely noise but contain a considerable amount of the response.

Such an effect may be modest in the data of Warzecha and Egelhaaf given that they interposed rest periods within each trial of their repeated 2.5 to 5 s motion stimuli but the actual adaptation rate will depend on the particular stimulus. Clearly, these types of issues will need to be addressed for continuous stimulation: long term changes in gain having been described even in the earliest recordings from motion sensitive insect neurones (Collett and Blest 1966). It has been demonstrated formally for the H1 neurone that consideration of epochs around 8 s (McCann 1974) is required to characterize adaptive changes.

This adaptation to image motion that changes not only the gain but also the temporal frequency tuning of neurones is indicated by responses to velocity impulses in H1 of flies (Zaagman et al. 1983; Maddess and Laughlin 1985; de Ruyter van Steveninick et al. 1986; Borst and Egelhaaf 1987) and those of visual interneurones in other insects (Maddess et al. 1991). When highly adapted the cells not only encode progressively higher image oscillation frequencies, but may also shift to encoding acceleration rather than velocity (Maddess and Laughlin 1985; Maddess et al. 1991; see also Shi and Horridge 1991).

The prospect of adapted optomotor neurones encoding something akin to acceleration is also foreshadowed by the comment of Warzecha and Egelhaaf that H1 responses contain higher temporal derivatives of the input (see also Egelhaaf and Reichardt 1987; Egelhaaf and Borst 1989). Nonlinear control systems generally have to deal with higher temporal derivatives (e.g. Dunstan and McRuer 1961). This may seem at odds with the data of Warzecha and Egelhaaf indicating that fly responses do not contain reliable information at temporal frequencies much above 30 Hz because, as optomotor neurones appear to shift towards encoding acceleration with adaptation, the high frequency components of the response are relatively larger (Maddess et al. 1991). For example, velocity impulse responses from some adapted butterfly optomotor neurones appear to encode information about image oscillation frequencies in the range 10 to 100 Hz. When unadapted the same neurones have most of their response power below 10 Hz. Overall, it would be surprising if flying insects did not make use of the 200 Hz bandwidth of their photoreceptors (Howard et al. 1984) to control their flight. Warzecha and Egelhaaf use conventional Fourier analyses where the average frequency content over the whole signal epoch is computed. A wavelet-like approach (e.g. Gabor 1946) might reveal significant short periods of high frequency response fluctuations. Such approaches permit short bursts of high frequency activity to be quantified where a normal Fourier approach, that looks at average frequency content, generates misleading results.

At the same time it should be recalled that Warzecha and Egelhaaf found that above 30 Hz the signal power was less than that of the noise under their test conditions. We did not examine the signal to noise ratio measured in highly adapted conditions where average responses appear to encode acceleration (Maddess and Laughlin 1985; Maddess et al. 1991). As stated at the outset animals do not get the chance to examine their average response to hundreds of presentations when navigating in the visual environment. Thus, our data should not be taken as refuting the findings of Warzecha and Egelhaaf, and more experiments are needed to determine the exact effects of adaptation and its significance to real-time behaviour.

So far our experiments on frequency response dynamics have been crude in that they have only used lengthy adaptation times but there is precedent for very rapid changes in tuning of visual systems. For example the contrast gain control system of vertebrate retinal ganglion cells regulates these cells' frequency response on a time scale of 15 ms (Victor 1988) giving them their transient character. This "contrast" gain control system is strongly spatial and temporal frequency dependent. Figures 1b,d of Warzecha and Egelhaaf indicate that H1's response is sometimes a nonlinear, and sometimes quite transient, function of the stimulus. Thus, a similar contrast gain control system may precede motion processing by H1, at times amplifying higher frequencies. Such rapid changes in frequency response may explain apparent discrepancies between the results of Harris et al. (1999), who examined impulse responses of motion sensitive neurones after full adaptation, and experiments with continuous velocity fluctuations. The presence of a gain control preceding motion computation as in Y-cells would lead to the prediction that the impulse responses would be biphasic, which is actually observed (Harris et al. 1999), and would show contrast dependent transients, which is observed in wallaby motion sensitive neurones (Ibbotson personal communication). For continuous stimuli the velocity impulse response would partially reflect the frequency response of the presynaptic units with rapid contrast adaptation.

3. Low image speeds and afterimage effects

So far we have considered the impact of adaptation and non-stationarity of the spike generating process upon potential methods for assessing real-time performance of optomotor neurones. Another effect may be relevant when average image slip speeds are low: the so-called afterimage-like effect (Maddess 1985, 1986). Perhaps the best demonstration of this phenomenon is obtained by briefly placing a low contrast stationary bar in the receptive field of an H1 neurone. If the receptive field is probed a second or more later with a thin moving line, an imprint in the sensitivity profile of the H1 cell receptive field is observed where the bar was. This is clearly not a light adaptation effect given that a dark adapting bar can produce a depression of sensitivity to a moving bright line, and that there is markedly different processing of ON and OFF responses (Maddess 1986). When a stationary grating is presented for 200 ms or more, a very deep modulation of the response can be observed once the grating begins to move. Afterimage effects for gratings moving at up to 100 °/s have also been demonstrated with gratings that had a spatial frequency of 0.1 c/° that are about optimal for flies (Maddess 1985). Hence the underlying mechanism is low pass with a corner frequency around 2 Hz (Fig. 2).

4. Summary and suggestions

Figure 2 summarizes the temporal frequency characteristics of a number of dynamic processes that might be considered in any real-time analysis of motion processing in the fly. The overall unadapted temporal frequency tuning curve for H1 responses to drifting grating stimuli (Fig. 2, "Tune", Maddess and Laughlin 1985) is presented to provide a reference. The relative modulation depth of sensitivity changes produced by the afterimage-like effect in response to slowly drifting gratings is also provided (Fig. 2, "After", Maddess 1985) together with the time constant describing the rate of the previously described gain change with adaptation to motion (Fig. 2, "Adapt", Maddess and Laughlin 1985). I have also plotted (Fig. 2, "Struct") a parameter describing the loss of structural invariance that was introduced by Mastebroek (1974). This parameter provides a measure of the magnitude of the temporal frequency dependent change in the spike process, i.e. a loss of stationarity, in response to flashed stimuli. Similar effects have been shown for drifting gratings (Gestri et al. 1980). Clearly a number of effects determine the dynamics of the H1 response and these operate within the band of frequencies of interest to the cell. Thus, as illustrated by figure 1, slowly changing adaptation can cause contamination the residuals from mean SIR models with signal rather than noise, which can affect estimates of noise structure and amplitude. Lack of stationarity can lead to differences in the suitability of measures of central-tendency, such as the mean, for different parts of a response.

Experiments by McCann (1974) where fly visual neurones were characterized by estimating Wiener kernels (e.g. Marmarelis and McCann 1973; James 1992) may suggest a way to characterize adapting real time responses. The Wiener kernel expansion provides improvements over the velocity impulse response method (Maddess and Laughlin 1985; Maddess et al. 1991) because in the Wiener expansion linear, quadratic, cubic and higher order response interactions are each quantified by separate kernels. The full nonlinear SIR (or the linear and nonlinear parts separately) can be easily computed from the kernels, complete with long term adaptive dynamics if desired. For example, McCann (1974) showed that the first and second order kernels computed with a memory length of 8 s formally captured the long term adaptive character of H1. Another benefit is that the stimuli used can be less repetitive and are thus more like natural visual stimuli. Interaction kernels, quantifying linear and nonlinear interactions between cells or between parts of a cell's receptive field, can also be computed.

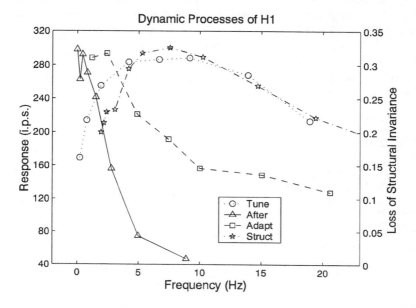

Fig. 2 Frequency dependent effects altering the response gain and dynamics of the fly H1 neurone. The following descriptions are labelled as in the figure legend. *Tune*: the overall unadapted tuning curve of H1 in response to image motion in the preferred direction, units (impulses per second) as for the left ordinate. *After*: the frequency tuning of the afterimage-like effect for afterimages induced by gratings moving in the preferred direction at the indicated contrast frequencies. The left ordinate units divided by 6 indicate the modulation depth in the H1 receptive field produced by the afterimage of a 0.1 c/° grating. The change in receptive field sensitivity is determined by moving a thin bar through the receptive field after presenting the moving grating, the thin bar producing a response of about 60 i.p.s. *Adapt*: the time constant of the change in gain of H1 neurones in response to motion of gratings in the preferred direction as a function of contrast frequency. The time constant in seconds has 30 times the units on the right ordinate. *Struct*: the "loss of structural invariance" described by Mastebroek (1974). This parameter indicates the magnitude of the change in the stochastic spike generation mechanism modulated by the neuronal response as a function of flicker frequency.

The Wiener method is not without problems, however, and methods modelling the changes in the velocity impulse response parametrically might be more parsimonious. Wiener models are non-parametric models where each point in every kernel is treated as a coefficient to be fitted. Thus, as the number and the dimension of the kernels is increased, there is an explosion in the number of coefficients to be estimated and so too in the number of data points required, since at least one datum per coefficient is needed. Parametric models can have many fewer coefficients. For example, Dubois (1993) demonstrated that the velocity impulse responses of butterfly optomotor neurones can be modelled by a third-order filter (i.e. having 3 coefficients) and that the dynamics can in turn be modelled by changes to just one stage (coefficient) of the filter. Interestingly, the changes to the third order filter provide the system with response dynamics characterized by a constant damping ratio.

In summary, better characterization of stimulus dependent changes in the spike generation process, and tests that assume prior knowledge (e.g. Bruckstein et al. 1983) in interpreting or decoding the spike signals should be considered in future.

References

Bialek W, Rieke F, de Ruyter van Steveninck RR, Warland D (1991) Reading a neural code. Science 252: 1854-1857

Borst A, Egelhaaf M (1987) Temporal modulation of luminance adapts time constant of fly movement detectors. Biol Cybern 56: 209-215

Bruckstein AM, Morf M, Zeevi YY (1983) Demodulation methods for an adaptive neural encoder model. Biol Cybern 49: 45-53

Collett TS, Blest AD (1966) Binocular, directionally selective neurones, possibly involved in the optomotor response of insects. Nature 212: 1330-3

de Ruyter van Steveninck RR, Zaagman WH, Mastebroek HAK (1986) Adaptation of transient responses of a motion-sensitive neuron in the visual system of the blowfly *Calliphora erythrocephala*. Biol Cybern 54: 223-236

Dubois R (1993) Visual processing of motion in the medulla of the butterfly *Papilio aegeus*. PhD, Australian National University

Dunstan G, McRuer D (1961) Analysis of nonlinear control systems. Wiley, New York

Egelhaaf M, Borst A (1989) Transient and steady-state response properties of movement detectors. J Opt Soc Am A 6: 116-127

Egelhaaf M, Reichardt W (1987) Dynamic response properties of movement detectors: theoretical analysis and electrophysiological investigation in the visual system of the fly. Biol Cybern 56: 69-87

Emerson RC, Bergen JR, Adelson EH (1992) Directionally selective complex cells and the computation of motion energy in cat visual cortex. Vision Res 32: 203-218

Gabor D (1946) Theory of communication. J IEE 93: 429-457

Gestri G, Masebroek HAK, Zaagman WH (1980) Stochastic constancy, variability and adaptation of spike generation: performance of a giant neuron in the visual system of the fly. Biol Cybern 38: 31-40

Giaschi D, Marlin RDS, Cynader M (1993) The time course of direction-selective adaptation in simple and complex cells in cat striate cortex. J Neurophysiol 70: 2024-2034

Harris RA, O'Carroll DC, Laughlin SB (1999) Adaptation and the temporal delay filter of fly motion detectors. Vision Res 39: 2603-2613

Howard J, Dubs A, Payne R (1984) The dynamics of phototransduction in insects. J Comp Physiol A 154: 707-718

Ibbotson MR, Clifford CW, Mark RF (1998) Adaptation to visual motion in directional neurons of the nucleus of the optic tract. J Neurophysiol 79: 1481-1493

Ibbotson MR, Maddess T (1994) Temporal frequency and binocularity govern adaptation of the human oculomotor system. Exp Brain Res 99: 148-154

Ibbotson MR, Maddess T, Dubois RA (1991) A system of insect neurons sensitive to horizontal and vertical image motion connects the medulla and midbrain. J Comp Physiol 169: 355-367

James AC (1992) Nonlinear operator network models of processing in the fly lamina. In: Pinter RB, Nabet V (eds) Nonlinear Vision. CRC Press, Boca Raton, pp 39-73

Kirschfeld K (1989) Automatic gain control in the movement detection of the fly. Naturwiss 76: 378-380

Lorenceau J (1987) Recovery from contrast adaptation: effects of spatial and temporal frequency. Vision Res 27: 2185-2191

Maddess T (1985) Adaptive processes affecting the response of the motion sensitive neuron H1. Proc Int Conf Cybern and Society, Tucson, pp 862-866

Maddess T (1986) Afterimage-like effects in the motion-sensitive neuron H1. Proc Roy Soc Lond B 228: 433-459

Maddess T, Dubois RA, Ibbotson M (1991) Response properties and adaptation of neurons sensitive to image motion in the butterfly *Papillio aegeus*. J Exp Biol 161: 171-199

Maddess T, Ibbotson MR (1992) Human ocular following responses are plastic: evidence for control by temporal frequency-dependent cortical adaptation. Exp Brain Res 91: 525-538

Maddess T, Laughlin SB (1985) Adaptation of the motion-sensitive neuron H1 is generated locally and governed by contrast frequency. Proc Roy Soc Lond B 225: 251-275

Maddess T, McCourt ME, Blakeslee B, Cunningham RB (1988) Factors governing the adaptation of cells in Area-17 of the cat visual cortex. Biol Cybern 59: 229-236

Maddess T, Vidyasagar TR (1992) Evidence that the adaptive gain control exhibited by neurons of the striate visual cortex is a co-operative network property. Proc Aus Conf Neural Net 3: 84-87

Marmarelis PZ, McCann GD (1973) Development and application of white-noise modeling techniques for studies of insect visual nervous system. Kybernetik 12: 74-89

Mastebroek H (1974) Stochastic structure of neural activity in the visual system of the blowfly. Doctorate, University of Groningen

McCann GD (1974) Nonlinear identification theory models for successive stages of visual nervous systems of flies. J Neurophysiol 37: 869-895

McKee SP, Silverman GH, Nakayama K (1986) Precise velocity discrimination despite random variations in temporal frequency and contrast. Vision Res 26: 609-619

Reichardt W (1961) Autocorrelation, a principle for the evaluation of sensory information by the central nervous system. In: Rosenblith WA (ed) Principles of sensory communication. Wiley, New York, pp 303-307

Shi J, Horridge GA (1991) The H1 neuron measures change in velocity irrespective of contrast frequency, mean velocity or velocity modulation frequency. Proc Roy Soc Lond B 331: 205-211

Vandenbussche E, Orban GA, Maes H (1986) Velocity discrimination in the cat. Vision Res 26: 1835-1849

Victor JD (1988) The dynamics of the cat retinal Y cell subunit. J Physiol 405: 289-320

Zaagman WH, Mastebroek HAK, de Ruyter van Steveninck R (1983) Adaptive strategies in fly vision: on their image-processing qualities. IEEE Trans Sys Man Cybern 13: 900-906

Part VI

Motion in the Natural Environment

Towards an Ecology of Motion Vision

Michael P. Eckert[1] and Jochen Zeil[2]

[1]Environmental Engineering Group, Faculty of Engineering, University of Technology, Sydney, Australia; [2]Visual Sciences Group, Research School of Biological Sciences, Australian National University, Canberra, Australia

Contents

1. Abstract

Natural motion signals, as a working definition, are those that are actually encountered by specific animals in the environment they normally operate in. The need to consider specific animals arises because the motion signals that are processed by a brain depend on the ethological and ecological context. Motion signals are determined by environmental motion, by the type and structure of locomotion of an animal, and by the visual topography of the world the animal operates in. We suggest that it is essential to consider natural motion signals in more detail, since they may reveal constraints that have shaped the evolution of motion detection and information processing mechanisms. The primary focus of this paper is to outline what needs to be considered and what is required to characterize the biologically relevant information content of the visual motion environment of an animal. In particular, we discuss the principal sources of image motion, critically assess the different ways of reconstructing, analysing and modelling natural motion signals, and briefly summarize current attempts to identify coding strategies, matched filters and optimization of neurones involved in processing visual information. We end with a survey of sensory and neural adaptations to show the multiple levels of processing at which motion filters have evolved under the influence of natural motion signals.

2. Natural motion signals, the visual environment, and behaviour

2.1 Environmental motion

The image motion experienced by an animal first manifests itself as image intensity variations at the photoreceptor array that are correlated in space and time. Correlated image intensity variations have a number of sources. For a stationary observer, they are caused by environmental motion: (i) wind-driven movements of vegetation, (ii) the movements of clouds and the passage of their shadows across the landscape, (iii) moving objects, and (iv) in the special case of fresh water or coastal habitats, the movement of water. To our knowledge, the statistical properties of environmental motion signals have not been analysed. A notable exception is the work by Fleishman (1986, 1988) who at least described the distribution of temporal frequencies in the natural habitat of lizards, which however is not sufficient to capture the distribution of true motion signals in the visual field of an animal. A few qualitative considerations suggest that environmental motion can generate strong and highly structured image motion patterns which animals need to cope with, in some situations as noise, and in others as useful cues. Wind induced movements of vegetation, for instance, consist of a large number of coupled oscillators, with probably a broad spectrum of frequencies, motion directions and amplitudes, that depend both on plant structure and wind pattern (e.g., Cutting 1982; Coutts and Grace 1995). For a stationary sit-and-wait hunter, plant movement patterns create a noisy background which makes the task of detecting moving prey quite daunting. For the prey in turn, the same background motion pattern may provide a means of camouflage, and for a herbivorous insect a way of identifying host plants. Compared to plant movements, water, wave and cloud movements can be much simpler in the sense that they tend to produce image motion that is either constant in direction or is rhythmically patterned. As a consequence, regular movement patterns such as these can be filtered out by a stationary observer, but they may also cause problems for animals that rely on vision for odometry or for determining egomotion parameters. Equally, the movements of animate, non-rigid bodies are also constrained, although in a more complicated way compared to clouds, both in terms of their paths through the world and in terms of "biological motion" (see Neri et al. 1998). Animals and their limbs tend to move along smooth paths and so do their images at the eye of an observer.

2.2 Behaviour and natural motion signals

2.2.1 Optic flow

We have so far considered sources of image motion from the viewpoint of a stationary observer. An actively moving or passively displaced observer experiences a distinct optic flow structure superimposed on the environmental motion field which depends on eye or body rotation, on the direction of translation, on the spatial layout of objects, and on contrast in the environment. The optic flow components that are elicited by rotations and by translation through the world are structured in a characteristic and invariant way (Gibson 1950; Koenderink 1986): the rotational components have the same magnitude and direction throughout most of the visual field and all objects in the environment contribute rotational flow components, independent of their distance from the observer. The translational components, in contrast, have directions radiating from the direction of heading and depend in their magnitude on the spatial layout of objects in the scene. Distant objects and objects in the direction of heading do not contribute motion vectors. Since most animals actively move through the world, a dominant part of the image motion they experience is produced and determined by their behaviour, that is both by the type of locomotion and by head and eye movements. For instance, a walking insect with firm contact to the ground is inherently more stable around the roll and the pitch axes than a flying insect and thus is rarely confronted with large image velocities in the vertical direction except in the direction of heading. In many animals which predominantly move forwards, the translational flow field, as far as motion vector directions are concerned, is for most of the time symmetrical to the longitudinal body axis. However, this is not always the case. Hovering insects and birds, for instance, are able to fly in any direction relative to their longitudinal body axis and consequently do not always experience rigidly oriented translational optic flow. The distribution of image motion directions at the retina of an animal is thus determined to a large extent by the type of movements it carries out, while the range of image motion speeds it experiences depends on the speed of locomotion and the density of objects in the environment.

2.2.2 The optomotor system and tracking

All visual tracking strategies – regardless of whether they involve eye, head or body movements – alter the velocity distribution (Eckert and Buchsbaum 1993a, b) and the structure of optic flow experienced by the nervous system in various ways (e.g., Lappe et al. 1998). The structure of eye movements thus has consequences for both coding and for the estimation of self-motion from optic flow (e.g., Perrone and Stone 1994). The optokinetic and vestibulo-ocular reflexes in vertebrates and the vestibular and optomotor systems in insects and crabs, for instance, remove most or all of the rotational optic flow from the retinal image. The smooth pursuit movements performed in foveal object tracking minimize retinal slip within the fovea only, but alter the distribution of image motion across the visual field. Tracking contours or objects close to the focus of expansion modifies the structure of translational optic flow and changes the position and conspicuousness of the pole of the flow field on the retina (Lappe et al. 1998). While most animals possess control systems that minimize the rotational component of image motion, some birds stabilize the retinal image (and not only part of it) even during translatory locomotion by linear head movements against the direction of translation (head bobbing, e.g., Frost 1978; Davies and Green 1988; Nalbach 1992). Depending on lifestyle, systems that subserve the visual stabilization of the retinal image can be very specialized and therefore are particularly instructive with regard to their functional significance. Waterstriders for instance live on the water surface and are displaced by wind and water movements. They can be considered to be "fliers in two-dimensional space" (Junger and Dahmen 1991). The insects compensate for rotations and drift by distinct behavioural programs that are driven separately by the rotational and the translational components of optic flow. When the bugs are simultaneously rotated and drifted by a flow gradient on the water surface, they continuously counteract rotation by rowing movements of one of their middle legs. Displacement through drift accumulates and the insects compensate for it periodically with saccadic jumps against the direction of drift. The jumps are executed by a fast and coincident action of the middle legs. The animals are able to extract the rotational and translational components of optic flow separately even when visual patterns are restricted to one side of an artificial river.

2.2.3 Structured movements and locomotion

While the optomotor system and foveal tracking serve to *compensate* for certain
image motion components, there are also many examples where eyes (or bodies)
are moved specifically to *produce* image motion. Particular patterns of locomotion
like search movements or regular flight paths may have evolved for many different reasons not directly related to vision, but they do have an influence on the
structure of image motion at an animal's eye. Many highly structured movements
and patterns of locomotion in insects may have been shaped at least in part by
image motion processing needs. In vertebrates there are also many different
patterns of locomotion, from hopping, to jumping, and running, which must influence to some degree what image motion the animal experiences. In some cases,

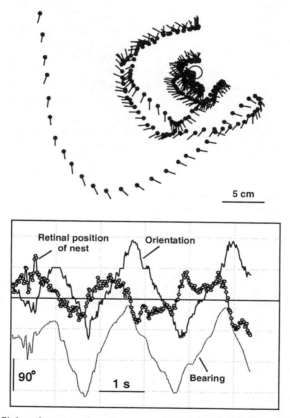

Fig. 1 Learning flight of a ground-nesting wasp (*Cerceris,* Sphecidae). The position and
orientation of the wasp are shown every 20 ms in the top diagram (dot marks the wasp's head and
the line attached to it the orientation of her longitudinal body axis). The nest hole in the ground is
marked by a circle. The lower diagram shows the time course of the wasp's bearing relative to
the nest (grey line), of the orientation of her body axis (black line) and of the retinal position of
the nest entrance (dotted line).

the function of movement patterns in visual information processing has been clearly demonstrated. Locusts and mantids for instance gauge the distance of a landing site before they jump by swaying their bodies from side to side, always perpendicular to their line of sight. That the insects generate and use motion parallax in their distance judgements can be shown by oscillating landing platforms in phase and out of phase with the animal's movements and by recording their subsequent jump distances (reviewed by Kral and Poteser 1997). Animals often adapt their behaviour to suit the visual task at hand: bees for instance change their flight direction relative to a pattern depending on the direction of contours that offer the strongest motion parallax cues (Lehrer and Srinivasan 1994).

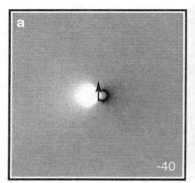 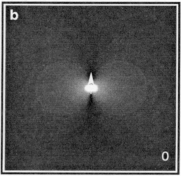

Fig. 2 Pattern of horizontal image motion components for pivoting and for translational movements. **a** Grey-level map (256 cm x 256 cm) of horizontal image motion components for a 5 cm translation in the direction of the black arrow and a rotation of 40° to the right. **b** Grey-level map of horizontal image motion components during a pure translation of 5 cm in the direction of the white arrow. Black = low velocity; white = high velocity.

The pivoting flight paths of some other insects probably also belong to this class of information processing behaviours. When wasps or bees depart from their nest or from a newly discovered food place for the first time, they perform a learning flight during which they acquire a visual representation of the goal environment (reviewed by Zeil et al. 1996). The insects back away from the goal in a series of increasing arcs that are roughly centred on the goal (Fig. 1). They constantly turn in such a way as to fixate the goal with frontal retina. These flight manoeuvres produce a particular image motion pattern at the retina with minimal image motion close to the goal on the ground and a gradient of increasing image speed with distance from the goal. Pivoting movements, like foveal tracking of objects at viewing directions perpendicular to the direction of travel, thus create a place (or object) centred parallax field (Fig. 2a) which differs from the parallax field produced by an animal moving along a straight path without rotation (Fig. 2b). Some ground-nesting wasps direct similar pivoting flight manoeuvres at novel objects (Voss and Zeil 1998, Fig. 3) and in two species of flies, males scrutinize females in this way (Collett and Land 1975a; Land 1993).

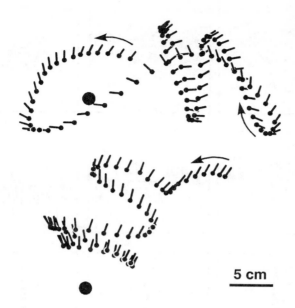

Fig. 3 Object-directed zig-zag flights of a wasp (*Odynerus*, Eumenidae). Two instances of pivoting flight in front of novel objects (a small metal cylinder marked by a black circle) are shown. Arrows indicate the direction of movement of the wasp. Other conventions as in Fig. 1.

There are many other cases of patterned locomotion. Male houseflies for instance patrol the airspace beneath indoor landmarks, like a lamp shade (Fig. 4a), much in the same way as empidid flies scan the water surface for drowning insect prey (Fig. 4b). The flies fly along straight paths which are linked by rapid changes in flight direction (Zeil 1986). As a consequence, rotational image flow is kept to a minimum and for most of the time the flies are experiencing the pure translational flowfield which may help them to judge their position relative to a guiding landmark and to detect moving objects by virtue of the fact that they create motion vectors that are usually different from those produced by translatory egomotion. Swarming midges, in contrast, fly along curved trajectories and, assuming that their longitudinal body axis is aligned with the flight path, would have to cope with a rather more complicated pattern of image motion at their retina (Fig. 4c). Insects that spend long periods of time in hovering flight, like male hoverflies at mating stations or guard bees (Collett and Land 1975b; Kelber and Zeil 1990), face an altogether different situation: the image motion field at their retina is mainly determined by wind-induced drift, displacement and rotational distur-bances which are all highly unpredictable.

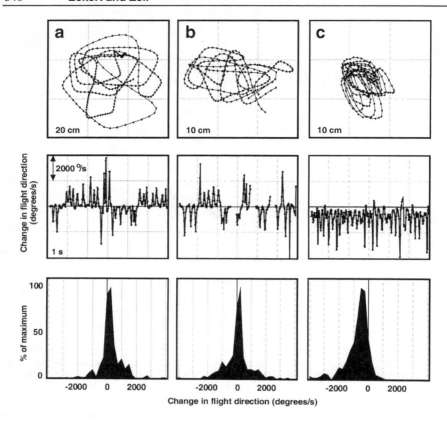

Fig. 4 Patterned locomotion in flies. Top panels show flight paths, centre panels the time course of changes in flight direction in °/s, and bottom panels frequency histograms of changes in flight direction. **a** A male *Fannia canicularis* (lesser housefly) patrolling the airspace below an indoor landmark hanging from the ceiling. The flight path is seen from below and positions are shown every 20 ms. **b** A dance fly (probably a male *Rhamphomyia*, Empididae) scanning the water surface of a puddle for drowning insect prey. Flight path as seen from above, 20 ms sampling. **c** The flight path of a swarming midge (*Chaoborus flavicans*, Chaoboridae) seen from the side (data from Wüst 1987). Positions shown every 20 ms.

2.3 Motion signals and lifestyle

Beside environmental motion and the type, speed and pattern of locomotion, there are also specific motion environments that depend on the spatial structure of the environment, the social context and the tasks that have to be solved. To exemplify how the specific biological context can lead to very different demands on visual motion processing, sometimes even within one species, let us consider two examples.

Female hoverflies spend most of their life feeding on nectar, flitting from flower to flower, hovering while feeding, and then searching for more food. The flight manoeuvres primarily involve hovering and linear flight, interspersed with rapid, saccade-like turns of the whole body (Collett and Land 1975a). The male, on the other hand, spends part of his life hovering and feeding as the female does, but much of the time he visually tracks, chases, and mates with females or female-like objects (Collett and Land 1975b, 1979). To track females and to hold hovering stations and return to them after a mating chase, requires flight manoeuvres and associated neural control systems that are different from those of females. Because they behave differently, male and female hoverflies thus are likely to have a different motion ecology and this in turn may exert different selective pressures on motion detection and processing mechanisms (O'Carroll et al. 1997; O'Carroll, this volume).

The semi-terrestrial crabs in our second example inhabit a world where most environmental motion is produced not by wind shaken plants but by conspecifics and by predators. Fiddler crabs live in an essentially flat world, the tropical mud-flats, and operate on the surface during low tide. Their eyes sit on mobile stalks, which allow both visual and non-visual reflexes to align the eye with the vertical or with the local visual horizon and to compensate for rotational optic flow around three axes (Barnes and Nalbach 1993). The flat world these crabs live in has some interesting properties with respect to image motion: translational flow is restricted to the ventral visual field because it is only here that objects are close enough to produce image motion during translatory locomotion. Large field image motion above the horizon indicates rotational flow for most of the time, because objects are likely to be far away (Nalbach and Nalbach 1987). An interesting exception to this rule occurs on windy days with rapidly moving clouds, but the main sources of environmental motion are normally predators and other crabs. Predatory birds that are running across the mudflats are seen by a crab above the visual horizon line because they are much larger than the crab observer. The movements of other crabs, in turn, are seen in a narrow band of visual space, just below the horizon (Zeil et al. 1986, 1989; Land and Layne 1995; Layne et al. 1997). By raising and waving their claws, male fiddler crabs produce visual motion signals in territorial and in mating interactions (Salmon and Hyatt 1983), which appear to be mimetic in the sense that motion above the visual horizon is normally associated with predatory birds (Christy 1995). The choreography of waving movements is species-specific which indicates that the crabs must have evolved the neural machinery to detect and discriminate quite complicated image motion patterns (Salmon et al. 1978; Zeil and Zanker 1997).

3. Describing natural motion signals: theory and methods of collection

Our short survey shows that the ecology of motion vision requires us to consider environmental motion, the structure of optic flow produced by sensor movements

– and thus the structure of behaviour – the layout of contours in the environment, and the biologically relevant information that is to be extracted from the distribution of natural motion signals. What then are the adequate levels of description for natural motion signals? We present a somewhat ad hoc hierarchy of descriptions, recognizing that the level of description depends on the level of analysis which is required. The first stages of visual processing are thought to be primarily concerned with efficient coding, which requires for analysis only a statistical description of images. Early to mid-level processing seems to involve various representations of different image contents which require both a statistical description of the image and a statistical description of the output of intermediate processing units (for instance of elementary motion detectors). Finally, processing higher up in the stream is concerned with extracting complex features in images and, for analysis, would require us to identify those features and examine their salience under natural conditions.

3.1 Statistics of time varying images

We may start by analysing the spatiotemporal statistics of image intensity at the photoreceptor array, as has been done for image sequences that were recorded from a human viewpoint (Eckert et al. 1992; Dong and Atick 1995a) or by using a spot sensor with a narrow field of view to mimic the time series likely to be experienced by a blowfly (van Hateren 1997). Such an analysis is typically used to address the "early vision problem" namely, how to represent and transmit as much of incoming signal as is needed for robust analysis at later stages. At this level, the issue is how the nervous system achieves efficient spatial and temporal frequency coding with neural channels that are limited in terms of bandwidth and dynamic range. Theoretical solutions predict gain adaptation of photoreceptor responses to handle changes in image contrast, and spatiotemporal filtering to protect against the effects of environmental and neural noise (Laughlin 1981; Srinivisan et al. 1981, van Hateren 1992; Dong and Atick 1995b). The formulation of the problem typically leads to a theoretical solution which is optimized to the first and second order statistics of the signal. The statistics of interest include first order statistics such as the probability distribution of image intensity and second order statistics such as the spatiotemporal correlation function (or power spectrum) or the probability distribution of image velocities (velocity distribution). The velocity distribution represents an alternative method of characterizing the second order statistics, but requires the reasonable assumption that image intensity variations result primarily from image motion across the visual field, in which case the spatiotemporal power spectrum can be computed from the velocity distribution and the spatial power spectrum (van Hateren 1992, 1993; Eckert and Buchsbaum 1993b; Dong and Atick 1995a). The advantage of the velocity distribution is that it may be easier to estimate in practice and that it can be more directly related to the velocity of observer motion.

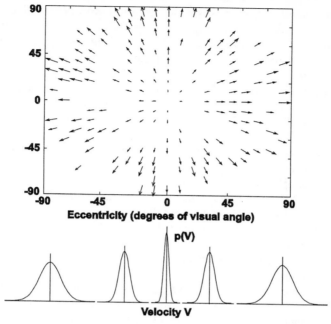

Fig. 5 Instantaneous velocity field for an observer translating directly ahead. Objects in the world are scattered at random distances from the observer. Below the velocity field, we show a typical set of eccentricity dependent velocity distributions which may arise when sampling the instantaneous velocity field through time.

For practical reasons, we know most about the spatial frequency spectrum of natural images since it depends only on scene structure and therefore can be analysed in still images. Temporal changes occur because of environmental motion or locomotion, so spatiotemporal statistics must, strictly speaking, be considered in the ethological context of an individual species, about which we typically have limited information. The key point to recognize is that animal behaviour actually influences the spatiotemporal, but not the spatial or the spectral statistics on the retina as long as they are the same everywhere in the scene. Velocity distributions in the same environment can be very different in terms of motion directions and amplitudes depending on the type of movement that is being carried out. Typically, a slowly moving animal will have a motion environment limited to low image velocities, and as a consequence, experience a velocity distribution with a small variance. Conversely, a fast flying animal in the same environment will have to face quite large velocities, and thus will experience a velocity distribution which is broader, with a large mean and variance. This relationship between speed of locomotion and image motion statistics is not simply an open loop process because animals interact with the world in a variety of ways, through flight speed adjustments, through patterned locomotion, or through eye movement control loops. However, as we show now, physical considerations allow simple functions to adequately describe the spatiotemporal statistics just as the $1/f$ model for the magnitude spectrum adequately describes the spatial frequency distribution of most natural images (reviewed by van der Schaaf and van Hateren 1996).

We approach our formulation of the spatiotemporal statistics of the environment by describing how the velocity distribution can be deduced from physical considerations of animal locomotion. We ignore environmental motion at this point.

Figure 5 illustrates an instantaneous velocity field which may arise in the visual field of an observer who travels forward at constant velocity through a scene in which objects are located at random distances from the observer. Below the velocity field, we illustrate typical retinal velocity distributions which might arise at different locations in the visual field. In the general case, the instantaneous velocity field depends on locomotion velocity, on direction of gaze relative to the direction of movement, and on the distance of objects from the animal. Since animals travel at different speeds and since objects lie at various distances, both are best represented as probability distributions. The complexity of the situation makes an analytical solution difficult except under a few simplifying assumptions. Van Hateren (1992) and Dong and Atick (1995a) assumed pure forward translational movement in a world in which objects are distributed uniformly in space, to arrive at a model for the distribution of retinal velocity, v, as a power function:

$$p(v) = c_v / (|v|+v_0)^n, \tag{1}$$

where v_0 is a constant which determines the mean and variance of the distribution, c_v is a normalization factor, and n characterizes how quickly the distribution falls off for large velocities. The parameter, v_0, can be used to specify the velocity "bandwidth" in the environment of a particular species, which for a flying insect is mainly determined by the average speed of flight. The parameter, n, largely depends on the distance distribution of objects in the world. If the probability of objects near the observer is large, then n will be smaller than if objects lie predominantly at great distances from the observer. An n near or slightly greater than 2 is reasonable for a cluttered world (van Hateren 1992; Dong and Atick 1995a). Note that a power law distribution is more sharply peaked and has a larger tail than a Gaussian distribution – it has a large kurtosis. The large kurtosis implies that quite high velocities may occur in some region of the visual field with a reasonably high probability. This can be an important factor in the design of robust detectors and estimators of motion fields, particularly if a theoretical optimization requires the assumption of Gaussian distributed statistics, which motion fields, in the global average sense, do not have. As an example, a robust estimate of the average velocity in a region of space is not obtained by a simple averaging of local velocity estimates, which would be optimal for a signal with a Gaussian distribution, but rather a non-linear combination of velocity estimates designed to reject outliers (Black and Rangarajan 1996).

The generality of the power law distribution in describing the situation in the real world has not yet been extensively investigated. Van Hateren (1997) showed that the time series collected with a head-mounted spot sensor in natural environments does behave according to the power law. However, considering freely moving animals, an area of concern is that the formulation assumes that the distribution of locomotion velocities and the distribution of distances in the world are uncorrelated and independent. As we have pointed out before, this assumption is, in fact, incorrect, because an animal may move more slowly when close to objects and more quickly when farther from objects in order to keep angular velocity across the receptor array within a particular range. This actually happens in bees that have to negotiate narrow passages and corridors (Srinivasan et al. 1996, 2000). Equally, an analysis of landing in bees shows that the insects slow down in proportion to their distance from a surface, apparently by keeping image velocity directly below them constant. Having image speed control flight speed may be a potent mechanism to negotiate a world full of obstacles (Martin and Franceschini 1994).

Until more detailed information is gathered regarding animal locomotion in a cluttered world, all we can say at the moment is that Equation 1 needs to be extended in several ways to approximate the natural situation (as pointed out by van Hateren 1992, 1993). First, the direction of image motion will need to be taken into account, since depending on the type of egomotion, the velocity distribution will differ for different directions of image motion. Considering how animals move, horizontal velocity components will be more dominant than vertical velocities for most of the time. In addition, there will be anisotropies both in azimuth and in elevation with respect to amplitudes and directions of image motion. Figure 6 illustrates two reasons for spatial anisotropies. Anisotropies in motion vector sizes result from the fact that forward motion generates smaller horizontal velocities in the direction of locomotion (black histograms marked by (a) in figure 6) compared to other directions of view (black histograms marked by (b) and (c)). At the same time, motion directions are also not uniformly distributed across the visual field, as can be seen in the two schematic views from the cockpit of a translating animal at the bottom of figure 6: vertical motion vectors are restricted to the part of the visual field that looks straight ahead, while at directions of view perpendicular to the direction of translation, horizontal motion dominates.

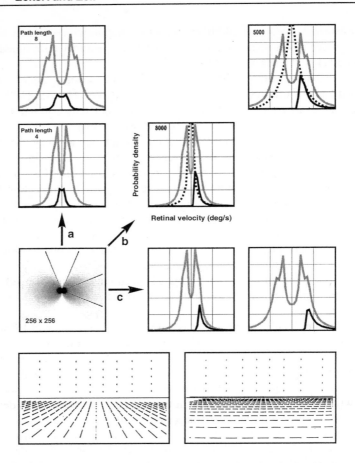

Fig. 6 This figure illustrates the primary factors causing the velocity distribution to be different in different parts of the visual field. The topography of motion amplitudes produced by a dense array of 256 x 256 "objects" during a pure translation in the centre is shown in the greylevel map on the lower left. Histograms show the frequency distributions of horizontal image motion vectors during a translation of 4 pixels/s (inner panels) and 8 pixels/s (outer panels). Binwidth is 0.5°/s. Grey curves show the velocity histograms over the whole array and black curves those for the three 45° sectors indicated in the greylevel map. Dashed curves are power law distributions with $v_0 = 1$, $n = 2$ (inner panel) and $v_0 = 50$, $n = 10$ (outer panel). Panels on the bottom show two views from the cockpit of a translating observer, one in the direction of heading (left) and the other perpendicular to it (right; modified after Voss 1995). For discussion see text.

There are a number of further issues to consider, such as the fact that occlusion of distant contours can alter the velocity distribution in quite dramatic ways. The higher the density of objects in the foreground, the rarer will be low image velocities and the velocity distribution will show a low velocity cut-off. This occlusion effect is apparent even in the rather simplistic simulation on which the velocity distributions in figure 6 are based and in which all contours within the array are assumed to be visible. The velocity distributions at 45° and 90° relative to the simulated direction of locomotion (black histograms (b) and (c)) are skewed with a steep low velocity cut-off because the simulated world is not large enough to contain distant contours which would contribute lower image velocities. For the same reason, the distribution over the whole array (grey histograms) has a trough at zero angular velocity. Occlusion and differences in depth distributions of contours again lead to anisotropies: in both walking and flying animals, the velocity distribution to the side may be different than above or below because the distribution of distances are likely to be different.

Furthermore, optomotor control loops or foveal tracking limit the velocity distribution even when the animal is not moving and the dominant image motion component is generated by environmental motion. Foveal tracking movements, irrespective of whether they operate during locomotion or on environmental motion only, result in an eccentricity dependent velocity distribution (Eckert and Buchsbaum 1993a, b). For a stationary observer, the eccentricity dependence in the velocity field arises because moving objects vary in size and because the probability that the image motion produced by moving objects outside the fovea differs from the image motion produced by tracking increases with angular distance from the fovea. A last point to note is that the distribution of oriented contours in natural scenes is not uniform (e.g., Switkes et al. 1978; van der Schaaf and van Hateren 1996; Coppola et al. 1998). Since vertical contours dominate on average, observer movements may preferentially elicit horizontal image motion components at the retina rather than vertical ones.

Despite these complicating factors, the velocity distribution can be linked to the spatiotemporal power spectrum in a fairly simple way. First, we must make an inspired guess about the spatial power spectrum of the world at each location in the retina. It is now notorious that the spatial frequency content of natural scenes has an amplitude spectrum of $1/f^n$, where n lies typically between 1 and 2, and seems to be relatively isotropic with respect to orientation (Burton and Moorhead 1987; Field 1987; van der Schaaf and van Hateren 1996; Ruderman 1997). While the scenes selected to establish this power law were anthropocentric, most natural images will have a similar frequency drop off, so it probably represents an acceptable model for virtually any species. Given a model for the spatial power spectrum, $P(k)$, and the velocity distribution, $p_v(x,v)$, it can be shown that the spatial velocity structure $S(x,k,v)$, is the product of the two (van Hateren 1992, 1993; Eckert and Buchsbaum 1993b; Dong and Atick 1995a; see also Dong, this volume):

$$S(x,k,v) = P(k)\, p_v(x,v) \,, \tag{2}$$

where $k = (k_v, k_h)$ is spatial frequency in cyc/° in vertical and horizontal directions, $x = (x_v, x_h)$ represents retinotopic spatial location, $v = (v_x, v_y)$ represents vertical and horizontal velocities, and $p_v(x,v)$ is a velocity distribution which depends on location, x, on the retina[1]. The spatiotemporal power spectrum is then found by integrating over all velocities, weighted by the velocity probability distribution, which satisfy the equation for constant velocity, $\delta(f\text{-}v\cdot k) = 0$. This gives an expression for the spatiotemporal power spectrum as:

$$S(x,k,f) = S(x,k) \int p_v(x,v)\, \delta(f\text{-}\, v\cdot k)\, dv \tag{3}$$

where f is temporal frequency in cyc/s, δ is the Dirac delta function, and the integration is over $v = (v_x, v_y)$. Equations (2) and (3) allow one to form an estimate of the spatiotemporal power spectrum simply by identifying the spatial power spectrum and velocity distribution at each location on the retina. Under the assumption of spatially stationary statistics, i.e., the spatiotemporal power spectrum does not vary with retinal location, an approximation to Equation 3 has been determined both theoretically (Dong and Atick 1995a) and empirically (Eckert et al. 1992) to be a separable function of the form:

$$S(k,f) = S_k(k)S_f(f). \tag{1}$$

Eckert et al. (1992) used $1/k^3$ for the spatial power spectrum and $1/f^2$ for the temporal power spectrum, though Dong and Atick (1995a) argue that the spatial power spectrum is quite flat, on the order of $1/k$, in some spatiotemporal regions. The primary utility of the second order statistics (Equations (1), (2), or (4)) is for the development of theories of early vision and we will return to assess them in that context below (see section 3).

3.2 Statistics of motion detector outputs

None of the statistical measures we have been discussing so far allows us to address the computational side of motion vision. Image motion is not directly represented in the activity of photoreceptors nor of visual interneurones at the first layer of visual processing. Visual motion information needs to be computed by comparing the input from at least two neighbouring locations in visual space. On this elementary level, the biological motion detectors as we know them do not faithfully represent local image velocity (Egelhaaf and Borst 1993). For a number of reasons, their output is dependent on scene properties like contrast and spatial frequency and their directional selectivity is very broadly tuned. A realistic way of analysing what the brain has to work with when extracting useful motion information from the retinal image would thus be at this level of elementary

---

[1] This corresponds to Equation (11) in Dong and Atick's paper and Equation (3) in Eckert and Buchsbaum's paper.

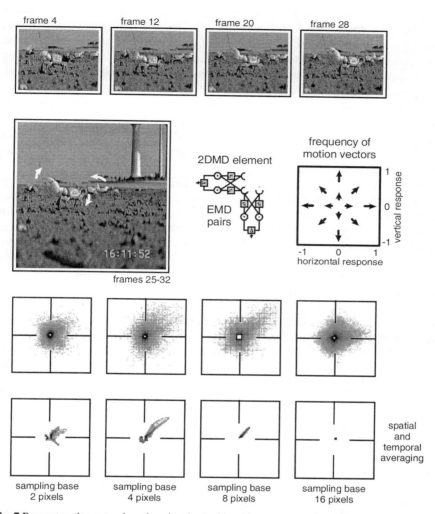

Fig. 7 Reconstructing natural motion signals. A video film sequence taken from the viewpoint of a fiddler crab shows several waving males moving about on a sandbank in Kuwait. Part of the sequence was used as input to a two-dimensional array of elementary motion detectors with sampling bases ranging from 2 pixels to 16 pixels (10 pixels = 0.6°). the output of the motion detector network is shown as two-dimensional frequency distributions of motion directions before (top row) and after spatial and temporal averaging (bottom row). Local spatial and temporal averaging is needed to extract the main component of image motion pattern in this crabworld scene (modified from Zeil and Zanker 1997).

motion detector activity. We suggest therefore that rather than examining the statistical characteristics of the input signal directly, looking at the outputs of elementary motion detectors would be a much more fruitful approach to understanding the visual ecology of motion vision (Zanker 1996; Zeil and Zanker 1997). With this approach it is possible to investigate neural and environmental constraints at the same time. For instance, some form of spatial and temporal integration is always needed to extract reliable information on the direction and the magnitude of motion vectors (Fig. 7). Integration, however, limits spatial and temporal resolution of motion signals which in turn is needed to process motion parallax information. Analysing the output of elementary motion detectors which are confronted with natural image motion, allows one to address the question how spatial and temporal integration need to be balanced in order to extract useful motion information in the presence of environmental and neuronal noise (Zeil and Zanker 1997).

Following such an approach, of course, the issue arises about the proper choice of models. Ideally it requires knowledge of either the computational structure of motion detection or of the response characteristics of motion sensitive cells, so that the elementary motion detectors can be adequately modelled. Fortunately, the computational structure and the constraints of the elementary process of motion detection are very well known at least in some biological systems (Borst and Egelhaaf 1993). There are several alternative schemes for implementing elementary motion detectors, and there is some debate on both their respective merits and on the most likely scheme implemented in biological systems. However, different detector mechanisms have been shown to have quite similar properties under certain conditions (Hildreth and Koch 1987; Borst and Egelhaaf 1993), although to our knowledge, it has not been systematically tested, how these different implementations compare in their performance if confronted with natural motion signals.

3.3 Higher order structure

Natural scene analysis and computational modelling of early stages of motion processing both lead to the question of how motion patterns can be extracted from the retinal image flow. A vast neurobiological and machine vision literature is concerned in one way or other with the problem of extracting higher order structure (Barron et al 1994, and reviews in this volume), but a number of critical issues have received relatively little attention: how motion patterns can be detected under natural conditions and with elementary motion detectors that do not – and probably cannot in principle - accurately represent the local image velocity (see Perrone, this volume). There is also a need to define motion events or "motion features" in the environment which are important for the survival of an animal, beyond optic flow processing for extraction of egomotion parameters. For animals operating in a simple environment, one might be able to specify "food events" or "predator events" as a set of spatiotemporal features and investigate their preservation at different levels of visual processing (c.f. Ewert 1980). It is worth noting, however, that many significant events in an animal's life may be difficult to describe in terms of motion vision alone, because not only are they a complex function of local motion features, they also may involve other visual cues such as colour and shape and non-visual cues such as behavioural context. The fusion of information from various sensory modalities clearly becomes more and more significant the higher the processing level one considers.

In summary, natural motion signals need to be characterized at several different levels, starting from a statistical characterization of the spatiotemporal power spectrum through the statistics of motion detector outputs and the analysis of neural filters that extract biologically relevant motion information, to defining complex spatiotemporal features which are important for the survival of the animal. The need for the different levels of characterization follows from the fact that the constraints at one level of visual processing may be quite different from the constraints at a higher level. For example, a simple description of second order statistics may be all that is required when modelling low level vision as a filter (or a set of filters) designed to protect the signal against neural and environmental noise. But this provides little insight into higher levels of motion processing which are likely to be optimized for more rigorous computational and ecological constraints to extract features of biological importance.

3.4 Reconstructing natural motion signals: an experimental approach

A complete description of the natural operating conditions for motion vision would ideally require us to place an appropriate imaging device close to an animal's eyes and have it move around with the animal. We would thus be in the position to reconstruct the natural motion signals the animal normally experiences. This approach is just about becoming feasible, given the right choice of animal and ecological setting. In many respects, however, there are serious limits in what we can do. We summarize and critically assess the state of the art below.

It is obviously easiest to analyse environmental motion, since observer movements need not be considered. The spatial and temporal power spectrum of windblown vegetation for instance can be derived from video sequences with standard methods and for the analysis of the velocity distribution the video films can be used as input to two-dimensional motion detector networks. In special cases, it is possible to reconstruct environmental motion from the viewpoint of an animal, as Zeil and Zanker (1997) have demonstrated for video films taken with a camera at the eye height of fiddler crabs. In such a well defined ethological context, rather specific questions can be asked as to what signals the animals have to glean from the noisy output of motion detectors (see Fig. 7).

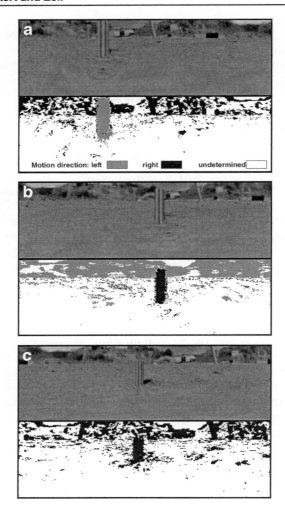

Fig. 8 Views from the cockpit of a learning wasp (top panels in a-c) and the spatial distributions of image motion directions the wasp generates during her learning flight (bottom panel in a-c). The situation is shown for a segment of flight shortly after lift-off from the nest entrance (**a**), about half-way through the learning flight (**b**) and shortly before the wasp departs (**c**). The location of the nest entrance in the ground can be seen in (a) as a small dark spot directly in front of the landmark (striped cylinder, of 2.2 cm diameter and 6.2 cm height, about 5 cm away from the nest entrance). The view was generated by first reconstructing the three-dimensional flight path and the orientation of the insect. The panorama was recorded by panning a video camera just above the nest entrance of the wasp, the foreground was cut away and replaced by a semi-natural plain stretching away to the distant panorama. With the knowledge of insect position and orientation it is then possible to simulate the (semi-) natural transformations of the scene the wasp had produced during her learning flight and to analyse the scene with a model of biological motion detectors. The reason for this involved procedure was that it is impossible in the field to account for translational flow, unless a camera is moved along the path flown by an insect which was not possible at the time (modified from Voss 1995).

It is clearly much more difficult to reconstruct motion signals for animals in unconstrained motion. The ideal way to go would be to have a camera on a freely behaving animal, like Passaglia et al. (1997) did with horseshoe crabs. The horseshoe crab however also reminds us of some serious, and possibly unsurmountable problems. The crabs are big enough to carry a camera, but most animals with interesting visual behaviour are too small for this. Big animals furthermore have a resolving power that is much superior to a video camera and in addition they are likely to move their eyes (but see Baddeley et al. 1997 for an attempt to capture image motion as seen from the viewpoint of a cat). Eye movements can be recorded, but so far only in consenting humans (Land 1994, and this volume). For those animals like insects that cannot move their eyes independently of their head or their body, a solution would be to use a robotic gantry to move a camera along the previously taken path. As long as flight paths are smooth and can be correctly reconstructed, the temporal limitations of the camera do not present problems for the reconstruction of the scene from the viewpoint of the animal, because any desired frame rate can be achieved simply by taking enough pictures along the flight path.

Up to now, controlled camera movements in the field were not achievable and researchers had to limit analysis to simple cases or use dirty tricks to reconstruct natural motion signals. Lambin (1987) and Voss and Zeil (1998) for instance simulated the global pattern of image motion in walking crickets and in flying wasps. A slightly more realistic procedure was developed by Voss (1995) who reconstructed the image motion in the visual field of a wasp during a learning flight on departure from her nest (Fig. 8) and by Kern and Warzecha (1998) who determined the optic flow experienced by flies moving through an artificial environment and replayed it to visual interneurones in subsequent electrophysiological experiments.

If an attempt is to be made using reconstruction methods to estimate the second order motion statistics for an animal, then the procedure is fairly straightforward. Provided an image sequence from the viewpoint of the animal has been obtained along a spatial path, and enough images along the path are collected to encompass the temporal bandwidth for the species of interest, then the second order statistics can be directly calculated from the image sequence, by computing the spatiotemporal frequency spectrum at each location on the retina. An alternative technique is to compute optic flow fields and to compute a histogram of velocities through time at every retinal location.

4. What to look for: coding and optimization

One of the goals of studying the visual ecology of motion vision is to understand the properties of neurones involved in image motion processing in light of the natural conditions in which they operate. The underlying assumption is that these neurones were shaped by natural selection and have evolved into efficient processing units that are adapted to provide reliable information for scene analysis and motor control and that ultimately contribute to the fitness of an animal. How can we know what influenced the design of motion sensitive neurones?

In the standard view, early vision mechanisms are thought to be designed so that information is effectively packaged to protect the signal against the deleterious effects of environmental and neuronal noise. This assumption is quite reasonable for many retinae, but allows one to proceed only a limited way along the processing stream, because coding in this context is typically not considered to have a particular computational purpose other than information preservation. An approach which is becoming more popular is to computationally model the first stages of visual processing, including prominent nonlinearities, and to investigate the advantages under natural operating conditions which accrue to the different possible implementations. Again, the processing units need to be modelled in a manner closely related to known physiology. This approach has the advantage that it does not require the formulation of elaborate theories for what is optimized. An example of this line of research is described by Field (1994). Finally, one can interpret visual optimization in terms of biologically relevant features. Such features may include the visibility of predators, prey, or even computationally complex features such as optic flow, and the interesting question is how neural processing mechanisms are optimized to detect such features.

4.1 Early vision: optimal filtering to remove environmental noise

One approach to optimization is to consider early visual neurones as filters which completely span the signal space of interest and remove environmental noise. This implies that neurones should be sensitive only to frequencies in which the signal has significant energy, and reject frequencies which possess predominantly environmental noise.

In spatiotemporal processing, the link between neural properties and lifestyle was first made by Autrum (1950), who observed that rapidly moving diurnal insects possess photoreceptors with significantly faster temporal responses than slower nocturnal insects. Srinivasan and Bernard (1975) subsequently showed theoretically that at high image velocities, spatial resolution is limited by photoreceptor dynamics which therefore ought to be matched to the motion vision ecology of a species. More recent work verifies Autrum's observation, even for insects within the same order (Howard et al 1984; de Souza and Ventura 1989; Laughlin and Weckström 1993). At low ambient light levels, visual processing is photon limited, so a slower temporal response is required to integrate individual photon-induced events and to maintain a reasonable signal to noise ratio. Because of the time needed by the visual system, flight speeds are also limited in order to avoid motion blur. Arguably, in this case, environmental limitations play the major role in determining the limits of flight behaviour. In daylight conditions, visual processing will no longer be primarily photon limited and other constraints become more important. Laughlin and Weckstrom (1993) for example argue that fast receptors are metabolically expensive, thus placing a large cost for significantly increasing temporal response speed. In this case, biophysical considerations represent the primary constraint, and one would expect very fast temporal responses to exist only in circumstances where strong evolutionary pressures favour fast flight (and the required fast temporal responses). Daylight provides a broad range of possible intermediate behaviours in terms of flight speed and temporal responses, which may not be predicted solely from physical or biophysical limitations, since the reaction times needed also depend on the details of lifestyle. If fast moving prey need to be detected, or a cluttered environment negotiated, the situation would be different than for an insect or a bird migrating at great height. However, based on arguments of efficiency, one would expect to find that response dynamics, velocity sensitivity and average flight speed to be closely linked (O'Carroll et al. 1996, 1997; O'Carroll, this volume).

The idea that early vision is concerned with the removal of environmental noise seems to hold rather well when matching flight behaviour to velocity sensitivity (at least for some species). However, the argument seems to fall apart at low velocities because most species have a bandpass velocity sensitivity when measured at typical ambient daylight levels. According to our assumption that filtering is optimized to remove environmental noise, there is no reason to attenuate the signal at low velocities. As we have seen, both the velocity distribution and the spatial power spectrum of natural images is a power function of the form $1/f^n$. Since the environmental noise levels can be assumed to be approximately white, the ratio of signal energy at each frequency to the noise level at each frequency is the highest at low frequencies. Indeed, the optimal "Wiener filter" which maximizes the overall signal to noise ratio for such a situation is a low pass filter, not a bandpass filter. Since the bandpass velocity response in biological systems cannot be explained in terms of protection against environmental noise, we have to consider constraints associated with noisy, bandwidth limited neural channels.

4.2 Early vision: coding to reduce redundancy

The concept of redundancy reduction as a first stage in neural processing was proposed by Barlow (1961). The goal of early vision in this theory is to protect the signal against distortion by neural noise and to code image information in a sparse way. This can be achieved by spatial and temporal lateral inhibition, a process which reduces signal variance in a reasonable manner. Srinivasan et al. (1982) developed and formalized this concept and suggested that the dynamic receptive field properties of the lamina monopolar cells in insects can be understood as a mechanism of predictive coding, whereby antagonistic surrounds provide a prediction of the signal in the excitatory centre of the receptive fields of neurones. Spatial and temporal correlation in the input signal is thus removed before further processing. However, predictive coding, as defined by standard Yule-Walker equations, does not consider variations in integration time or variations in centre size to protect the signal against distortions from both environmental and neural noise.

Van Hateren (1992) developed a more general theory which predicted spatiotemporal frequency transfer functions to optimize the signal to noise ratio in early visual processing. He asked how different levels of internal (neuronal) and external (environmental) noise affect the spatiotemporal response properties of the lamina interneurones of insects. He found that at high ambient light levels, that is conditions of low environmental noise, neuronal noise will dominate because of the limited dynamic range of neurones. In this case, a bandpass velocity response will maximize the signal to noise ratio by reducing redundancy, and thereby reducing the dynamic range required to represent the signal. The signal energy at low frequencies is so high in this situation that all of it is not needed to retain the signal integrity. At dawn, at dusk or at night, the environmental noise dominates and the neuronal noise is not a significant factor. In this situation, increased low pass filtering is required to increase the signal to noise ratio, and there will be little or no spatial or temporal inhibition. The theory presented by van Hateren (1992) thus considers both optimal noise reduction and optimal redundancy reduction.

Unfortunately, all of the approaches to vision which rely on signal processing theory have limitations. First, they assume that higher order processing needs do not influence the coding strategies in early vision and second, and depending on the optimization, usually imply that the degree of importance for a particular velocity is only determined by the frequency of occurrence of that velocity in the environment. We now briefly discuss these assumptions in the next two sections.

4.3 Mid-level vision: computational modelling of early visual units

Most of the approaches for modelling early vision based on redundancy reduction rely solely on optimization with respect to image statistics, with some additional constraints such as limited channel capacity. However, it is clear that what has been lacking are investigations associated with the advantages and disadvantages of different representations in what we will refer to as "mid-level vision" – basically, an examination of the output statistics of early visual processing units rather than the statistical characteristics of the image itself. An example of this level of description is Field's (1994) investigation into the advantages of an octave bandwidth representation of images. Rather than focus on the energy compaction properties of Gabor transforms, he considered the effect this representation has on the image statistics, comparing the statistics of the raw image with the statistics of the octave bandwidth filter bank representation. He shows that this representation significantly increases the kurtosis of the probability distribution function so that "unusual" image components are highlighted. He argues that this provides significant advantages for feature detectors at higher levels of processing. A corresponding approach would seem to be particularly fruitful for motion processing simply because the nonlinearities of motion processing units limit the level of insight that can be obtained by using analytical techniques.

If higher order processing needs have an influence on the best coding strategy at earlier stages of processing ("early vision") then for motion vision the first question is how to achieve a robust directional selectivity and a linear representation of image velocity. Correlation-type elementary motion detectors (EMDs) can only deliver this information after some spatial and temporal integration and after the signals from EMDs with different preferred directions are appropriately combined. The choice of mechanisms and the level of accuracy required, however, depend on the task. The best coding strategy may be different for such high level functions as motion parallax processing, odometry or image stabilization. Motion parallax processing and odometry for instance require true image velocity information which may not necessarily be needed in closed-loop image stabilization. Spatial integration is a potential problem for motion parallax processing, but not for odometry or the optomotor response. It is also worth remembering, that a large part of motion signal extraction operates in a closed sensory-motor loop. Motion detector properties thus have evolved and are maintained within the performance and stability constraints of feedback systems. One advantage of the bandpass-filter properties of biological elementary motion detectors for instance is that their high frequency roll-off prevents the feedback system involved in image stabilization from becoming unstable (Warzecha and Egelhaaf 1996). A similar argument was made by Nalbach (1989) who argued that the fastest of the three parallel motion detection channels in crabs has to habituate rapidly, in order to prevent feedback oscillations.

4.4 Feature based optimization

The efficient coding approach considers neural design as driven by the need to preserve information with minimal cost. Yet an animal must solve problems and detect certain features if it is to survive in the world and there are many examples of neurones that can be interpreted as "feature detectors", tuned to biologically relevant events. Feature detectors do not necessarily reflect the average signal distributions in the environment. Efficient coding and feature detection are not at odds as long as biologically relevant features are represented. However, efficient coding is normally modelled strictly in terms of spatiotemporal statistics rather than biologically relevant features and the widespread use of second order statistics for analysing early vision may therefore be misleading. Usually, early vision is analysed in terms of identifying a filter structure which preserves and protects the visual signal. The analysis typically includes constraints of linearity for the filter with optimization in the mean squared sense, resulting in a linear filter which depends only on the second order statistics. A statistical description models the average characteristics of a signal, with the implied assumption that the importance of a particular velocity range for instance is related to its frequency of occurrence. This conjecture seems reasonable in the context of egomotion control since there is little doubt that neural control of eye, head and body movements require a velocity bandwidth of motion sensitive pathways which is related to that generated by egomotion. However, when considering all aspects of motion processing, this assumption may overstate or understate the importance of certain velocities because it ignores the possibility that crucial events in an animal's life are rare. For instance, the appearance of a predator, may have a spatiotemporal pattern which has a low probability with respect to the average image statistics. Detection of such a rare pattern will nevertheless have a significant impact on the survival of the animal, and thus, represents a case where the average statistics may not adequately characterize the selective pressure on neuronal signal processing. The question then becomes whether the velocity distribution of biologically relevant events such as the movement of food or the movement of predators is contained within the range of image velocities caused by egomotion.

Let us take Lettvin et al.'s (1959) somewhat schematic description of a frog's life at face value. The frog then is an animal in which the velocity structure of biologically relevant events does not match the frequency of occurrence of those velocities in the environment. "His eyes do not move, as do ours, to follow prey, attend suspicious events, or search for things of interest. If his body changes its position with respect to gravity or the whole visual world is rotated about him, then he shows compensatory eye movements. These movements enter his hunting and evading habits only, e.g., as he sits on a rocking lily pad. Thus his eyes are actively stabilized ... He will starve to death surrounded by food if it is not moving. His choice of food is determined only by size and movement. He will leap to capture any object the size of an insect or worm, providing it moves like one".

This description illustrates several characteristics of the frog's motion world. The frog is a sit-and-wait hunter who swims in short bouts, hops or moves with a slow crawl, so that in terms of the fraction of total time it experiences comparatively little image flow generated by locomotion (Dieringer et al. 1983). Passive displacements on a rocking lily pad are largely compensated for by vestibular and optokinetic reflexes, whereby to stabilize the image on the retina, vision only needs to cover the residual image velocities not compensated for by vestibular reflexes. On the other hand, prey insects, worms and predators move and some of them probably do so much faster than the average velocity of whole image motion across the retina. The visual behaviour of toads can be considered to be a reflection of this situation: prey catching and escape is elicited predominantly by high image velocities (Fig. 9a) in a range where the gain of the optomotor response is rather low. The range of image velocities produced by prey and predators may thus lie in the low probability regions of the velocity distribution induced by egomotion on the retina. As a result, a blind application of the velocity distribution as a measure of importance in the frog's or the toads world would completely miss the range of velocities of biologically relevant events. In essence, these amphibians may show us how to use motion to separate out the presence of significant events, but they also remind us that the probability of experiencing a particular velocity is not commensurate with biological importance. For the toad a crucial discriminator of whether image motion signals prey, predators, or egomotion is where in the visual field it occurs and whether it is small field or large field image motion. Neither spatiotemporal statistics nor the velocity distribution capture this important fact.

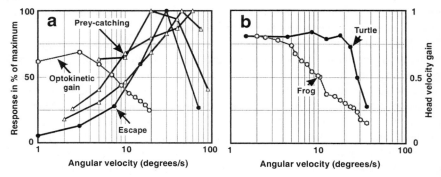

Fig. 9 a The probability of prey-catching behaviour (triangles) and of escape behaviour (dots) in toads over the image velocity of prey or predator dummies (data from Ewert 1969; Ewert and Rehn 1969; Ewert et al. 1979; Burghagen and Ewert 1983). The three curves with triangles show the results of three different experiments. Open circles and grey line show the optokinetic gain of the toad (data from Manteuffel et al. 1986). **b** The gain of the optomotor system in frog and turtle over pattern velocity (data from Dieringer et al. 1983).

The difficulty of analysing feature based optimization, from the experimental point of view, is that it requires assessing the spatiotemporal content of biologically relevant events and quantifying their "importance" to the species. The problem is not so much the ethological analysis, but in many cases it may be practically impossible to specify the cost functions for different events. Costs and benefits may be fairly easy to measure in the case of predator avoidance. But the details of egomotion parameter extraction or the efficiency with which environmental motion is filtered out may be difficult to assess in this respect.

Ryan (1990) has illustrated the many ways in which signals in the world of animals have influenced the evolution of sensory processing mechanisms, especially in the context of mating systems. Ryan did not speak specifically about motion vision, but there is reason to believe that in addition to the uses of image motion for orientation and navigation, visual motion is also a biological signal. If so, visual motion processing would be subject to multiple selective pressures, which result from a complex interaction of functions and environmental influences (Endler 1992). As an example, motion pattern analysis is needed to extract information about the observers own motion, and to identify other moving organisms in order to predict their direction of motion and intent. A particular set of movements may consist of a mating ritual for a conspecific, while different spatiotemporal patterns will indicate movement of a predator, and other patterns indicate movement of prey (e.g., Fleishman 1988). The task then is to determine the relative importance of all these influences, to characterize them in terms of the underlying visual processing and feature extraction mechanisms, and to identify how the web of influences results in an optimized receiver system. Unfortunately, while it is pleasant to speculate about such an approach, the reality is that the motion environments of most animals are so complex, and the unknowns about the importance of different features so numerous, that a feature based approach may not be feasible except in a few fortunate cases.

Note that the term "feature" could be replaced with the term "computational problem" if one wishes to follow the approach of Marr (1982). An animal must solve a set of problems necessary for its survival. The solution to each of these problems has a cost and a benefit, and each must be dealt with simultaneously in the visual processing chain. In his computational approach to vision, Marr suggested that one should specify a problem, design algorithms which solve the problem, and investigate different implementations of these algorithms. The primary difficulty with attempting to elucidate visual processing in animals in this way is the expectation that we can correctly specify and isolate the problems and constraints which are present for an organism. Unfortunately, this is often not the case. Both the problem and the constraints might be difficult to identify correctly and comprehensively. In addition, if one attempts to investigate the optimization of low and mid-levels of visual processing, one is presented with the fact that different computational problems may have different survival benefits to the organism, so that an inefficient biological solution to one problem may be acceptable if it allows the more efficient solution of other more important computational problems. As an example of the difficulty of all approaches involving understanding visual processing in terms of optimization, how do we decide whether the main selective pressure on early visual processing is redundancy reduction, reliability or sparseness of representation, or the image representation which allows efficient feature extraction?

5. Examples of visual motion processing matched to natural motion signals

There are many reasons to expect that motion detection and processing mechanisms in animals have been shaped by the general constraints imposed by neural hardware, by the regularities of image transformations produced by the closed sensory-motor loop of a moving observer and by the visual environments in which detection and processing takes place. The question is how well we can describe these constraints. As we have seen, the problem we are facing at the moment is that we have virtually no quantitative data on the natural operating conditions of motion vision. As the selective survey of known motion detection and processing mechanisms below will show, adaptations to both the properties of the visual environment and to the closed loop conditions of vision, are found on many levels of motion processing, involving the properties of single neurones and those of whole behavioural systems. But for the time being, we are left with rather indirect evidence as regards the functional significance under natural operating conditions of most of these adaptations.

5.1 Adaptation to environmental motion

A few scattered examples suggest that the properties of motion sensitive neurones have been shaped or are being shaped by environmental motion. One such case has been documented in a comparative study of the startle response in crabs by Tomsic et al. (1993), in which it turned out that crabs that live in areas with high environmental motion (in between vegetation) habituate faster to motion than crabs that live in rocky habitats. The second case involves a study of prey detection and of visual signals in lizards by Fleishman (1986, 1988). When confronted with prey objects in the presence of background motion, lizards appear to respond to particular object motion patterns while at the same time habituating to other common patterns of motion (Fleishman 1986). Lizards use vertical head bobbing movements as a way of signalling to conspecifics. The dynamics of these head movements appear to be designed to contrast with the movements of windblown vegetation in the environment of the lizards and thus help receivers to discriminate the signal from environmental motion (Fleishman 1988).

5.2 Adaptation to the structure of optic flow

In insects, the front end of the visual system, namely the retinal sampling array, appears to reflect the structure of optic flow. In many flies, for instance, the resolution of compound eyes is best in the forward pointing part of the retina which views the normal   direction of heading. Resolution decreases from the front towards the back and it has been suggested that this distribution of sampling distances across the retina matches the structure of optic flow vectors during normal flight (Land 1989). On a higher level of visual processing, there are many examples from a variety of animals of motion sensitive neurones that in one or the other of their properties are clearly matched to the structure of optic flow. The neurones are often found in distinct neuropils or brain areas devoted to analyse the egomotion components of retinal image flow. In the accessory optic tract in birds, for instance, neurones respond to large field image motion patterns in a way that suggests that their prime function is to sense particular optic flow components (Frost 1993, 1997; Wylie et al. 1998). A dominant feature is that some of the neurones receive input from opposite sides of the 360° visual field which makes them sensitive to coherent image motion that signals rotational disturbances.

Similar neural interactions have been suggested to underly the separation of rotational and translational optic flow components in the eye movement control system of crabs (Kern et al. 1993; Blanke et al. 1997). An instructive example of a brain area devoted to optic flow analysis is the lobula plate in insects. The neuropil contains about 60 different motion sensitive neurones which integrate image motion over large parts of the visual field. They are neatly arranged in horizontal and vertical motion detectors that respond to the rotational movements of insects in flight (Hausen 1993). The distribution of directional selectivity in the receptive fields of these movement sensitive neurones are matched in surprising detail to the structure of optic flow around the principal rotational axes of a flying animal and even seems to reflect the average distribution of contours in the environment (Krapp and Hengstenberg 1997; Krapp et al. 1998; Dahmen et al., this volume).

5.3 Adaptation to environments and lifestyle

Examples of visual motion processing being shaped by the specific environmental situation an animal finds itself in are comparatively rare while there are many cases in which neurone properties can be related to the lifestyle of animals, including their mode of locomotion, their tracking strategies and the specific tasks they have to solve.

A clear case of neural filters that make specific use of environmental topography can be found in animals that live in a flat world, like on the water surface, on plain open country or on tropical mudflats. In shore crabs, waterstriders, and also in vertebrates like the rabbit, the sensitivity to large-field motion around the yaw axis is not distributed equally throughout the visual field, but optomotor sensitivity reaches a narrow maximum just above the horizon (Kunze 1963; Dubois and Collewijn 1979; Nalbach and Nalbach 1987; Dahmen and Junger 1988; Zeil et al. 1989). This regionalization of motion sensitivity does not coincide with the high resolution equatorial acute zone in these animals and it is absent in animals that inhabit spatially more complex environments. It is thus thought to constitute a motion filter matched to a specific constraint in a flat world, namely the fact that image motion above the horizon will always signify rotational optic flow because objects are likely to be far away. Restricting optomotor sensitivity to the part of the visual field viewing distant objects therefore helps in separating the rotational and the translational components of optic flow (Nalbach and Nalbach 1987).

In considering the general effect of lifestyle on motion processing, a way to examine how locomotion or moving objects influence the image motion normally experienced by an animal is to first define the characteristic velocity as the speed required for a point object to travel across a receptive field (in degrees) in one integration time (in seconds). One is then in a position to study the link between flight behaviour and velocity sensitivity (Srinivasan and Bernard 1975; van Hateren 1992, 1993; Land 1997). Essentially, the characteristic velocity is a parameter that provides a measure of the speed above which there will begin to be significant spatial blurring due to motion. Note that the characteristic velocity refers to retinal velocity, not velocity of flight, but the distinction is not usually significant for animals without eye movements as long as the environment can be assumed to be a "cluttered" world. For situations in which this assumption is violated, such as for high flying birds, the link between retinal velocity and velocity of flight will, of course, depend on the height of the flight path above the ground. The implication of the characteristic velocity is that the visual system of fast flying animals which have a large characteristic velocity should be designed to ensure little loss of spatial resolution for quite high flight speeds, while hoverers and slow fliers which may have small characteristic velocities, would be predicted to loose spatial resolution at high speeds (Srinivasan and Bernard 1975). As we have pointed out before, one may have to consider in addition how the characteristic velocity changes with retinal topography.

A recent comparative electrophysiological analysis of wide-field directionally selective motion sensitive interneurones in the lobula plate of different insects indeed shows that the spatial and temporal frequency characteristics of these neurones are species- and in one case also sex-specific (O'Carroll et al. 1996, 1997). The authors measured the spatiotemporal contrast sensitivity curves of wide-field motion sensitive neurones for a number of insects with different flight behaviour. They found that the neurones in hovering species tend to have lower characteristic velocities (with 65-70°/s), compared to fast flying, non-hovering species (160-170°/s). Although some of these differences may be due to the differences in eye size of the species investigated and due to the fact that at least flies have fast, non-visual information on unintended rotations through their halteres, there is the strong possibility that elementary motion detectors, like photoreceptors (Laughlin and Weckström 1993) are differently tuned depending on lifestyle. The spatial and temporal filters that feed into motion detectors thus appear to be matched to the velocity distribution most frequently encountered by a particular species. Although physical constraints suggest that most energy in the velocity distribution lies at small velocities and that energy will decrease at higher velocities, faster moving insects will certainly experience a much broader velocity distribution, and so could be expected to be sensitive to higher velocities than insects which move comparatively slowly.

A similar relation to lifestyle may also underly the differences that have been found in the oculomotor system in vertebrates. Dieringer et al. (1983) described species differences in the dynamical properties of the optomotor control system in frogs and turtles that can be interpreted along the same lines. Both animals move rather slowly and, as expected, their closed loop gaze stabilization works for only quite low angular velocities. However, the velocity bandwidths for gaze stabilizsation are significantly different in the two animals, with the turtle being able to compensate for higher image velocities than a frog can (Fig. 9b). A frog crawl is still slower than a turtle walk, and the gaze stabilization of the two species may reflect this fact although we have no information about the characteristic velocity of these animals. Head movements also differ in the two species: a turtle moves its head while orienting and exploring much more frequently than a frog, and therefore may require input from the optokinetic system to help compensate for rotational image motion. An alternative way of interpreting this difference is the need of the turtle who may catch prey on the move, to have a well stabilized retinal image in order to increase signal to noise ratio. A comparative study of fish shows similar specializations of the eye movement control systems related to lifestyle (Dieringer et al. 1992). Bottom dwelling fish tend to possess low-gain optokinetic stabilizing systems while the gain in freely swimming fish is generally high.

The image motion normally experienced by many animals is heavily influenced by eye movements and by the type of control systems that stabilize gaze. The weights given to visual and non-visual information can be quite different depending on habitat and lifestyle. A comparative study of crabs for instance has shown that swimming crabs rely more on their statocyst signals to drive compensatory eye movements, while crabs living on rocks or on mudflats make in addition heavy use of vision, whereby vision can overide non-visual input. In rock crabs, stabilizing eye movements are also elicited by proprioceptive input from the legs (reviewed by Nalbach 1990).

All animals possess both visual and non-visual image stabilization reflexes, but they differ in the extent to which they are foveate and able to track targets. Stabilizing the retinal image, over the full field and/or locally through foveal smooth pursuit, means that even large velocity motion in the scene will be mapped to low image velocity as long as image velocity is correctly estimated by the optomotor system. The characteristic velocities of humans, found in psychophysical measurements is between 1 and 2°/s (Kelly 1979), suggesting that motion processing is optimized to deal with much lower velocities than in insects. However, the velocity bandwidth for humans is quite large, so that even large velocities can be detected. The characteristic velocity in foveate animals is thus likely to be optimized to accommodate the retinal velocity distribution after gaze stabilization rather than the raw velocity distribution. One would expect, therefore, that motion sensitivity reflects the image stabilization strategies used by an animal. As an example, smooth pursuit tracking in humans results in a narrow velocity distribution (low average velocity) at the fovea and a large velocity distribution in the periphery (Eckert and Buchsbaum 1993a, b). This implies that the velocity of peak sensitivity should increase from the fovea to the peripheral visual field, and indeed, this observation holds true for humans and other primates with similar image stabilization systems (McKee and Nakayama 1984; van de Grind et al. 1986; Johnston and Wright 1986).

Again in insects, we find adaptations related to tracking tasks, albeit of a different kind. In many cases, most notably in flies, the visual system is strongly sexually dimorphic. In houseflies and blowflies male specific visual interneurones have been identified in the third optic ganglion, the lobula, which receive input from a forward and upward looking part of the male eye in which facets are enlarged and inter-ommatidial angles are small (Hausen and Strausfeld 1980).

These neurones respond preferentially to small moving objects (Gilbert and Strausfeld 1991) and are thus likely to mediate the chasing behaviour of male flies (Land and Collett 1974).

6. Outlook

We have identified a need to analyse natural motion signals in order to understand the evolution of motion detection and motion pattern extraction mechanisms. Methods of reconstruction and analysis of the image motion experienced by freely moving animals in their natural environment are just about becoming available and will help us to tackle problems ranging from image motion statistics, through coding in motion vision, to motion pattern extraction. As we have argued, it will be important to work as closely as possible to specific lifestyles and habitats, to be able to identify and characterize quantitatively the biologically relevant information content of natural scenes and to assess the importance of differences in the velocity distribution across the visual field. However, although research on a number of issues in motion vision is likely to benefit from knowledge of natural motion signals, the signals themselves are not the only selective force that has shaped the evolution of biological motion detectors. There are bound to be many constraints not directly imposed by information processing needs per se which had had a say in the design of motion processing mechanisms and which cannot necessarily be identified by studying visual ecology alone.

Acknowledgements

We are grateful to Hans van Hateren, Mandyam Srinivasan, Eric Warrant, and Johannes Zanker for their helpful and critical comments on earlier versions of the manuscript. Jochen Zeil acknowledges financial support from HFSP 84/97.

References

Autrum H (1950) Die Belichtungspotentiale und das Sehen der Insekten (Untersuchungen an *Calliphora* und *Dixippus*) Z Vergl Physiol 32: 176-227
Baddeley R, Abbott LF, Booth MCA, Sengpiel F, Freeman T, Wakeman EA, Rolls ET (1997) Responses of neurons in primary and inferior temporal visual cortices to natural scenes. Proc Roy Soc Lond B 264: 1775-1783
Barlow HB (1961) Possible principles underlying the transformation of sensory messages. In: Rosenblith WA (ed) Sensory communication. MIT Press, Cambridge, pp 217-234
Barnes WJP, Nalbach H-O (1993) Eye movements in freely moving crabs: their sensory basis and possible role in flow-field analysis. Comp Biochem Physiol 104A: 675-693
Barron JL, Fleet DJ, Beauchemin SS (1994) Performance of optical flow techniques. Int J Comp Vision 12: 43-77

Black MJ, Rangarajan A. (1996) On the unification of line processes, outlier rejection, and robust statistics with applications in early vision. Int J Comp Vision 19: 57-92.

Blanke H, Nalbach H-O, Varjú D (1997) Whole-field integration, not detailed analysis, is used by the crab optokinetic system to separate rotation and translation in optic flow. J Comp Physiol A 181: 383-392

Borst A, Egelhaaf M (1993) Detecting visual motion: Theory and models. In: Miles FA, Wallman J (eds) Visual motion and its role in the stabilisation of gaze. Elsevier, Amsterdam, pp 3-27

Burghagen H, Ewert J-P (1983) Influence of the background for discriminating object motion from self-induced motion in toads *Bufo bufo* (L.). J Comp Physiol 152: 241-249

Burton GJ, Moorhead IR (1987) Color and spatial structure in natural scenes. Appl Optics 26: 157-170

Christy JH (1995) Mimicry, mate choice, and the sensory trap hypothesis. Am Nat 146: 171-181

Collett TS, Land MF (1975a) Visual control of flight behaviour in the hoverfly, *Syritta pipiens* L. J Comp Physiol 99: 1-66

Collett TS, Land MF (1975b) Visual spatial memory in a hoverfly. J Comp Physiol 100:59-84

Collett TS, Land MF (1979) How hoverflies compute interception courses. J Comp Physiol 125: 191-204

Coppola DM, Purves HR, McCoy AN, Purves D (1998) The distribution of oriented contours in the real world. Proc Natl Acad Sci 95: 4002-4006

Coutts MP, Grace J (1995) Wind and trees. Cambridge University Press, Cambridge

Cutting JE (1982) Blowing in the wind: perceiving structure in trees and bushes. Cognition 12: 25-44

Dahmen HJ, Junger W (1988) Adaptation to the watersurface: structural and functional specialisation of the Gerrid eye. In: Elsner N, Barth FG (eds) Sense Organs. Proc 16th Göttingen Neurobiol Conf. Thieme Verlag, Stuttgart, p 233

Davies MNO, Green PR (1988) Head-bobbing during walking, running and flying: relative motion perception in the pigeon. J Exp Biol 138: 71-91

Dieringer N, Cochran SL, Precht W (1983) Differences in the central organization of gaze stabilizing reflexes between frog and turtle. J Comp Physiol 153: 495-508

Dieringer N, Reichenberger I, Graf W (1992) Differences in optokinetic and vestibular ocular reflex performance in Teleosts and their relationship to different life styles. Brain Behav Evol 39: 289-304

Dong DW, Atick JJ (1995a) Statistics of natural time-varying images. Network: Comp Neural Syst 6: 345-358

Dong DW, Atick JJ (1995b) Temporal decorrelation: A theory of lagged and nonlagged responses in the lateral geniculate nucleus. Network: Comp Neural Syst 6: 159-178

Dubois MFW, Collewijn H (1979) The optokinetic reactions of the rabbit: Relation to the visual streak. Vision Res 19: 9-17

Eckert MP, Buchsbaum G (1993a) Effect of tracking strategies on the velocity structure of two-dimensional image sequences. J Opt Soc Am A10: 1993-1996

Eckert MP, Buchsbaum G (1993b) Efficient coding of natural time varying images in the early visual system. Phil Trans Roy Soc Lond B 339: 385-395

Eckert MP, Buchsbaum G, Watson AB (1992) Separability of spatiotemporal spectra of image sequences. IEEE Trans Pattern Anal Machine Intell 14: 1210-1213

Endler JA (1992) Signals, signal conditions, and the direction of evolution. Am Nat 139: S125-153

Egelhaaf M, Borst A (1993) Movement detection in arthropods. In: Miles FA, Wallman J (eds) Visual motion and its role in the stabilisation of gaze. Elsevier, Amsterdam, pp 53-77

Ewert J-P (1969) Quantitative Analyse von Reiz-Reaktionsbeziehungen bei visuellem Auslösen der Beutefang-Wendereaktion der Erdkröte (*Bufo bufo* L.). Pflügers Arch 308: 225-243

Ewert J-P (1980) Neuroethology. Springer Verlag, Berlin

Ewert J-P, Rehn B (1969) Quantitative Analyse der Reiz-Reaktions-Beziehungen bei visuellem Auslösen des Fluchtverhaltens der Wechselkröte (*Bufo viridis* Laur.). Behaviour 35: 212-233

Ewert J-P, Arend B, Becker V, Borchers H-W (1979) Invariants in configurational prey selection by *Bufo bufo* (L.). Brain Behav Evol 16: 38-51

Field DJ (1987) Relations between the statistics of natural images and the response properties of cortical cells. J Opt Soc Am A 4: 2379-2394

Field DJ (1994) What is the goal of sensory coding? Neural Computation 6: 559-601

Fleishman LJ (1986) Motion detection in the presence and absence of background motion in an *Anolis* lizard. J Comp Physiol A159: 711-720

Fleishman LJ (1988) Sensory and environmental influences on display form in *Anolis auratus*, a grass anole from Panama. Behav Ecol Sociobiol 22: 309-316

Frost BJ (1978) The optokinetic basis of head-bobbing in the pigeon. J Exp Biol 74: 187-195

Frost BJ (1993) Subcortical analysis of visual motion: Relative motion, figure-ground discrimination and induced optic flow. In: Miles FA, Wallman J (eds) Visual motion and its role in the stabilisation of gaze. Elsevier, Amsterdam, pp 159-175

Frost BJ, Sun H (1997) Visual motion processing for figure/ground segregation, collision avoidance, and optic flow analysis in the pigeon. In: Venkatesh S, Srinivasan MV (eds) From living eyes to seeing machines. Oxford Univesity Press, Oxford, pp 80-103

Gibson JJ (1950) The perception of the visual world. Houghton Mifflin, Boston

Gilbert C, Strausfeld NJ (1991) The functional organisation of male-specific visual neurons in flies. J Comp Physiol A 169: 395-411

van de Grind WA, Koenderink JJ, van Doorn AJ (1986) The distribution of human motion detector properties in the monocular visual field. Vision Res 26: 797-810

van Hateren JH (1992) Theoretical predictions of spatiotemporal receptive fields of fly LMC's, and experimental validation. J Comp Physiol A 171: 157-170

van Hateren JH (1993) Three modes of spatiotemporal preprocessing by eyes. J Comp Physiol A 172: 583-591

van Hateren JH (1997) Processing of natural time series of intensities by the visual system of the blowfly. Vision Res 37: 3407-3416

Hausen K (1993) The decoding of retinal image flow in insects. In: Miles FA, Wallman J (eds) Visual motion and its role in the stabilisation of gaze. Elsevier, Amsterdam, pp 203-235

Hausen K, Strausfeld NJ (1980) Sexually dimorphic interneuron arrangements in the fly visual system. Proc Roy Soc Lond B 208: 57-71

Hildreth EC, Koch C (1987) The analysis of motion: From computational theory to neural mechanisms. Ann Rev Neurosci 10: 477-533

Howard J, Dubs A, Payne R (1984) The dynamics of phototransduction in insects. J Comp Physiol A 154: 707-718

Junger W, Dahmen HJ (1991) Response to self-motion in waterstriders: visual discrimination between rotation and translation. J Comp Physiol A 169: 641-646

Johnston A, Wright MJ (1986) Matching velocity in central and peripheral vision. Vision Res 26: 1099-1109

Kelber A, Zeil J (1990) A robust procedure for visual stabilisation of hovering flight position in guard bees of *Trigona* (Tetragonisca) *angustula* (Apidae, Meliponinae). J Comp Physiol A 167: 569-577

Kelly DH (1979) Motion and vision. II. Stabilized spatio-temporal threshold surface. J Opt Soc Am 69: 1340-1349

Kern R, Nalbach H-O, Varjú D (1993) Interactions of local movement detectors enhance the detection of rotation. Optokinetic experiments with the rock crab, *Pachygrapsus marmoratus*. Visual Neurosci 10: 643-646

Kern R, Warzecha A-K (1998) Coding of motion as seen out of the cockpit of a behaving fly. In: Elsner N, Wehner R (eds) New Neuroethology on the Move. Proc 26th Göttingen Neurobiol Conf. Thieme Verlag Stuttgart, p 126

Koenderink JJ (1986) Optic flow. Vision Res 26: 161-180

Kral K, Poteser M (1997) Motion parallax as a source of distance information in locusts and mantids. J Insect Behav 10: 145-163

Krapp HG, Hengstenberg R (1997) Estimation of self-motion by optic flow processing in single visual interneurons. Nature 384: 463-466

Krapp HG, Hengstenberg B, Hengstenberg R (1998) Dendritic structure and receptive field organisation of optic flow processing interneurons in the fly. J Neurophysiol 79: 1902-1917

Kunze P (1963) Der Einfluss der Grösse bewegter Felder auf den optokinetischen Augennystagmus der Winkerkrabbe (*Uca pugnax*). Ergeb Biol 26: 55-62

Lambin M (1987) A method for identifying the nearby spatial cues used by animals during transverse orientation. Behav Processes 14:1-10

Land MF (1989) Variations in the structure and design of compound eyes. In: Stavenga DG, Hardie RC (eds) Facets of vision. Springer, Berlin, pp 90-111

Land MF (1993) The visual control of courtship behaviour in the fly *Poecilobothrus nobilitans*. J Comp Physiol A 173: 595-603

Land MF (1995) The functions of eye movements in animals remote from man. In: Findlay JM, Walker R, Kentridge RW (eds) Eye movement research. Elsevier, Amsterdam, pp 63-76

Land MF (1997) Visual acuity in insects. Ann Rev Entomol 42: 147-177

Land MF, Collett TS (1974) Chasing behaviour of houseflies (*Fannia canicularis*). J Comp Physiol 89: 331-357

Land MF, Layne J (1995) The visual control of behaviour in fiddler crabs: I. Resolution, thresholds and the role of the horizon. J Comp Physiol A 177: 81-90

Land MF, Lee DN (1994) Where we look when we steer. Nature 369: 742-744

Lappe M, Pekel M, Hoffmann K-P (1998) Optokinetic eye movements elicited by radial optic flow in the Macaque monkey. J Neurophysiol 79: 1461-1480

Laughlin SB (1981) A simple coding procedure enhances a neuron's information capacity. Z Naturforsch 36: 910-912

Laughlin SB, Weckstrom M (1993) Fast and slow photoreceptors - a comparative study of the functional diversity of coding and conductances in the Diptera. J Comp Physiol A 172: 593-609

Layne J, Land MF, Zeil J (1997) Fiddler crabs use the visual horizon to distinguish predators from conspecifics: A review of the evidence. J Mar Biol UK 77: 43-54

Lehrer M, Srinivasan MV (1994) Active vision in honeybees: task-oriented suppression of an innate behaviour. Vision Res 34: 511-516

Lettvin JY, Maturana HR, McCulloch WS, Pitts WH (1959) What the frog's eye tells the frog's brain. Proc of the Inst Radio Engineers 47: 1940-1951

Manteuffel G, Kopp J, Himstedt W (1986) Amphibian optokinetic afternystagmus: properties and comparative analysis in various species. Brain Behav Evol 28: 186-197

Martin N, Franceschini N (1994) Obstacle avoidance and speed control in a mobile vehicle equipped with a compound eye. In: Masaki I (ed) Intelligent vehicles. MIT Press, Cambridge, pp 381-386

Marr D (1982) Vision. Freeman and Company, New York

McKee SP, Nakayama K (1984) The detection of motion in the peripheral visual field. Vision Res 24: 25-32

Nalbach H-O (1989) Three temporal frequency channels constitute the dynamics of the optokinetic system of the crab, *Carcinus maenas* (L.). Biol Cybern 61: 59-70

Nalbach H-O (1990) Multisensory control of eyestalk orientation in decapod crustaceans: An ecological approach. J Crust Biol 10: 382-399

Nalbach H-O (1992) Translational head movements of pigeons in response to a rotating pattern: characteristics and tool to analyse mechanisms underlying detection of rotational and translational optical flow. Exp Brain Res 92: 27-38

Nalbach H-O, Nalbach G (1987) Distribution of optokinetic sensitivity over the eye of crabs: its relation to habitat and possible role in flow-field analysis. J Comp Physiol A 160: 127-135

Neri P, Morone MC, Burr DC (1998) Seeing biological motion. Nature 395: 894-896

O'Carroll DC, Bidwell NJ, Laughlin SB, Warrant EJ (1996) Insect motion detectors matched to visual ecology. Nature 382: 63-66

O'Carroll DC, Laughlin SB, Bidwell NJ, Harris EJ (1997) Spatio-temporal properties of motion detectors matched to low image velocities in hovering insects. Vision Res 37: 3427-3439

Passaglia C, Dodge F, Herzog E, Jackson S, Barlow R (1997) Deciphering a neural code for vision. Proc Natl Acad Sci 94: 12649-12654

Perrone JA, Stone LS (1994) A model of self-motion estimation within primate extrastriate visual cortex. Vision Res 34: 2917-2938

Ruderman D (1997) Origins of scaling in natural images. Vision Res 23: 3385-3398

Ryan MJ (1990) Sexual selection, sensory systems and sensory exploitation. Oxford Surveys of Evol Biol 7: 157-195

Salmon M, Hyatt G, McCarthy K, Costlow JD (1978) Display specificity and reproductive isolation in the fiddler crabs, *Uca panacea* and *U. pugilator*. Z Tierpsychol 48: 251-276

Salmon M, Hyatt GW (1983) Communication. In: Vernberg FJ, Vernberg WG (eds) The Biology of Crustacea 7: Behavior and ecology. Academic Press, New York, pp 1-40

van der Schaaf A, van Hateren H (1996) Modelling the power spectra of natural images: statistics and information. Vision Res 36: 2759-2770

de Souza JM, Ventura DF (1989) Comparative study of temporal summation and response form in hymenopteran photoreceptors. J Comp Physiol A 165: 237-245

Srinivasan MV, Bernard G (1975) The effect of motion on visual acuity of the compound eye: a theoretical analysis. Vision Res 15: 515-525

Srinivasan MV, Laughlin SB, Dubs A (1982) Predictive coding: a fresh view of inhibition in the retina. Proc Roy Soc Lond B 216: 427-459

Srinivasan MV, Zhang SW, Lehrer M, Collett TS (1996) Honeybee navigation *en route* to the goal: visual flight control and odometry. J Exp Biol 199: 237-244

Srinivasan MV, Zhang SW, Chahl JS, Barth E, Venkatesh S (2000) How honeybees make grazing landings on flat surfaces. Biol Cybern (in press)

Switkes B, Mayer MJ, Sloan JA (1978) Spatial frequency analysis of the visual environment: Anisotropy and the carpentered environment hypothesis. Vision Res 18: 1393-1399

Tomsic D, Massoni V, Maldonado H (1993) Habituation to a danger stimulus in two semi-terrestrial crabs: ontogenic, ecological and opioid modulation correlates. J Comp Physiol A 173: 621-633

Voss R (1995) Information durch Eigenbewegung: Rekonstruktion und Analyse des Bildflusses am Auge fliegender Insekten. Doctoral Thesis, Universität Tübingen

Voss R, Zeil J (1998) Active vision in insects: An analysis of object-directed zig-zag flights in a ground-nesting wasp (*Odynerus spinipes*, Eumenidae). J Comp Physiol A 182: 377-387

Warzecha A-K, Egelhaaf M (1996) Intrinsic properties of biological motion detectors prevent the optomotor control system from getting unstable. Phil Trans Roy Soc Lond B 351: 1579-1591

Wüst R (1987) Studien zum Verhalten schwärmender Mücken am Beispiel einer Chaoboridenart (*Chaoborus spec.*). Diploma Thesis, Universität Tübingen

Wylie DRW, Bischof WF, Frost BJ (1998) Common reference frame for neural coding of translational and rotational optic flow. Nature 392: 278-282

Zanker JM (1996) Looking at the output of two-dimensional motion detector arrays. Invest Ophthalm Vis Sci 37: 743

Zeil J (1986) The territorial flight of male houseflies (*Fannia canicularis* L.). Behav Ecol Sociobiol 19: 213-219

Zeil J, Kelber A, Voss R (1996) Structure and function of learning flights in bees and wasps. J Exp Biol 199: 245-252

Zeil J, Nalbach G, Nalbach H-O (1986) Eyes, eye stalks, and the visual world of semi-terrestrial crabs. J Comp Physiol A 159: 801-811

Zeil J, Nalbach G, Nalbach H-O (1989) Spatial vision in a flat world: Optical and neural adaptations in Arthropods. In: Singh RN, Strausfel NJ (eds) Neurobiology of sensory systems. Plenum Press, New York, pp 123-137

Zeil J, Zanker JM (1997) A glimpse into crabworld. Vision Res 37: 3417-3426

Spatiotemporal Inseparability of Natural Images and Visual Sensitivities

Dawei W. Dong

Center for Complex Systems and Brain Sciences, Florida Atlantic University, Boca Raton, USA

1. Abstract

The visual system is concerned with the perception of objects in a dynamic world. A significant fact about natural time-varying images is that they do not change randomly over space-time; instead image intensities at different times and/or spatial positions are highly correlated. We measured the spatiotemporal correlation function – equivalently the power spectrum – of natural images and we find that it is non-separable, i.e., coupled in space and time, and exhibits a very interesting scaling behaviour. This behaviour is shown to be related to the motion in the images and the power spectrum is naturally separable into a spatial term and a velocity term. The same kind of spatiotemporal coupling and scaling exists in visual sensitivity measured in physiological and psychophysical experiments. By assuming that the visual system is optimized to process information of natural images, a quantitative relationship can be derived between the power spectrum of natural images and the visual sensitivity. This reveals some interesting aspects of motion vision.

2. Statistics of natural time-varying images

Interest in properties of time-varying images dates back to the early days of development of the television (Kretzmer 1952). The statistical properties of static images have been studied for many years (Burton and Moorhead 1987; Field 1987; Tolhurst et al. 1992; Hancock et al. 1992; Ruderman and Bialek 1994). The statistical properties of time-varying images, on the other hand, has been studied more carefully in recent years (Dong and Atick 1995a). We will briefly summarize the results first and later-on will verify our assertion that the power spectrum is dominated by image motion and then will further relate the power spectrum to motion vision.

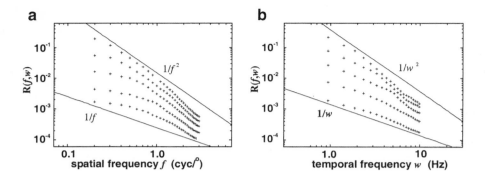

Fig. 1 Measured spatial and temporal power spectra of natural time-varying images. **a** The temporal frequency increases from 1.4, 2.3, 3.8, 6, to 10 Hz as we go from the highest to the lowest curve. **b** The spatial frequency increases from 0.3, 0.5, 0.8, 1.3, to 2.1 cyc/° as we go from the highest to the lowest curve. Also shown are the lines representing the power-laws $1/f^2$, $1/f$ (a) and $1/w^2$, $1/w$ (b), for reference.

We collected many samples of natural time-varying images from recordings of a moving camera and analysed their spatial-temporal statistics. We measured the spatiotemporal correlation function – or the power spectrum – for an ensemble of more than a thousand segments of motion pictures, and we find significant regularities. Figure 1 illustrates the spatial and temporal scaling behaviour found in natural images (adapted from Dong and Atick 1995a). It shows that the natural time-varying images do not change randomly over space-time; instead, image intensities at different times and/or spatial positions are highly correlated. Had natural scenes been random in space and time, i.e. white noise, we would have observed a flat power spectrum in both domains, i.e. the power lines would lie horizontally. The measurement indicates otherwise. Natural scenes have more power at low frequencies and this power decreases as spatial and/or temporal frequency increases. For a given temporal frequency, the data shows that the power spectrum decreases roughly as a reciprocal power of spatial frequency, f:

$$R \sim 1/f^a. \tag{1}$$

Similarly, for a given spatial frequency, the power spectrum decreases roughly as a reciprocal power of temporal frequency, w:

$$R \sim 1/w^b. \tag{2}$$

Both the a and b are positive numbers. In figure 1, on the left, $1/f^2$ and $1/f$, and on the right, $1/w^2$ and $1/w$ are plotted for reference; in the double logarithmic plot, they are straight lines.

A straightforward inspection of figure 1 shows that the power spectrum cannot be separated into pure spatial and pure temporal parts, space and time are coupled in a non-trivial way. A more careful examination of the power spectrum showed that spatial and temporal power spectra of natural images are intertwined in a special way related to relative motions of objects and observer. To see this we have replotted the power spectrum in figure 2a as a function of spatial frequency f but for fixed w/f ratio. One can see clearly that the curves for different w/f ratios are just a horizontal shift from each other and all of them follow a very precise power law, i.e. a straight line in log-log plot:

$$1/f^{m+1}. \tag{3}$$

In fact, if we multiply the spectrum by a power of f, i.e. if we plot $f^{m+1}R(f,w)$ as a function of w/f then all curves coincide very well, as shown in figure 2b, which means that

$$R(f,w) = (1/f^{m+1})\, F(w/f). \tag{4}$$

This exhibits a very interesting scaling behaviour in which the power spectrum is non-separable, i.e. coupled in space and time; but is separable into two functions of the spatial frequency f and the ratio of the temporal and spatial frequencies w/f, respectively.

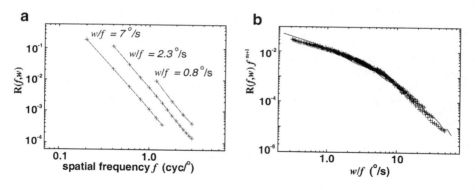

Fig. 2 Scaling behaviour of spatiotemporal power spectrum. **a** The power spectrum is plotted for three velocities – ratios of temporal and spatial frequencies – 0.8, 2.3 and 7°/s. **b** The power spectrum is replotted as a function of w/f after multiplication by f^{m+1} - all the data points fall on a single curve. The solid curve is the velocity distribution P(v) multiplied by the constant C_s (see equation (7)).

3. Spatiotemporal inseparability and motion of visual scene

The special property shown in figure 2 and equation (4) tells a lot about the origin of the spatiotemporal power spectrum. Assuming that the dominant contribution to the spatiotemporal variability in the signal is relative motion, one can derive that

$$R(f,w) = (1/f) \, R_s(f) \, P(w/f) \tag{5}$$

in which $R_s(f)$ is the spatial power spectrum of the static objects and $P(v)$ is the probability distribution of one-dimensional image velocity v due to the relative motion of the objects to the observer (Dong and Atick 1995a).

By measuring the static power spectrum for this collection of images (frames treated as snapshots), we find

$$R_s(f) = C_s / f^m \tag{6}$$

with m = 2.3 for some constant C_s - this is precisely the same m that we get from figure 2. The measured static power spectrum is shown in figure 3a, which is in general agreement with other earlier measurements on static images (Burton and Moorhead 1987; Field 1987; Ruderman and Bialek 1994). In most of the measurements, m ~ 2. Substituting equation (6) into (5), we arrive at

$$R(f,w) = (C_s / f^{m+1}) \, P(w/f). \tag{7}$$

This indicates that all the data points $R(f,w) \, f^{m+1}/C_s$ should fall on a curve which has a direct physical interpretation: the velocity density distribution of the images.

To verify the above claim, we measured the velocity distribution of the same collection of images by calculating the optical flow fields between frames using the following method: for a small area of image in one frame, find its translated one in the next frame (i.e., the one with the least-square-difference) thus get the large movement in number of pixels, then calculate the sub-pixel small movements by

$$v_x \, \partial I(x,y)/\partial x + v_y \, \partial I(x,y)/\partial y = - \, \partial I(x,y)/\partial t \, . \tag{8}$$

The one-dimensional velocity density distribution is calculated by accumulating over all possible spatial and temporal locations (for references to similar methods, e.g., Jain and Jain 1981; Horn and Schunk 1981). The measured $P(v)$ is shown in figure 3b. It is interesting to see that in certain regime $P(v) \sim 1/v^2$, confirming some similar velocity distributions proposed earlier (van Hateren 1993; Dong and Atick 1995a).

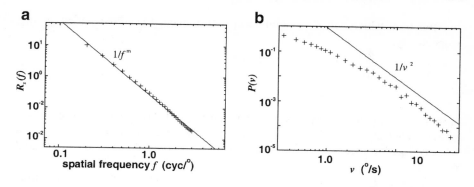

Fig. 3 Spatial power spectrum and velocity distribution. **a** Measured spatial power spectrum of snap shot images. It shows that $R_s(f) \sim 1/f^m$ is a good approximation to the spectrum (the solid line). In our measurement m=2..3. **b** Measured velocity density distribution P(v). For reference, the $1/v^2$ is plotted (the solid line).

The P(v) curve multiplied by the constant C_s is plotted with $R(f,w) f^{m+1}$ in figure 2b (m and C_s are determined from the static power spectrum). It confirms the equation (7).

4. Visual sensitivities to moving stimuli

Given the measured spatiotemporal power spectrum, it is not difficult to predict the filter which is optimized for transmitting information from nature scenes. Do the measurements of visual systems' responses to moving stimuli agree with the theoretical prediction?

We have derived earlier (Dong and Atick 1995b) that the optimal coding requires decorrelation when the signal-to-noise ratio is high and requires smoothing where noise is significant. We derived the following relationship for the visual sensitivity K and the power spectrum R in the presence of noise power N:

$$K = \frac{(1/R)^{1/2}}{(1 + N/R)^{3/2}} \tag{9}$$

This relationship is compared with experimental data from both cat and human. In those experiments, the stimuli are gratings of spatial frequency f moving at velocity v (which is also characterized as changing at temporal frequency $w = fv$).

As shown in the previous section, the spatiotemporal power spectrum of natural images has a special form which is not separable in space and time but separable in space and velocity, i.e., equation (7) can be rewritten as the product of two functions which only depend on the spatial frequency f and the velocity $v=w/f$, respectively:

$$R(f,v) = \frac{C_s}{f^{m+1}} P(v) \ . \tag{10}$$

Assuming white noise power N, the equation (9) and (10) predict: for any fixed spatial frequency f, the contrast sensitivity K only depends on the velocity v; and for any fixed velocity v, the contrast sensitivity K only depends on the spatial frequency f.

In general the $P(v)$ is not a simple power-law function. But there is a regime of interest the $P(v)$ approximates a universal power-law $1/v^2$ (see Fig. 3b). Included in this regime is the region of low spatial and intermediate temporal frequencies, which is where the experiments on the LGN temporal tuning properties are done (see references in Dong and Atick 1995b). It is easy to show that our theoretical prediction in this case is

$$K(v) = \frac{(v^2 C_f)^{1/2}}{(1+v^2 C_f)^{3/2}} \ , \tag{11}$$

in which C_f only depends on f and is a constant for the contrast sensitivity curve of different velocities. This is verified with experimental data from cat LGN (Troy 1983a) where single cell recordings were made for contrast sensitivity to a grating of fixed spatial frequency moving at various velocities (Fig. 4a). It is clear that the agreement is very good.

Similarly, for fixed velocity v, the theory predicts (assuming $m \sim 2$):

$$K(f) = \frac{(f^3 C_v)^{1/2}}{(1+f^3 C_v)^{3/2}} \ , \tag{12}$$

in which C_v only depends on v and is a constant for a contrast sensitivity curve of spatial frequency. Not only so, for different velocity v, the curves in the log-log plot are only parallel shifts from each other along the spatial frequency axis. Again, as shown in figure 4b, the theory agrees with experimental data (Kelly 1979) very well. In fact, except at very high and very low spatial or temporal frequencies, the entire spatiotemporal contrast sensitivity surface can be represented by one curve over a single variable which is the spatial frequency scaled by velocity density distribution function (Dong 1997).

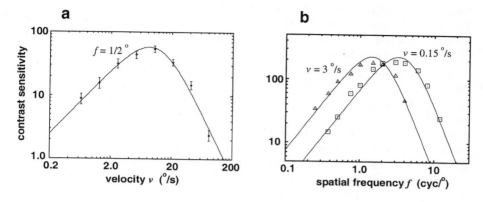

Fig. 4 Comparison between predicted contrast sensitivity and experimental data. **a** Predicted contrast sensitivity curve for different velocity (solid curve) and cat physiological data (diamond symbols and error bars from twenty-seven X LGN cells of nine cats measured by Troy 1983a). **b** Predicted spatiotemporal contrast sensitivity curves (solid curves) and human psychophysical data, triangle symbols and square symbols are measured by Kelly (1979) for two velocities – w/f ratios (0.15 and 3°/s).

5. Discussion

We should point out that while our predictions show that, in general, the human visual sensitivity is strongly space-time coupled, we do predict a regime where decoupling is a good approximation. This is based on the fact that in certain regime the velocity density distribution approximates a universal power-law $\sim 1/v^2$ thus the power spectrum of natural images is separable into spatial and temporal parts. In a previous work we have used this decoupling to model response properties of cat LGN cells and where we have shown that these can be accounted for by the theoretical prediction based on the power spectrum in that regime (Dong and Atick 1995b). The cat LGN data used in current paper were generated from moving gratings and different from the data we modelled earlier which were generated from temporal modulation of receptive field centres. The striking agreements in both cases give very strong support of the theory that the visual system is optimized to process spatiotemporal information of natural images.

In general, the spatial and temporal parts of R(f,w) are not separable and hence receptive fields of neurones cannot be fully characterized in space independently of time. The scaling behaviour that we illustrated suggests a natural way for dealing with this coupling. More precisely, it suggests that a better way to examine spatiotemporal tuning data from real visual system is to plot the data not as a function of f and w separately but as a function of w/f. This is a natural representation for motion vision. In this representation we expect that vision will exhibit more universal behaviour as we have shown for the human psychophysical data. We should point out that this scaling behaviour is expected to break down for very high temporal and spatial frequency where the effect of the temporal and spatial modulation function of the eye (Campbell and Gubisch 1966; Schnapf and Baylor 1987) cannot be ignored. Also this scaling behaviour might not show up in single cells since in real visual system neurones could have different spatiotemporal receptive fields. For example there are different cell types, and even the same type of cells could tune to different spatial frequencies in LGN (Troy 1983b). So an ecological theory of relating natural image properties with visual processes need to make distinctions about single cells and ensembles (Atick 1992) and different types of cell ensembles can participate in overall efficient coding (Van Essen and Anderson 1990; Li 1992; Eckert and Buchsbaum 1993).

Finally, we want to comment on the asymptotic behaviours of our measured power spectrum. While the detailed form of the function $F(w/f)$ depends on the image velocity distribution $P(v)$ which in turn depends on the distributions of motion velocities and spatial distances of visual objects relative to observer, there is an interesting asymptotic regime where the form of $P(v)$ is $\sim 1/v^2$. This gives rise to the asymptotic behaviour at relative low spatial frequency and relative high temporal frequency:

$$R(f,w) \sim \frac{1}{f^{m-1}w^2} \tag{13}$$

On the other end of this, i.e. relative high spatial frequency and relative low temporal frequency:

$$R(f,w) \sim \frac{1}{f^{m+1}} \tag{14}$$

In general, from equation (10), whenever $P(v)$ can be approximated as $\sim 1/v^a$, the power spectrum can be approximated as

$$R(f,w) \sim \frac{1}{f^{m+1-a}w^a} \tag{15}$$

Our measured power spectrum does not include a regime of

$$R(f,w) \sim \frac{1}{f^m w^2} \tag{16}$$

with $m \geq 2$, such as what was observed by Eckert et al. (1992). On the other hand, another simple relationship is implied in the coupled spatiotemporal behaviour: when temporal power spectrum is measured for a single spatial point (van Hateren 1997) which moves randomly across static image, it is not difficult to show that

$$R(w) \sim \frac{1}{w^{m-1}} \tag{17}$$

in which the temporal power index is simply the spatial power index minus one. This behaviour can also be derived by integrating the spatiotemporal power spectrum over all spatial frequencies. For more discussion about the origins of scaling in natural images see Ruderman (1997).

References

Atick JJ (1992) Could information theory provide an ecological theory of sensory processing? Network: Comp Neural Syst 3: 213-251

Burton GJ, Moorhead IR (1987) Color and spatial structure in natural scenes. Appl Optics. 26: 157-170

Campbell FW, Gubisch RW (1966) Optical quality of the human eye. J Physiol 186: 558-578

Dong DW (1997) Spatiotemporal coupling and scaling of natural images and human visual sensitivities. In: Mozer MC, Jordan MI, Petsche T (eds) Advances in neural information processing systems 9. MIT Press, Cambridge, pp 859-865

Dong DW, Atick JJ (1995a) Statistics of natural time-varying images. Network: Comp Neural Syst 6: 345-358

Dong DW, Atick JJ (1995b) Temporal decorrelation: a theory of lagged and nonlagged responses in the lateral geniculate nucleus. Network: Comp Neural Sys 6: 159-178

Eckert MP, Buchsbaum G (1993) Efficient coding of natural time varying images in the early visual system. Phil Trans Roy Soc Lond B 339: 385-39

Eckert MP, Buchsbaum G, Watson AB (1992) Separability of spatiotemporal spectra of image sequences. IEEE Trans Pattern Anal Machine Intell 14: 1210-1213

van Essen DC, Anderson CC (1990) Information processing strategies and pathways in the primate retina and visual cortex. In: Zotnetzer SF, Davis JL, Lau C (eds) Introduction to neural and electronic networks. Academic Press, Orlando, pp 43-72

Field DJ (1987) Relations between the statistics of natural images and the response properties of cortical cells. J Opt Soc Am A 4: 2379-2394

Hancock PJB, Baddeley RJ, Smith LS (1992) The principal components of natural images Network: Comp Neural Syst 3: 61-70

van Hateren JH (1993) Spatiotemporal contrast sensitivity of early vision. Vision Res 33: 257-267

van Hateren JH (1997) Processing of natural time series of intensities by the visual system of the blowfly. Vision Res 37: 3407-3416

Horn BKP, Schunk BG (1981) Determining optical flow. Artif Intell 17: 185

Jain JR, Jain AK (1981) Displacement measurement and its application in interframe image coding. IEEE Trans Commun 29: 1799-1808

Kelly DH, (1979) Motion and vision II. Stabilized spatio-temporal threshold surface. J Opt Soc Am 69: 1340-1349

Kretzmer ER (1952) Statistics of television signals. The Bell System Technical Journal 751-763

Li Z (1992) Different retinal ganglion cells have different functional goals. Int J Neural Systems 3: 237-248

Ruderman DL (1997) Origins of scaling in natural images. Vision Res 37: 3385-3398

Ruderman DL, Bialek W (1994) Statistics of natural images: scaling in the woods. Phys Rev Let 73: 814-817

Schnapf JL, Baylor DA (1987) How photoreceptor cells respond to light. Scient Am 256: 40-47

Tolhurst DJ, Tadmor Y, Chao T (1992) Amplitude spectra of natural images. Opthal Physiol Optics 12: 229-232

Troy JB (1983a) Spatio-temporal interaction in neurons of the cats dorsal lateral geniculate nucleus. J Physiol 344: 419-423

Troy JB (1983b) Spatio contrast sensitivities of X and Y type neurons in the cats dorsal lateral geniculate nucleus. J Physiol 344: 399-417

Motion Adaptation and Evidence for Parallel Processing in the Lobula Plate of the Bee-Fly *Bombylius major*

David C. O'Carroll

Department of Zoology, University of Washington, Seattle, USA

1. Introduction

As different insects fly about their world, they experience an enormous range of ambient image speed, from near stationary in hovering flies or hawkmoths (Farina et al. 1994) to 5,000 °/s or more during aerial pursuit by male flies (Land and Collett 1974). The optimum strategy for sampling and analysing complex patterns of optic flow must depend (and to a large extent be limited by) the behaviour of the animal (see also Eckert and Zeil, this volume). This notion is supported by comparative analysis of photoreceptor kinetics (Laughlin and Weckström 1993) which show a remarkable diversity of coding strategies within the Dipteran flies, consistent with differences in flight speed. We have recently described a correlation between spatiotemporal properties of motion detecting neurones in the optic lobes of flying insects and their flight behaviour (O'Carroll et al. 1996, 1997) suggesting that this principle also holds at the level of "central" processing. Most species that we have studied have "fast" photoreceptors, able to encode even rapidly moving images at the level of the retina. Yet specialized hovering insects such as hawkmoths and syrphid flies have motion detectors with maximal sensitivity at much lower (10-20 fold) image velocities than other insects such as butterflies or bees (O'Carroll et al. 1996). Detecting low velocities reliably may be crucial for stabilizing precise hovering flight by these species, but usually incurs a cost, either in the form of very large eyes, or reduced sensitivity of motion detectors to high pattern speeds.

The degree to which we might expect to see an optimum tuning of spatiotemporal properties to the general demands of behaviour is limited in some cases by the *diversity* of behaviour that we see within a species. Many insect hoverers are also extremely agile and engage in spectacular flight at high speed. How do they cope with the conflicting demands of motion detection at both high and low speeds? One mechanism proposed to extend the range of velocities over which motion detectors can respond reliably is adaptation of the time constant of the elementary motion detector (EMD) delay mechanism (Clifford et al. 1997). This would shift the optimum tuning of the motion detector from low to high velocities. In the blowfly, prolonged exposure to high image velocities indeed results in "motion adaptation", manifested by a decrease in absolute speed sensitivity and an accompanying increase in the relative sensitivity to changes about the adapting speed (Maddess and Laughlin 1985). Similar changes have been recorded physiologically in the motion pathway of the wallaby and observed psychophysically in humans (Clifford and Langley 1996; Ibbotson et al. 1998). We recently showed, however, that motion adaptation in certain flies does not seem to result in substantial change in the delay filter properties or spatial tuning (Harris et al. 1999) and hence does not extend the range of velocities to which the EMD responds.

As an alternative, our previous work (O'Carroll et al. 1997) suggested that motion detecting neurones in the bee-fly, *Bombylius*, which I study here, achieve sensitivity to a broad range of velocities by an unusual mechanism. Many properties of these neurones are consistent with input from EMDs of the correlation type. Yet temporal tuning curves for neurones in this species show evidence of two "humps" (O'Carroll et al. 1997) rather than a single clear optimum as predicted by the classic Reichardt correlation model (Buchner 1984; Borst and Bahde 1986; Egelhaaf et al. 1989). These broad tuning curves are consistent with a motion detector mechanism employing two or more delay filters, operating in parallel. Here I present a detailed analysis of temporal tuning in *Bombylius*. By investigating the effect of contrast on the unadapted response tuning, I present evidence to support the hypothesis that the motion sensitive neurones of this species combine outputs from at least two parallel delay mechanisms in the motion detection pathway. These provide an inherent tuning to a large range of image speeds. Using selective motion adaptation experiments, I further show that, in contradiction to the mechanism proposed by Clifford et al. (1997), motion adaptation seems to operate by selectively reducing the contribution of the slower pathway and thus sharpening temporal tuning at high image speeds, without extending the dynamic range.

2. Methods

I recorded intracellularly from wide-field, direction selective cells in the lobula plate of *Bombylius major* captured from the wild and stimulated with sinusoidal gratings presented on a Tektronix 608 CRT display under computer control, at a frame rate of 200 or 300 Hz. For details of the recording methods and display system see O'Carroll et al. (1997). Recordings were from a class of cells with equatorial receptive fields which respond selectively to front to back (progressive) motion. We previously showed that these cells are neuroanatomically similar to blowfly HS cells (O'Carroll et al. 1997). I recorded from such cells in 27 flies with long enough intracellular recordings for the experiments presented here (>30 minutes) in 15 female flies. Like blowfly HS cells, they respond to wide-field motion with a graded response, depolarizing up to 12 mV on stimulation with a pattern moving from front to back and hyperpolarizing by a similar amount to motion in the opposite direction (O'Carroll et al. 1997).

3. Results

3.1 Unadapted temporal frequency tuning

To investigate the effect of contrast on temporal frequency tuning, and to provide a basis for comparison with adapted response tuning, I estimated the unadapted temporal frequency tuning using two alternative methods, contrast steps and contrast ramps. In the contrast step experiments, the cell is adapted to a blank screen of mean luminance for several seconds, then presented with a grating moving against this background at a fixed "test" contrast, with no change in mean luminance. Response (change in membrane potential) is then plotted as a function of the temporal frequency (speed) of the test grating. In contrast ramp experiments, the test pattern increases in contrast over a period of 10 s, allowing the contrast sensitivity (the inverse of the contrast at which a threshold response level is exceeded) to be estimated. For details of this method see O'Carroll et al. (1997).

Figure 1 shows the unadapted temporal frequency tuning recorded from HS cells in *Bombylius* using both methods. Contrast sensitivity reaches a maximum just above 30 (threshold contrast 0.03) but is shown normalized to 1 in this plot to permit comparison with the contrast step data. In addition to a maximum close to 8 Hz, both curves show a secondary "shoulder" at 1.5 Hz and some evidence of a further shoulder at even lower frequencies. The secondary shoulders are much more apparent in the contrast sensitivity curve, which is also flatter across the middle part of the range of temporal frequencies studied, to the extent that sensitivity is more than 80% maximal (threshold contrast < 4%) across an entire log unit of temporal frequencies from just above 1 to above 10 Hz. Given that the contrast of the natural scenes that these flies would experience during natural behaviour is very high (van Hateren 1997), high contrast sensitivity and broad tuning would permit these motion detectors to provide significant responses over a huge range of pattern speeds.

Why does the contrast sensitivity data suggest broader response tuning than the contrast step data? Likely explanations lie in several confounding problems with the contrast step method. Firstly, motion adaptation causes the response to decline during presentation in a manner dependent on the stimulus temporal frequency (Borst and Egelhaaf 1987), becoming more pronounced as temporal frequency or contrast increases (Maddess and Laughlin 1985; Borst and Egelhaaf

1987). In order to plot "unadapted" responses, I consider only the initial part of the response following the presentation of the moving pattern. Unfortunately, models predict that it can take several hundred milliseconds for responses of a Reichardt correlator to reach a steady state (Egelhaaf and Borst 1989). At high temporal frequencies the initial response exceeds the steady state level, while the opposite is true at low temporal frequencies. Considering only the transient response, the resulting curve is thus biased towards higher temporal frequencies. On the other hand, my initial experiments (not shown) showed that in the "steady state" condition, motion adaptation depresses responses at high temporal frequencies, so that tuning is biased towards lower temporal frequencies. I minimized both effects by using a moderate contrast stimulus (Michelson contrast of 0.1 or 0.15), a short presentation (300 ms) and discarding the initial 100 ms of the response to exclude transients. Nonetheless, the resulting curve remains a partial trade-off between these various effects. Finally, the non-linear correlation operation in the EMDs produces an approximately quadratic response increase with increasing contrast (Hausen and Egelhaaf 1989; see also O'Carroll et al. 1997 for an example). At higher contrasts, this quadratic relationship saturates, leading to increasingly flattened tuning curves (data not shown). At the low contrasts used, however, this non-linear operation might exaggerate the effect of small differences in sensitivity in the "flattened" region of the curves.

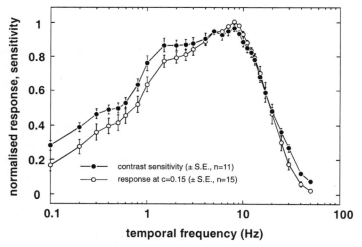

Fig. 1 Temporal frequency tuning of unadapted HS cells to sinusoidal patterns at a spatial frequency of 0.08 cyc/°. Data shown are either the response measured in a 200 ms window, 100 ms after presentation of a moderate contrast grating (c=0.15) or the contrast sensitivity estimated by the contrast ramp method (see text). Both methods reveal an optimum at 8 Hz and a secondary peak or "shoulder" at 1.5 Hz.

3.2 Effect of contrast on spatiotemporal tuning

Figure 2 shows the spatiotemporal response tuning measured for a single HS cell by presenting contrast steps at 200 combinations of spatial and temporal frequency and at 2 different basic contrasts, 0.1 and 0.4. In each case, data are normalized with respect to the maximum response in each (contour intervals at 0.1 normalized response units). The spatiotemporal response "surface" at low contrast (0.1) has a clear optimum at a position consistent with the lower "shoulder" of figure 1, with the upper peak of figure 1 showing up as a plateau region or "saddle" in the spatiotemporal data. Although the curve is noticeably flattened at high contrast (0.4), possibly due to saturation of the response (as mentioned above), the optimum clearly shifts to the higher temporal frequency, as seen in figure 1. This higher contrast data set again reveals a change in slope at very low temporal frequencies, in the vicinity of 0.5 Hz, consistent with a third, low frequency component. All three peaks or plateaux are centred at the same spatial frequency of 0.12 cyc/°, so they can not be attributed to EMDs with different sampling distances.

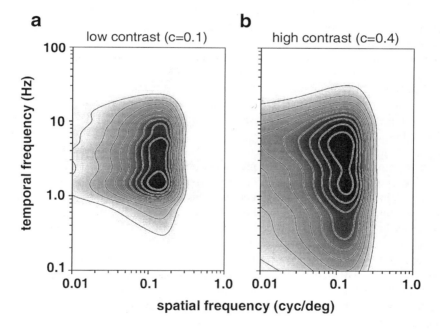

Fig. 2 Spatiotemporal response surfaces for an HS cell to drifting gratings with two different contrasts. **a** At low contrast the response optimum is at 1.5 Hz. **b** At high contrast the temporal response is much broader, and clearly peaks at higher temporal frequency, with a secondary plateau corresponding to the optimum in a and with a further shoulder at 0.5 Hz.

3.3 Motion adaptation and temporal tuning

I further tested temporal frequency tuning before and after motion adaptation, in order to see if adaptation might act differently on the parallel pathways implied by the data in figures 1 and 2. Figure 3 illustrates the "test-adapt-test" experimental protocol used to investigate the effect of motion adaptation. After adapting cells to a blank screen of mean luminance for 10 s, a moderate contrast (0.3), high spatial frequency (0.1 cyc/°) grating appeared, drifting in the preferred direction at the test temporal frequency for 1 s (the initial test stimulus). This test stimulus was followed by 3 or 4 s exposure to one of two adapting stimuli (see below) before the test grating was presented a second time. Responses were averaged before and after adaptation during the first 300 ms of each test presentation (as indicated), after discarding the initial 100 ms "transient" period (as in unadapted contrast step experiments described earlier). Because motion and flicker have both been shown previously to adapt the motion pathway of other flies (Borst and Egelhaaf 1987), I investigated adaptation in response to two adapting stimuli. In one experiment (Fig. 3a) the adapting pattern was a low spatial frequency (0.02 cyc/°) sinusoidal grating with high contrast (0.9), and high temporal frequency (20 Hz), drifted in the preferred direction ("motion adaptation"). In a second experiment (Fig. 3b), the same grating was "counterphased" (each stripe in the grating reverses contrast over time in a sinusoidal manner) at high temporal frequency (20 Hz). Both stimuli produced powerful adaptation, as evident from the depressed response to the second test presentation (Figs. 3a and b). Because the test pattern in figure 3 had a low temporal frequency (1 Hz) which did not itself produce strong adaptation, the response can be seen to recover towards the end of the second presentation in both sequences.

Counterphasing gratings have the advantage over wide-field flicker or motion stimuli in that they do not produce a large net response from a cell with a large receptive field, such as an HS neurone (see Fig. 3), because any local motion energy is in equal and opposite directions. This allows the response to the second test presentation to be made without a large motion after-effect, such as we observed in our previous study of adaptation in hoverfly HS cells (Harris et al. 1999). By comparison, adaptation by motion in the preferred direction produces pronounced hyperpolarization relative to the resting potential immediately following the end of the adaptation period and the response to the second test stimulus "rides" on this motion after-effect (Fig. 3a). There will inevitably be some confusion between the time courses of this motion after-effect and any recovery or transient behaviour in the response. However, because the flicker or local motion cues produced by a counterphasing grating will be weak in the vicinity of the "zero crossings" (the grey transition regions between dark and light stripes) some of the underlying EMDs in the receptive field of the HS cell may be more strongly adapted than others by a simple counterphasing pattern. To overcome this problem, I made the adapting grating jump through 90° of phase in alternate directions after every 2 complete cycles of the counterphase. This ensured that the extent of adaptation was evenly applied to each part of the receptive field of the neurone. While each jump produced a brief transient motion response, these were frequent and in opposite directions and did not appear to result in a large motion after-effect (Fig. 3b).

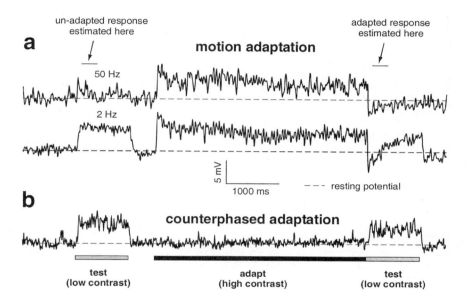

Fig. 3 Raw responses from an HS cell illustrating the test-adapt-test protocol used to determine the response tuning following adaptation. Following adaptation to a blank screen of mean luminance for 10 s, a low contrast test pattern appeared and moved for 1 s. 500 ms later, a high contrast, low spatial frequency and high temporal frequency adapting pattern appears, either drifting in the preferred direction for 4 s (**a**) or counterphasing (**b**), as described in the text. At the end of this adaptation period, the test pattern appeared again, and the response was estimated by averaging the membrane potential between 100 ms and 300 ms after the onset of each test period, as indicated.

Figure 4a shows the temporal tuning of HS cells, before and after motion adaptation with either counterphased or drifting gratings. The data shown are normalized with respect to the maximum response before adaptation. Several effects are obvious: Firstly, both classes of adapting stimulus produce large response depression compared with unadapted levels. Secondly, high velocity motion produces more powerful adaptation than the counterphased pattern at the same contrast, temporal and spatial frequency. Thirdly, the motion after-effect described above is powerful enough that "responses" to some test patterns are below resting potential. This is most evident at very high test temporal frequencies (above 20 Hz) which produced little or no response before motion adaptation. In further control experiments (not illustrated) where no second test stimulus was presented, I measured post-adaptation potentials as large as 3 mV below the un-stimulated resting potential.

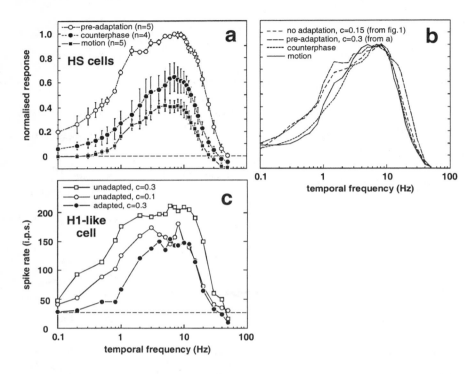

Fig. 4 Effects of motion adaptation on temporal frequency tuning. **a** Data for 5 HS cells according to the method described in figure 3. The response shown is either unadapted (none) or following adaptation for 3 s with either a counterphasing grating or motion in the preferred direction of a grating with low spatial frequency and high temporal frequency (i.e. high velocity). See figure 3 for details of the adaptation procedure. The response of the cells is shown normalized with respect to the maximum response obtained in the no adaptation condition. **b** The same data after normalization with respect to the maximum in each case and to the minimum (in order to remove the effects of the motion after-effect described in figure 3). Contrast step data from figure 1 are also shown aftern the same normalization. The adapted responses show pronounced narrowing in tuning, particularly at lower temporal frequency. **c** Results of a similar experiment on an H1-like cell. The motion test-adapt-test sequence illustrated by figure 3a was repeated at two different contrasts. At lower contrast, the unadapted response generates similar responses at middle and high temporal frequency to the high contrast stimulus following adaptation. The adapted response, however, is again much narrower and weaker at low temporal frequency.

Figure 4b shows the data from figure 4a after re-normalization with respect to the maximum and minimum response levels. Changes in the qualitative shape of the temporal tuning following even powerful adapting stimuli are very subtle.

In these experiments, the curve obtained with the pre-adaptation test stimulus is flattened compared with the unadapted data from figure 1 (reproduced on this plot for comparison). Adaptation was so powerful that I needed to use a high test contrast (0.3) in order to get a large enough response following adaptation to provide a useful curve. As in figure 2b, such high contrast patterns drive the unadapted neurone towards its maximum response level at intermediate temporal frequencies and the observed "flattening" is likely to be due to response saturation. Both adapted and unadapted responses are in remarkably close agreement in the upper part of the temporal frequency range (above 3 Hz) but diverge greatly below this frequency. The adapted curve is also much more symmetrical about the optimum than the unadapted curve, and the two humps at 1.5 and 0.5 Hz are no longer obvious.

This finding was confirmed by a series of experiments from a further cell of a different physiological class, with similar response properties (spiking response and a unilateral preference for regressive, wide-field motion) to the well studied H1 neurone in the lobula plate of the blowfly (Hausen and Egelhaaf 1989). The responses are illustrated in figure 4c. In this case, the motion adapting protocol (Fig. 3a) was used to construct tuning curves at two different test contrasts, 0.1 and 0.3. The post-adaptation response at a test contrast of 0.1 was too weak to provide a useful tuning curve. The higher contrast stimulus, however, produced a similar response at the optimum temporal frequency (8 Hz) to that of the pre-adaptation response to low contrast stimulation, allowing direct comparison of these curves without normalization. Neither set of stimuli (open and filled circles in Fig. 4c) generated strongly saturated responses, as illustrated by the unadapted response at high contrast (squares) which gave peak responses 20% higher.

The response tuning before and after adaptation in this H1-like cell was similar to that in the HS cells (Figs. 1, 2, 3 and 4a), with an overall optimum at the higher peak (8 Hz) and a pronounced shoulder at lower temporal frequency. At lower contrasts, the tuning shows more bias towards the low temporal frequency peak, as in figure 2a. Following adaptation, the response is noticeably weaker at low temporal frequencies, despite remaining very similar at high frequencies.

4. Discussion

4.1 Asymmetry in temporal frequency tuning

Both male and female *Bombylius* experience a large range of image speeds during natural behaviour. In addition to hovering, both sexes move rapidly from flower to flower while feeding from nectar. Females also make rapid transitions from hovering to rapid forward and backward flight while laying eggs. Males also experience high image speeds during territorial pursuit of conspecifics (O'Carroll et al. 1997). In keeping with this unusually varied behaviour, temporal sensitivity in *Bombylius* motion detectors is unusually broad and biased towards low temporal frequencies compared with that which we have published for other insects, including flies (O'Carroll et al. 1996, 1997). The unadapted temporal frequency tuning obtained by both methods (Fig. 1) is asymmetric, with a strong bias towards lower frequencies. The sharp roll-off in sensitivity that starts at 8 Hz is unlikely to be due to low-pass temporal filtering by the photoreceptors, which have corner frequencies above 50 Hz in typical diurnal flies (Laughlin and Weckstrom 1993), suggesting that temporal frequency tuning is determined primarily by the properties of the delay filter in the EMDs. Yet a motion detector with a single, simple delay mechanism, such as first order low-pass filter, should produce a temporal frequency tuning curve that is symmetrical about the optimum when plotted on a logarithmic temporal frequency axis (Buchner 1984; Harris et al. 1999). It is noteworthy that temporal frequency tuning in neurones from several other insect species (bees, moths and butterflies) obtained using similar methods are much more symmetrical about the optimum than I see here (O'Carroll et al. 1996, 1997).

The most obvious possibility to explain the asymmetric, humped shape of unadapted temporal tuning curves in *Bombylius* is that the peaks and plateaux represent the outputs of EMDs with two or even three delay filters operating in parallel, with different time constants. Nalbach (1989) has proposed a similar mechanism for optokinetic responses in crabs. These could exist as separate, parallel elementary motion detectors EMDs which are then integrated by the wide-field units from which we record, or possibly as different delay filters operating within a single EMD. The peak at 1.5 Hz (and possibly that at 0.5 Hz) would thus represent contributions from "slow" channels – delay filters with long time constants, while that at 8 Hz would be from a "fast" channel with a shorter time constant. The summed output of the fast and slow channels produces the humped appearance of the tuning and the asymmetric bias towards low temporal frequencies.

The fact that the shape of the supra-threshold temporal frequency tuning changes with increasing contrast (Fig. 2) suggests that the slow and fast pathways exist as separate, parallel EMDs. This is further supported by the finding that the contrast sensitivity is similar at the two peaks (Fig. 1). This result is explicable if we assume that the two delay pathways contribute inputs to the HS cell with different contrast gain characteristics. Thus, while the contrast sensitivity of the slower component may be very close to that of the fast, the contribution from it appears to saturate more readily, so that the fast component dominates responses at higher contrasts.

4.2 Effect of motion adaptation on temporal tuning

My data show that motion adaptation influences temporal tuning in two ways. Firstly, there is a clear reduction in response to low contrast patterns, although higher contrast patterns still evoke high response levels. This suggests that contrast sensitivity of the neurones is reduced by motion adaptation. Secondly, this sensitivity (or contrast gain) reduction is greatest at low temporal frequencies, leading to narrower, more symmetrical tuning curves. This change in the shape of the adapted temporal tuning curve is distinct from the effect of increasing contrast on the unadapted neurone observed earlier (Fig. 2). The post-adaptation response is noticeably narrower despite the use of a higher contrast stimulus, whereas I noted previously that response saturation at higher contrast produces broader tuning at both higher and lower temporal frequencies in the unadapted response (Figs. 2, 4b and c). Given the similarity in the post- and pre-adaptation tuning at temporal frequencies above 5 Hz, I suggest that this is most likely the result of a reduction in the relative contributions of "slow" components of the motion detector delay filter mechanism.

Taken together, these results represent convincing evidence that the temporal tuning that we see in the HS cells results from the summation of the output of at least two and possibly three delay mechanisms operating in parallel and possibly in different EMDs.

4.3 Motion adaptation and velocity contrast discrimination

What are the functional consequences of these findings for the natural behaviour of these flies? I have already mentioned that a motion detector employing several parallel delay mechanisms would benefit from a broad response tuning. The overall sensitivity to low temporal frequencies resulting from what may be the "slow" component is extraordinary: at 0.1 Hz, *Bombylius* HS cells retain 30% of their maximum contrast sensitivity. This sensitivity level extends to beyond 20 Hz. This would "pre-adapt" the system to respond strongly to motion at a correspondingly large range of speeds. However, one price paid for such broad tuning is that the response will be similar across a large range of pattern speeds. Thus while this system could detect a large range of speeds, it would not be able to differentiate between them reliably on the basis of the output of such neurones. If we assume that the function of these HS cells and other wide-field cells of the lobula plate is not only to detect, but to compare and analyse different classes of motion encountered during natural behaviour, this "flat" response is not ideal.

In *Bombylius*, this problem may be overcome by the motion adaptation mechanism. By reducing the relative contribution of "slow" channels when the animal experiences high velocities, adaptation causes a narrower response tuning, more typical of the predictions of the correlator model for a single delay EMD (Buchner 1984). High speed responsiveness is not altered by this mechanism, but the response/temporal frequency curve has a much steeper slope at speeds below the optimum. Thus the sensitivity to changes in image speed – "velocity contrast" – will be enhanced, as previously observed by Maddess and Laughlin (1985) following adaptation of the neurone H1 of the blowfly.

4.4 How does adaptation aid motion coding during natural behaviour?

If we consider some key characteristics of the natural scenes for which these eyes evolved, the interplay between adaptation and the contrast saturation of the slow EMD pathway becomes explicable. Natural images contain a broad range of spatial frequencies, with power (which is proportional to the square of contrast) declining as frequency increases (Burton and Moorhead 1987; Field 1987). High spatial frequencies are passed poorly by the optics of the eye (Buchner 1984) further reducing their contrast in the image presented to the motion detectors. As the fly moves through or rotates within its world, this spatial spectrum would be transformed to an equivalent temporal frequency spectrum. The highest temporal frequencies will always be generated by the lowest contrast components of the image and vice versa. Because the spatial structure of natural images are so stereotyped, it is the animals' behaviour that determines the spatiotemporal distribution of energy to be detected by the motion pathway. At the very lowest speeds, only the highest spatial frequency components of the image would generate high enough temporal frequencies to be detectable. These are the very components that are present with the lowest contrast. At higher speeds, even the lowest (and thus higher contrast) components of the image would start to generate detectable temporal frequencies.

From this argument, it follows that in addition to being "tuned" to low temporal frequencies, a motion detector optimized for detecting low speed motion of natural images requires high contrast sensitivity and a high gain, since it will be stimulated by relatively "weak" components of the image. Conversely, a motion detector optimized for higher image speeds need not be so sensitive. Indeed, lower sensitivity might prevent saturation at high speeds as higher contrast (lower spatial frequency) components of the image start to generate optimal temporal frequencies.

In *Bombylius* high contrast sensitivity at low temporal frequencies is provided by the "slow" channel, allowing useful responses with high gain at even very low velocities. The contribution from this slow channel would be prone to saturation as velocity increases. The overall response of the neurone to higher image speeds would thus be determined principally by the faster channel, which will be stimulated by higher spatial frequency (and thus lower contrast) image components. My data suggest that this channel saturates less readily, even at moderate contrasts, although it also has high contrast sensitivity.

In natural scenes, average contrasts are very high (Tolhurst et al. 1992), compared with the low contrasts required to evoke significant responses from *Bombylius* motion detectors. At intermediate and high speeds the "slow" pathway would thus contribute a constant, saturated input to the overall response. This redundant signal would be undesirable because it will lead to compression of the effective range of membrane potentials that an HS neurone can generate as speed increases. Motion adaptation may play a key role in reducing this redundant response component via several mechanisms. The first, as already described, is a reduction in the relative contribution of the "slow" pathway to the overall response, and the accompanying reduction in contrast sensitivity. The second is the addition of a standing potential which opposes the partially redundant excitatory signal – as implied by the large (up to 3 mV) hyperpolarization seen following adaptation. This motion after-effect has also been observed previously in the responses of the neurone H1 in the blowfly lobula plate (Srinivasan and Dvorak 1979) and is similar to the well-described "waterfall illusion" observed in human psychophysics. The third is an overall reduction in the magnitude of the response at all frequencies (i.e. a reduction in the overall "gain" of the system), by a mechanism which is not clear from my experiments.

References

Borst A, Bahde S (1986) What kind of movement detector is triggering the landing response of the housefly? Biol Cybern 55: 56-69

Borst A, Egelhaaf M (1987) Temporal modulation of luminance adapts time constant of fly movement detectors. Biol Cybern 56: 209-215

Buchner E (1984) Behavioural analysis of spatial vision in insects. In: Ali MA (ed), Photoreception and vision in invertebrates. Plenum, New York, pp 561-621

Burton GJ, Moorhead IR (1987) Color and spatial structure in natural scenes. Appl Optics 26: 157-170

Clifford CWG, Langley K (1996) Psychophysics of motion adaptation parallels insect electrophysiology. Curr Biol 6: 1340-1342

Clifford CWG, Ibbotson MR, Langley K (1997) An adaptive Reichardt detector model of motion adaptation in insects and mammals. Visual Neurosc 14: 741-749

Egelhaaf M, Borst A (1989) Transient and steady-state response properties of movement detectors. J Opt Soc Am A 6: 116-127

Egelhaaf M, Borst A, Reichardt W (1989) Computational structure of a biological motion-detection system as revealed by local detector analysis in the fly's nervous system. J Opt Soc Am A 6: 1070-1087

Farina WM, Varjú D, Zhou Y (1994) The regulation of distance to dummy flowers during hovering fligth of the hawkmoth Macroglossum stellatarum. J Comp Physiol A 174: 239-247

Field DJ (1987) Relations between the statistics of natural images and the response properties of cortical cells. J Opt Soc Am A 4: 2379-2394

Harris RA, O'Carroll, DC, Laughlin SB (1999) Adaptation and the temporal delay filter of fly motion detectors. Vision Res 39: 2603-2613

van Hateren JH (1997) Processing of natural time series of intensities by the visual system of the blowfly. Vision Res 37: 3407-3416

Hausen K, Egelhaaf M (1989) Neural mechanisms of visual course control in insects. In Stavenga DG, Hardie R (eds) Facets of vision. Springer Verlag, Berlin, pp 360-390

Ibbotson MR, Clifford CWG, Mark RF (1998) Adaptation to visual motion in directional neurons of the nucleus of the optic tract. J Neurophysiol 79: 1481-1493

Land MF, Collett TS (1974) Chasing behaviour of houseflies (Fannia canicularis): a description and analysis. J Comp Physiol A 89: 331-357

Laughlin SB, Weckström M (1993) Fast and slow photoreceptors - a comparative study of the functional diversity of coding and conductances in the Diptera. J Comp Physiol A 172: 593-609

Maddess T, Laughlin SB (1985) Adaptation of the motion-sensitive neuron H1 is generated locally and governed by contrast frequency. Proc Roy Soc Lond B 225: 251-275

Nalbach HJ (1989) Three temporal frequency channels constitute the dynamics of the optokinetic system of the crab. Biol Cybern 61: 59-70

O'Carroll DC, Bidwell NJ, Laughlin SB, Warrant EJ (1996) Insect motion detectors matched to visual ecology. Nature 382: 63-66

O'Carroll DC, Laughlin SB, Bidwell NJ, Harris RA (1997) Spatio-temporal properties of motion detectors matched to low image velocities in hovering insects. Vision Res 37: 3427-3439

Srinivasan MV, Dvorak DR (1979) The waterfall illusion in an insect visual system. Vision Res 19: 1435-1437

Tolhurst DJ, Tadmor Y, Chao T (1992) Amplitude spectra of natural images. Ophthal Physiol Optics 12: 229-232

Index

Printing: Weihert-Druck GmbH, Darmstadt
Binding: Buchbinderei Schäffer, Grünstadt